WITHDRAWN

Tablet and Capsule Machine Instrumentation

Tablet and Capsule Machine Instrumentation

Edited by

Peter Ridgway Watt
MSc, PhD, CChem, FRSC, CPhys, FInstP
Formerly Instrument Services Co-ordinator
Beecham Pharmaceuticals Research Division
Brockham Park, UK

and

N Anthony Armstrong
BPharm, PhD, FRPharmS, FCPP
Formerly Senior Lecturer in Pharmaceutics
Welsh School of Pharmacy
Cardiff University, Cardiff, UK

London • Chicago   Pharmaceutical Press

Published by the Pharmaceutical Press
An imprint of RPS Publishing

1 Lambeth High Street, London SE1 7JN, UK
100 South Atkinson Road, Suite 200, Grayslake, IL 60030–7820, USA

© Peter Ridgway Watt and N Anthony Armstrong 2008

(PhP) is a trade mark of RPS Publishing

RPS Publishing is the publishing organisation of the Royal Pharmaceutical Society of Great Britain

First published 2008

Typeset by J&L Composition, Filey, North Yorkshire
Printed in Great Britain by TJ International, Padstow, Cornwall

ISBN 978 0 85369 657 5

All rights reserved. No part of this publication may be reproduced, stored in a retrieval system, or transmitted in any form or by any means, without the prior written permission of the copyright holder.

The publisher makes no representation, express or implied, with regard to the accuracy of the information contained in this book and cannot accept any legal responsibility or liability for any errors or omissions that may be made.

The right of Peter Ridgway Watt and N Anthony Armstrong to be identified as the authors of this work has been asserted by them in accordance with the Copyright, Designs and Patents Act, 1988.

A catalogue record for this book is available from the British Library

Dedication

I first met Peter Ridgway Watt about 30 years ago when we were both speakers at a very early conference on instrumented tablet presses. We quickly found that we had many interests in common. In 1988, Peter brought out his textbook on instrumentation, *Tablet Machine Instrumentation in Pharmaceutics,* and we collaborated several times in organising short courses on the topic. It was at one of the most recent of these that Peter and I decided that a revision of his textbook was called for, to be written partly by us, but inviting experts in certain areas to contribute chapters on selected topics. Peter threw himself into the task, but his health began to fail, and he died on 12 February 2007, only five days after the text of the one remaining chapter had been received.

This book is dedicated to Peter Ridgway Watt, an inspiring colleague and a good friend.

N Anthony Armstrong
Harpenden, UK

February 2007

Contents

Preface xi
Contributors xiii

1 Introduction — 1
N Anthony Armstrong

Introduction 1
A brief overview of instrumented systems 2
Units of measurement 4
The instrumentation of tablet presses and capsule-filling equipment 6
References 8
Further reading 9

2 The measurement of force — 11
Peter Ridgway Watt

Introduction 11
Strain measurement 12
Strain gauges 13
Siting strain gauges 22
The Wheatstone bridge circuit 32
Load cells 38
Miscellaneous methods 46
References 49

3 The installation of strain gauges — 51
Anton Chittey

Introduction 51
Health and safety considerations 51
Surface preparation 51
Bonding with adhesive 53
Leadwire attachment 57
Protection of the installation 58
Inspection and testing 58
Specialist applications 59

Tools and installation accessories 61
Professional assistance 63
References 64
Further reading 64

4 The measurement of displacement
Peter Ridgway Watt

65

Introduction 65
Displacement transducers with analogue output 65
Displacement transducers with a digital output 76
Dynamic measuring devices 79
Miscellaneous methods of displacement measurement 82
References 84

5 Power supplies and data acquisition
Peter Ridgway Watt

87

Introduction 87
Gauge excitation level 87
The power supply unit 88
Mains noise 89
Battery power 91
Power supply to and data acquisition from tablet presses 91
Signal display 95
References 97

6 Instrumented tablet presses
N Anthony Armstrong and Peter Ridgway Watt

99

Introduction 99
The eccentric press 99
Rotary tablet presses 111
The measurement of displacement in tablet presses 119
Measurement of ejection forces 127
Measurement of punch pull-up and pull-down forces 129
Measurement of punch face adhesive forces 131
Instrumentation packages 132
References 136

7 Calibration of transducer systems
Peter Ridgway Watt

139

Introduction 139
Force 140

Displacement 142
Calibration problems 145

8 Data handling
Alister P Ridgway Watt

Introduction 147
Sampling system theory 147
Electronics sub-systems 153
Embedded systems 157
Computer interfacing 157
Software 162
Data backup 163
Further reading 165

147

9 Applications of tablet press instrumentation
N Anthony Armstrong

Introduction 167
Punch displacement–time profiles 168
Force–time profiles 176
Force–porosity relationships 182
The force–displacement curve 187
Punch velocity 190
Die wall stress 196
Applications of press instrumentation to lubrication studies 201
References 202
Further reading 205

167

10 The instrumentation of capsule-filling machinery
N Anthony Armstrong

Introduction 207
Capsule-filling equipment 208
Instrumentation of dosating disk capsule-filling machines 209
Instrumentation of dosating nozzle capsule-filling machines 214
References 220
Further reading 222

207

11 Automatic control of tablet presses in a production environment
Harry S Thacker

Introduction 223
The force–weight relationship 224
Monitoring systems 225

223

Control systems 225
Reject systems 226
Weighing systems 227
Computer-controlled systems 228
References 239
Further reading 240

Appendix: Suppliers of materials and services 241
Index 249

Preface

WHEN WILLIAM BROCKEDON patented the notion of 'shaping pills, lozenges and black lead by pressure in dies', he could hardly have imagined the extent to which this apparently simple idea would grow. It was largely this invention that extended the industrial revolution to the preparation of medicines, giving rise to the pharmaceutical industry as it now exists. Individual pharmacies would no longer need to make up small quantities of medicines themselves, large-scale production in a relatively small number of manufacturing sites was now feasible, and mechanical engineering methods could be applied to the process.

In Brockedon's original invention (Figure P1), the upper punch was removed so that powder could be loaded into the die. The punch was replaced and was then struck with a mallet to compress the charge between the faces of the two punches. It would have been possible to make a few tablets in a minute.

At the present time, there are rotary tablet presses with many sets of punches and dies that are capable of making compressed tablets at a rate of up to one million in an hour. Yet for more than 100 years, the satisfactory operation of the process was dependent on the skill and experience of the men who ran the machines. They might evaluate a tablet by breaking it in half and listening to the snap, but they did not have the facility to measure what was happening in accurate detail.

Since the 1960s, the situation has changed dramatically. We have reached a point where we

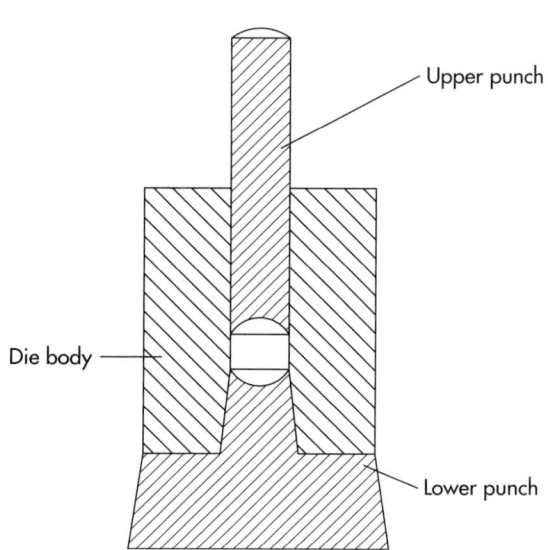

Figure P1 Schematic view of Brockedon's original punch and die assembly.

are in a position to measure many variables before, during and after the compaction event, and to use the constant stream of information to control the press automatically. In this book, we have described a selection of measuring devices that have been developed in the general field of engineering instrumentation, and we have shown how some of them have been applied in our particular area of interest. Readers might be concerned that many of the references quoted are of some considerable age, but in fact there has been little published work on new measuring systems for several decades. The most significant advances have been in the field of electronics, and the application of computer techniques to data acquisition and processing, but measuring devices such as strain gauges and displacement transducers have not changed greatly since the 1980s.

As for the equipment described in these pages, we have assumed little prior knowledge on the part of the reader and have attempted to define any new terms as they appear. Many tablet press manufacturers offer machines that are already fitted with measuring devices and data processing systems. Nevertheless, it is still necessary to understand the essential principles of press instrumentation, the importance of transducer selection, siting, and calibration, and to have an appreciation of what a particular instrumentation technique can and – equally important – cannot do. It is our hope that these pages will help to promote such understanding.

Of course, the idea that research progresses smoothly from one stage to the next is a myth, usually supported by papers and publications that conveniently omit all mention of the dead ends and disasters that happen in real life. We have, therefore, included a few anecdotes from our own experience, which confirm the hypothesis that if something can go wrong, it will!

N Anthony Armstrong and
Peter Ridgway Watt

February 2007

Contributors

N Anthony Armstrong
Harpenden, UK
Formerly Senior Lecturer in Pharmaceutics, Welsh School of Pharmacy, Cardiff University, Cardiff, UK

Peter Ridgway Watt
Formerly Instrument Services Co-ordinator, Beecham Pharmaceuticals Research Division, Brockham Park, UK

Anton Chittey
Technical Support Engineer, Vishay Measurements Group UK Ltd, Basingstoke, UK

Alister P Ridgway Watt
Technical Director, QBI Ltd, Walton on Thames, UK

Harry S Thacker
Ormskirk, UK; formerly of Manesty Machines, Knowsley, UK

1

Introduction

N Anthony Armstrong

Introduction

The year 1843 saw the publication of British Patent Number 9977. It was issued to William Brockedon, an English inventor, and its object was that of 'shaping pills, lozenges and black lead by pressure in dies'. This marked the introduction of the dosage form now known as the tablet. Brockedon did not set out to invent a dosage form. His original aim was to reconstitute the powdered graphite left as a waste product when natural Cumberland graphite was sawn into narrow strips for pencil 'leads'. However, he later realised that his invention could be applied to the production of single-dose units of medicinally active compounds.

The introduction of the tablet marked the impact of the Industrial Revolution in the production of medicines and opened up a whole range of new possibilities for the pharmaceutical industry. Compared with earlier dosage forms such as the pill, it offered a stable, convenient form that was capable of being mass produced by machines. Furthermore, with appropriate formulation, a range of different types of tablet could be produced, including those to be swallowed intact, sucked, held within the buccal pouch or under the tongue, dissolved or dispersed in water before ingestion, or so formulated that the active ingredient is released in a controlled manner. So popular has the tablet become that it has been estimated that of the 600 million National Health Service prescriptions written per annum in the UK, over 65% are for tablets. There are 336 monographs for tablets in the 2005 edition of the *British Pharmacopoeia*.

The original Brockedon press consisted of a die and two punches, force being applied by a blow from a hammer. Mechanised versions of this device soon followed, either eccentric presses with one die and one set of punches or rotary presses with many sets of tooling. A modern rotary press can turn out approximately one million tablets every hour, rejecting any that are unsatisfactory. Such presses are often designed to operate without continuous human supervision, and to achieve this aim, highly sophisticated control systems are required. However, all tablet presses involve compression of a particulate solid contained in a die between two punches, which is essentially Brockedon's invention.

The capsule originated at about the same time as the tablet. The first recorded patent was granted in 1834 to two Frenchmen, Dublanc and Mothes. This was a single piece unit that today is usually referred to as a soft-shell capsule, the contents of which are almost invariably liquid or semisolid. The hard-shell capsule was invented a few years later in 1846 by another Frenchman, Lehuby. Such capsules consist of two parts, the body and the shell, and are usually made from gelatin. The fill is almost always a particulate solid, and the filling process usually involves the application of a compressive force. Hard-shell capsules also proved to be a popular dosage form, and there are 64 monographs for hard-shell capsules in the 2005 edition of the *British Pharmacopoeia*.

Research into the formulation and manufacture of tablets, and to a lesser extent that of hard-shell capsule fills, soon followed but suffered from a major handicap. Many tablet properties – thickness, crushing strength, resistance to abrasion, disintegration time, release of active ingredient – are dependent on the pressure that has been applied to the tablet during manufacture. If the means of accurately measuring the applied pressure are lacking, it follows that meaningful studies are impossible.

Measuring the force applied to a tablet in a press was not easy, given the constraints of early twentieth century technology. Even using the relatively simple presses of that era, the compression event lasted only a fraction of a second, and hence the measurement system had to react to the change in pressure sufficiently rapidly. Mechanical devices, owing to their inherent inertia, were not appropriate for this purpose. Such devices are suitable for measuring pressure during a longer-lasting event (e.g. compression in a hydraulic press), but this is unrealistically slow in terms of tablet manufacture.

It is instructive to consider how the pressure in a tablet press arises. As the punch faces approach each other, the volume containing the particulate solid decreases. When the solid is in contact with the faces of both punches, then pressure exerted by one punch will be transmitted through the solid mass and will be detected at the other punch. The magnitude of the pressure is thus a function of the distance separating the punch faces.

Many presses have some form of mechanical indication of pressure. For example, the Manesty F3 press has an eccentric cam graduated with a linear scale. The reading on this scale is related to the depth of penetration of the upper punch into the die. It takes no account of lower punch position and, therefore, is not a measure of the distance separating the punch faces. The relationship between punch separation and pressure is not linear, and it must be borne in mind that the relationship between pressure and punch face separation differs for different solids. Consequently, though the graduated scale gives a useful reference point, it is not a device for actually measuring pressure.

The major step that enabled compression pressure in a tablet press to be directly measured was the independent discovery by Simmons and by Ruge in 1938 that wires of small diameter could be bonded to a structure to measure surface strain. Since strain is proportional to force, this marked the invention of the strain gauge as a device for measuring force. The strain gauge was developed considerably during World War Two, primarily in the aircraft industry. Its application to tablet presses soon followed. The construction and mode of operation of the strain gauge is described in Chapter 2. However, its essential characteristic, namely representing force in terms of an electrical signal, means that force in the die of a tablet press can be directly measured *in situ* with the press operating at its normal rate of production.

The first report of the use of strain gauges in a study of tablet preparation was made by Brake at Purdue University in 1951. This report was in the form of a Master's thesis that unfortunately was never published as a conventional scientific paper. A year later, the first in a series of papers entitled 'The physics of tablet compression' was published by T. Higuchi and others at the University of Wisconsin. In one of the earlier papers in the series, the term 'instrumented tablet machine' was used for the first time. The importance of this series, publication of which continued until 1968, cannot be overemphasised and it can be said to have initiated the systematic study of the tabletting process and of tablet properties.

Further important steps in the development of instrumented tablet presses and capsule-filling equipment are given in Table 1.1. The instrumented tablet press, with its output often linked to a computer, is now a widely used research tool. In the pharmaceutical production environment, many presses are routinely fitted with some form of instrumentation during construction.

A brief overview of instrumented systems

The basic components of an instrumented system are shown in Figure 1.1.

All instrumentation systems have several essential attributes:

- a transducer of appropriate sensitivity
- a suitable site for fixing the transducer to the equipment

Table 1.1 Historical milestones in the instrumentation of tablet presses and capsule-filling machinery

1951	Utilisation of strain gauges in tablet preparation by Brake, Purdue University, USA
1952–1968	'The physics of tablet compression' a series of papers by T Higuchi et al., University of Wisconsin, USA
1954	First use of the term 'instrumented tablet machine' by Higuchi et al. (1954)
1967	The instrumentation of a rotary tablet press reported by Knoechel et al. (1967), Upjohn, Kalamazoo, USA
1971	The first reported linking of an instrumented tablet press to a computer by de Blaey and Polderman (1971), University of Leiden, Netherlands
1972	The first report of a tablet press simulator (Rees et al., 1972, Sandoz, Switzerland)
1972–1977	Instrumentation of capsule filling machinery (Cole and May (1972), Merck, Sharp and Dohme, Hoddesden, UK: Small and Augsburger (1977), University of Maryland, USA)
1980	Linkage of a microcomputer to an instrumented tablet press (Armstrong and Abourida, 1980, Cardiff University, UK)
1982	Simulated capsule filling machinery (Jolliffe et al., 1982, Chelsea College, University of London, UK)

- a power supply and a means of getting that power to the transducer
- a means of getting the output away from the transducer
- amplification circuitry
- a method of observing and/or recording the signals from the transducer
- a method of calibration.

The main parameters of interest in the instrumentation of tablet presses and capsule-filling equipment are force, distance and time, with the first two often being measured in relation to the last. Measurement is carried out by means of transducers. A transducer is a device that permits the measurement of one physical parameter (input) by presenting it as another (output). An everyday example of a transducer is a conventional thermometer, in which temperature is measured in terms of the volume of a liquid. Proportionality must be established between the input of the transducer and its output – in this case between temperature and the liquid volume. In other words, the transducer must be calibrated.

Almost all the transducers used in instrumented tablet presses have electrical outputs of some sort, which by appropriate circuitry can be changed into signals based on voltage These, in turn, perhaps after transformation into digital form, can be measured, stored and manipulated.

Numerous parameters involved in the tabletting process can be measured, though some are more difficult to measure than others. For example, with a rotary tablet press fitted with force and displacement transducers on upper and lower punches, it is possible to measure all the parameters described in Table 1.2.

Most of these involve force (pressure) and movement. Since these parameters will have been recorded with respect to time, it is possible to measure the duration of events in the compression cycle. The rate of change can also be measured; for example, punch speed can be derived from knowledge of punch movement with respect to time. It is also possible to record one of these parameters as a function of another. Examples of what can be measured are given in Table 1.3, and their significance will be discussed later in this book.

If the primary objective for using an instrumented press or capsule-filling equipment is fundamental research or to optimise a new formulation, it may be useful to measure as many of these parameters as possible. Conversely, if the aim is to control a production machine, then fewer need to be monitored. It must be borne in mind that instrumentation can be expensive, both in terms of equipment costs and the costs of skilled personnel to use it, maintain it and to interpret its output. Hence a 'let's measure everything' approach can be unnecessarily costly. As in all scientific work, careful consideration of the objectives of the work and the benefits that may be achieved must be undertaken as an initial step.

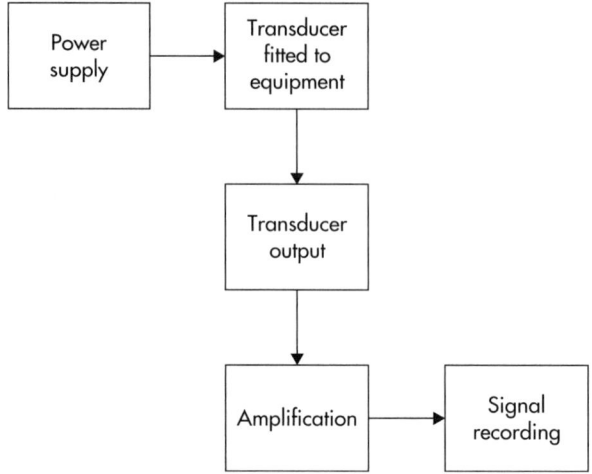

Figure 1.1 The basic components of an instrumented system.

Table 1.2 Parameters that can be measured using a rotary tablet press fitted with force and displacement transducers on upper and lower punches

Parameter	Units
Upper punch precompression force	Force (N)
Lower punch precompression force	Force (N)
Upper punch compression force	Force (N)
Lower punch compression force	Force (N)
Ejection force	Force (N)
Upper punch pull-up force	Force (N)
Lower punch pull-down force	Force (N)
Die wall force	Force (N)
Upper punch movement	Distance (m)
Lower punch movement	Distance (m)
Punch or die temperature	Temperature (°C)

Furthermore it is vitally important to be confident that the collected information is a measure of the intended parameter, and not an artefact introduced by the measuring device or its attachment, an error in data collection or manipulation, or some uncontrolled feature of the overall system.

Units of measurement

Units of measurement can often be the source of confusion, though this would be reduced if SI units were invariably used. Wherever possible, units outside the SI system should be replaced by SI units and their multiples and sub-multiples formed by attaching SI prefixes. In the SI system, there are seven basic units from which all others can be derived. These base quantities, together with their units and symbols, are shown in Table 1.4. Such variables as displacement, time and temperature can, therefore, be referred in principle to the base units of the SI system. Variables, such as force, that are not among the seven fundamentals must be derived from combinations of the latter.

In practice, all the base units are not equally accessible for everyday use. It is, therefore, normal to approach them through the use of derived units, and the derivation of some of these is shown below.

Base units

The base unit of length, the metre, is defined in terms of time and the speed of light, which is

Table 1.3 Parameters that can be derived from data obtained from a tablet press fitted with force and displacement transducers on upper and lower punches

Upper punch	Lower punch	Upper and lower punches
Punch speed (m s^{-1})	Punch speed (m s^{-1})	Ratio of peak forces
Peak force (N)	Peak force (N)	Distance between punch faces (m)
Punch penetration (m)	Punch displacement (m)	Tablet thickness (m)
Work of compression (N m)		Tablet density (kg m^{-3})
Work of expansion (N m)		Porosity
	Ejection force (N)	
	Work of ejection (N m)	
Area under force–time curve (N s)	Area under force–time curve (N s)	
Rise time (s)	Rise time (s)	
Stress rate (N s^{-1})	Stress rate (N s^{-1})	
	Ejection displacement (m)	

Table 1.4 Basic units in the SI system of measurement

Base quantity	Unit	Symbol
Length	metre	m
Mass	kilogram	kg
Time	second	s
Electric current	ampere	A
Thermodynamic temperature	kelvin	K
Luminous intensity	candela	cd
Amount of substance	mole	mol

299 792 458 m s^{-1}. Thus, the metre is the length of the path travelled by light in a vacuum during a time interval of 1/(299 792 458) of a second. Secondary sources are lasers in the visible and near infrared spectrum, and physical objects are calibrated by direct comparison with these lasers.

The base unit of mass is the kilogram, and this is the only unit of the seven that is currently represented by a physical object. The international prototype of the kilogram is a cylinder made of a platinum–iridium alloy kept at the International Bureau of Weights and Measures at Sèvres near Paris. Replicas are kept at various national metrology laboratories such as National Physical Laboratory in the UK and the National Bureau of Standards in the USA.

The SI unit of thermodynamic temperature is the kelvin (K). The kelvin is defined as the fraction 1/(273.16) of the thermodynamic temperature of the triple point of water.

The SI unit of time is the second, which is defined as 9 192 631 770 periods of the radiation derived from an energy level transition of the caesium atom. As such, it is independent of astronomical observations on which previous definitions of time depended. The international atomic time is maintained by the International Bureau of Weights and Measures from data contributed by time-keeping laboratories around the world. A quartz clock movement, kept at a reasonably constant temperature, can maintain its rate to approximately one part per million, equivalent to 1 s in about 12 days.

Derived units

The SI unit of force is the newton (N), and is defined as the force that imparts an acceleration of one metre per second every second (1 m s^{-2}) to a body having a mass of one kilogram.

The SI unit of pressure is the pascal (Pa), which represents one newton per square metre (1 N m^{-2}). The pascal is an inconveniently small unit for practical purposes. For example, atmospheric pressure is approximately 10^5 Pa.

The SI unit of energy or work is the joule (J), which is the work done by a force of one newton when the point at which that force is applied is displaced by one metre in the direction of the force.

The SI unit of power is the watt (W), and one watt is the power that gives rise to the production of energy at the rate of one joule per second.

Velocity is the rate of change of position of a body in a particular direction with respect to time. Since both a magnitude and a direction are implied in this definition, velocity is a vector. The rate of change of position is known as speed if only the magnitude is specified, and hence this is a scalar quantity.

Force is the most important parameter that is measured in instrumented tablet presses and capsule-filling equipment, though often the term 'pressure' is used. In some texts, the terms 'force' and 'pressure' seem to be used interchangeably, as if they were both measurements of the same thing. This is incorrect, since pressure is force per unit area. In some cases, such as when flat-faced tablet punches are used, the area over which the force is applied can be easily measured, and so if the force is known, then the pressure can be readily calculated. However, if the area of contact is not known, or if the force is not equally distributed over the whole surface of contact as, for example, with concave-faced punches, then calculation of the pressure is more complex.

Table 1.5 shows the wide variety of units, both SI and otherwise, that have been used in recent years in scientific papers describing the relationship between applied force or pressure and the crushing or tensile strength of the resultant tablets. Comparison of data from sources that use different units of measurement is difficult, and the value of using a standard system such as SI is apparent.

The instrumentation of tablet presses and capsule-filling equipment

Instrumentation techniques that can be applied to tablet presses and capsule-filling equipment are summarised here but are described in more detail later in this book.

Eccentric tablet presses

Much of the earliest work on instrumented tablet presses was carried out on eccentric presses. The upper punch is readily accessible so that force transducers can be easily fitted, and there is no problem in getting the electrical supply to the transducers and their signals out from them. It is usually considered desirable to mount the force transducers as near to the point of action as possible (i.e. on the punches). This implies that if the tablet diameter or shape is changed, another set of instrumented punches must be provided. An alternative approach is to mount the force transducers on the punch holder or eccentric arm, an arrangement that

Table 1.5 Examples of units that have been used to describe force, pressure, tablet crushing strength and tablet tensile strength in papers on tablet research in recent years

Abscissa Parameter	Unit	Ordinate Parameter	Unit
Force	kg	Crushing strength (hardness)	kg
	lb		Strong-Cobb units
	kN		N
	N		kp
Pressure	kg cm^{-2}	Tensile strength	kg cm^{-2}
	Pa		Pa
	MPa		MPa
	lb in^{-2}		

can accommodate changes of punch. It is usually possible to mount transducers directly on to the lower punch, though a popular alternative is to use a load cell fitted into a modified punch holder.

There is also adequate room to mount displacement transducers on an eccentric press, but the siting of these may cause problems owing to distortion of the press itself during the compaction event.

Rotary presses

The essential action of a rotary press – compression of a particulate solid in a die between two punches – is the same as that of an eccentric press. The main problem in fitting instrumentation to a rotary press is that the active parts of the press, the punches and dies, are moving in two horizontal dimensions, as well as the vertical movement of the punches into and out of the die. Hence, if the transducers are to be directly attached to the punches, fixed links between the power supply and the transducers and between the transducers and the output devices are impracticable. There are two approaches. Firstly, the transducers may be fitted to static parts of the press such as the tie bar, compression roll bearings, etc. The disadvantage of this approach is that these parts are distant from the punches, and intervening components such as bearings or linkages may introduce errors. However, Schmidt and Koch (1991) showed that in practice these errors were not significant and siting the force transducers distant from the punches gave a satisfactory outcome.

Secondly, a non-continuous link may be employed to get power to the transducers and their signals out. Radio-telemetry, slip-rings and optical devices have been used. Such systems usually preclude the use of a full set of punches and dies.

Ejection forces can be measured in a rotary press by fitting force transducers to the ejection ramp. The measurement of punch displacement is somewhat more difficult, owing to the difficulty of mounting the transducers close to the punches. However, modified punches are available. It has been shown that, provided allowance is made for press and punch deformation, patterns of punch movement in rotary presses follow predicted paths more fully than those of eccentric presses, and it has been suggested that punch position in a rotary can be 'assumed' rather than measured (Oates and Mitchell, 1990).

Compaction simulators

Since patterns of punch movement differ from press to press, it is an attractive proposition to have a machine that can simulate any type of press. The tablet-press simulator is essentially a hydraulic press, movement of the platens of which can be made to follow a predetermined path with respect to time. This path is designed to imitate the patterns of punch movement of a specific press operating at a specific speed. The die is usually filled by hand with a weighed quantity of solid. Therefore, only small quantities of raw material are needed. However, tablet-press simulators are extremely expensive. Much of the expense arises from the need to move relatively large amounts of hydraulic fluid rapidly and precisely.

A cheaper alternative to the simulator is a motorised hydraulic press, though this has two limitations. The punch speed is constant (which is not the case in tablet presses) and it is much slower than the punch speeds used in most presses. However, it is noteworthy that many workers with a simulator also opt for a constant punch speed, often referred to as a 'saw tooth' profile, even though, presumably, they have the option of a more complex speed profile.

Capsule-filling machinery

It is surprising how little work has been carried out on the instrumentation of capsule-filling machinery, despite the popularity of the capsule as a dosage form, and the fact that in much of this equipment the same two parameters of force and movement are important. There are potentially two main problems. The forces are much lower than in tablet presses, being at most a few hundred newtons rather than tens of kilonewtons. Hence a more sensitive measuring system is needed. Secondly, for reasons of signal stability, transducers must be fixed to a 'massive' component of the machine; otherwise distortion will ensue. These positions are readily available on a tablet press, but are not so abundant on capsule-filling machinery. A further complication is that there are two distinct types of capsule-filling equipment, dosating tube and dosating disk, the filling mechanisms of which differ. Solutions to instrumentation challenges in one type might not be applicable to the other. Both dosating tube and dosating disk equipment have been simulated.

Instrumentation and computers

With the availability of cheap computing power, the use of computers for the acquisition, storage and manipulation of compression data is a natural progression. Virtually all transducers used in tablet-press instrumentation give out electrical signals that can be converted by appropriate circuitry to a voltage. However, the transducer gives out an analogue signal, which must be converted to a digital signal before it can be processed by the computer.

It must be stressed that it is perfectly possible to have an instrumented press without a computer. Also the availability of suitable software must be considered. The use of spreadsheets such as Excel can be invaluable here.

Methods of interfacing a computer to a tablet press or capsule-filling equipment are described in Chapter 8.

Instrumentation packages

Only a few years ago, if one wanted to instrument a tablet press, it was necessary to fit the transducers to the press oneself, and select suitable amplification and signal-conditioning equipment. This is no longer the case. Many production presses are available with instrumentation built in, primarily for the purpose of automatic weight control leading to automated press operation.

Also available are instrumentation packages capable of being fitted to a press. These typically comprise the transducers, power source, amplifiers, a computer interface and a computer for data capture, storage and manipulation. Setting up is simplified by a 'menu' display on the computer screen. Computer software is available to transform the data received from the transducers into parameters used for characterising the compaction process. Care must be taken that the definitions of such parameters are correct. It is the authors' experience that these parameters are sometimes incorrectly defined, and potential users must satisfy themselves on this score.

References

Armstrong NA, Abourida NMAH (1980). Compression data registration and manipulation by microcomputer. *J Pharm Pharmacol* 32: 86P.

Cole GC, May G (1972). Instrumentation of a hard shell encapsulation machine. *J Pharm Pharmacol* 24: 122P.

de Blaey CJ, Polderman J (1971). Compression of pharmaceuticals. 2: Registration and determination of force–displacement curves using a small digital computer. *Pharm Weekblad* 106: 57–65.

Higuchi T, Nelson E, Busse LW (1954). The physics of tablet compression. 3: Design and construction of an instrumented tabletting machine. *J Am Pharm Assoc Sci Ed* 43: 344–348.

Jolliffe IG, Newton JM, Cooper D (1982). The design and use of an instrumented mG2 capsule filling machine simulator. *J Pharm Pharmacol* 34: 230–235.

Knoechel EL, Sperry CC, Ross HE, Lintner CJ (1967). Instrumented rotary tablet machines. 1: Design, construction and performance as pharmaceutical research and development tools. *J Pharm Sci* 56: 109–115.

Oates RJ, Mitchell AG (1990). Comparison of calculated and experimentally determined punch displacement on a rotary tablet press using both Manesty and IPT punches. *J Pharm Pharmacol* 42: 388–396.

Rees JE, Hersey JA, Cole ET (1972). Simulation device for preliminary tablet compression studies. *J Pharm Sci* 61: 1313–1315.

Schmidt PC, Koch H (1991). Single punch instrumentation with piezoelectric transducer compared with a strain gauge on the level arm used for compression force–time curves. *Pharm Ind* 53: 508–511.

Small LE, Augsburger LL (1977). Instrumentation of an automatic capsule filling machine. *J Pharm Sci* 66: 504–509.

Further reading

Armstrong NA (2004). Instrumented capsule filling machines and simulators. In Podczeck F, Jones BE (eds), *Pharmaceutical Capsules*, 2nd edn. London: Pharmaceutical Press, pp. 139–155.

Celik M (1992). Overview of compaction data analysis techniques. *Drug Dev Ind Pharm* 18: 767–810.

Celik M, Marshall K (1989). Use of a compaction simulator in tabletting research *Drug Dev Ind Pharm* 15: 759–800.

Celik M, Ruegger CE (1996). Overview of tabletting technology. 1: Tablet presses and instrumentation. *Pharm Tech* 20: 20–67.

Hoblitzell JR, Rhodes CT (1990). Instrumented tablet press studies on the effect of some formulation and processing variables on the compaction process. *Drug Dev Ind Pharm* 16: 469–507.

Wray PE (1992). The physics of tablet compaction revisited. *Drug Dev Ind Pharm* 18: 627–658.

2

The measurement of force

Peter Ridgway Watt

Introduction

There are two fundamental approaches to the measurement of force: we can describe these as direct and indirect.

The direct approach is exemplified by the use of a two-pan balance or domestic scales to weigh an object of unknown mass. Placing the object on one pan deflects the balance beam; known masses are then added to the other pan until the system is returned to a state of zero deflection. The restoring force may also be generated by electrical means, but the general principle remains the same. As the deflection is returned to zero for each operation, the system is largely independent of the elastic properties of its components.

Weighing by a version of this technique was for many years the most precise operation that could be carried out in a chemical laboratory, though it was a notably time-consuming procedure. In the double-weighing procedure suggested by Gauss, the unknown mass was first weighed as carefully as possible on one side of the balance. Then it was transferred to the other pan and weighed again, thus compensating for any asymmetry of the beam. The declared mass was taken as an average of the two readings. At present, the National Physical Laboratory precision balance is capable of comparing kilogram masses to the nearest microgram, or one part in a thousand million, though the operation still calls for a great deal of patience, care and time. In spite of their achievable accuracy, direct methods of this kind are clearly unsuited to continuous or near-continuous measurement. Force balancing calls for a system that can be restored to zero deflection for each measurement, and this effectively rules out its application to tablet presses and associated machines, where the forces are large and vary rapidly. However, the balance method is very important for the static calibration of gauges and transducers, as we shall see later. Force balance systems were analysed in 1975 in a monograph by Neubert, and in principle have hardly varied since then.

The indirect approach is that in which forces are inferred by the effect that they have on deformable objects. This method is exemplified by the spring balance, where the gravitational force applied by an unknown mass extends a spring whose extension is read on an arbitrary scale to represent the mass. Within certain limits, the deformation of an elastic solid is accurately proportional to the applied force, so if we can assess the deformation we can infer the force that produced it.

Carrying out such a measurement on a working machine need not affect its normal operation, provided that appropriate systems are used. Indeed, a basic principle of good instrumentation is that it causes only minimal disturbance to the characteristics being measured. Nevertheless, machines such as tablet presses are constructed as more or less rigid supports for their operating elements, and in order to infer the forces acting within them, it is necessary to measure very small dimensional changes, or 'surface strains'.

The measurement of surface strain is normally achieved by the use of 'strain gauges' attached to the component under examination, although many other methods exist, and we shall see more of these in succeeding pages. A review of the techniques available for force measurement was published in 1982 by Erdem, and although this particular article is now over 20 years old, once again we can say that the essential principles have not changed materially since then.

Strain measurement

Stress and strain

We have already noted that mechanical stresses produce dimensional changes in solid materials. Stress is conventionally measured in terms of applied force per unit area, and it may take various forms. Thus, 'tensile' stress is a force that tends to elongate or stretch a specimen, while 'compressive' stress produces a corresponding shortening of the specimen; stress may also result in 'torsion', when the specimen is twisted, or 'shear'. Shear stresses are always in the plane of the area being considered and are at right angles to compressive or tensile stresses.

Strain is defined as the proportional dimensional change resulting from the applied force. For example, if a wire were to be stretched until its length increased by 1% of its original value, it would be said to exhibit 1% strain. Strain is often signified by the Greek letter ε. In most practical applications, very much smaller levels of strain are to be expected, and the usual working unit is the 'microstrain', $\mu\varepsilon$, or proportional part per million. Both stresses and strains are vector quantities, with direction as well as magnitude, and this has to be borne in mind when measuring systems are devised.

Hooke's law

Hooke's law tells us that the extension of a spring or an elastic solid is proportional to the applied load, up to the so-called 'proportional limit'. For many steels, this corresponds to an extension of approximately 0.1% from the original length. If such a load is removed, the solid will return to its original dimensions; in other words, it exhibits 'elastic recovery'. Beyond the proportional limit, there may still be some recovery but the relation between force and extension ceases to be linear. At and beyond the 'elastic limit', the material no longer returns to its original dimensions when the applied force is removed, some deformation having become permanent. For most engineering materials, the proportional limit and the elastic limit are virtually the same. Finally, as the load increases to the 'yield point', often in the region of 0.2% elongation, there may be a sudden increase in strain without any further increase in stress. In other words, catastrophic failure ensues (Figure 2.1).

Young's modulus

Within the proportional limit in tension or compression, the ratio of stress to strain is nearly constant and is defined as the modulus of elasticity, or Young's modulus, which is often signified by E. It is important to know the E value of a component that is to be gauged, as it enables us to calculate the likely dimensional change that will result from a given force. If a specimen of length L and cross-sectional area A is subjected in tension to a force W, thereby producing an extension e, then Young's modulus is given by $WL/A\,e$.

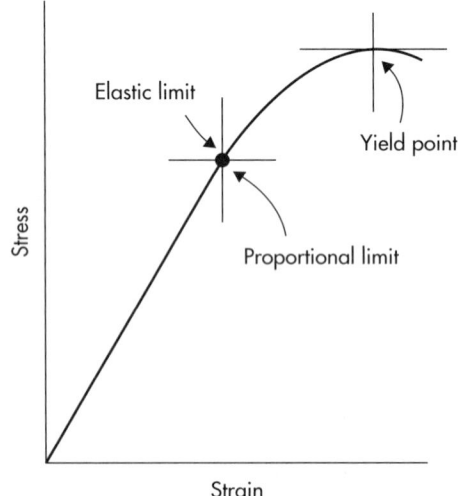

Figure 2.1 Hooke's law: the progressive extension of a metal specimen under tension from the proportional region to ultimate failure.

The approximate value of E for mid-range steels is $210\,\text{GN}\,\text{m}^{-2}$, while that for aluminium is $100\,\text{GN}\,\text{m}^{-2}$, though in both examples, the values of E are not precisely constant since they are affected by changes in temperature. As a rule, higher temperatures produce lower values of E.

Poisson ratio

When a length of solid material, such as a metal rod or wire, is placed under tension, its length increases, and at the same time its diameter almost always decreases. There are some anomalous materials, known as 'auxetics', that expand under tension, but most of these are polymer foams that fall outside our current areas of interest.

The ratio of lateral contraction per unit breadth to longitudinal extension per unit length is defined as 'Poisson ratio', a dimensionless value often signified by the Greek letter nu (ν). If the volume of the metal remained constant, it would be a simple matter to calculate the change in diameter for a given change in length. In practice, the original volume is not preserved, and different materials exhibit a range of values for the Poisson ratio. Typical examples may be found in most engineering reference books, but the reader should be warned that the published values, although similar, do not always agree precisely for all materials. The Poisson ratio for zinc, for example, has been given variously as 0.21 and 0.331; that for cast iron as 0.25 and 0.21. If the precise figure might be critical, it is always advisable to obtain a value from the metal suppliers.

Gauge factor

As a result of changes both in its length and in its diameter, the electrical resistance of a stretched wire increases, and measuring that resistance can, therefore, provide information about the applied force. The proportional change in resistance for a given proportional change in length is called the 'gauge factor' for that material, often indicated by the symbol K, and is a measure of its strain sensitivity.

Interestingly enough, it is not possible to predict the gauge factor directly from Poisson ratio alone, since most resistive alloys show additional changes in their bulk resistivity when stressed, and this effect increases their strain sensitivity to some extent. Without such an additional effect, the gauge factor for a typical alloy would be around 1.6, whereas in practice most commercial foil gauges have a factor slightly greater than 2.0. So an elongation of, say, ten parts per million (ppm) would result in a resistance change of approximately 20 ppm.

Strain gauges

Various devices have been developed for the assessment of strain in structures and mechanisms, but the term 'strain gauge' when used without qualification is generally taken to imply the electrical resistance strain gauge, in which small changes in length produce accurately and reproducibly related changes in the resistance of a conductive element.

The development of electrical resistance strain gauges

The effect of strain on the resistance of a wire element was observed by Lord Kelvin as early as 1856, and the principle was employed in the early 1930s by Charles Kearns to measure strains in propeller blades, using carbon composite resistors. In the period 1937–8, Simmons and Ruge independently experimented with fine resistance wires as the basis for a strain-gauging device. Later, following developments in instrumentation during World War Two, it became possible to manufacture wire-based strain gauges as a commercial proposition. For convenience in measurement, it was considered desirable to use wires with a resistance of at least 100 ohms (Ω), though there was a practical limit to the fineness of the wire that could be made and handled without damage. The first gauge elements were, therefore, made from a comparatively long thin wire that could achieve the required resistance. This was wound into a flat grid pattern so that the resulting gauge could be of reasonably small dimensions. In these early examples, the wires were usually held by adhesive between two layers of insulating paper. However, they were difficult to make and apply; they were not particularly reproducible, and their thickness meant that they were not ideally adapted to following small movements of the surfaces to which they were attached.

All this changed in the 1950s with the development of photochemical etching techniques for the manufacture of printed electronic circuits. Nowadays, gauges are mass produced from very thin metallic foil on an insulating backing sheet. During manufacture, the foil is coated with a film of light-sensitive 'resist', which is then exposed to light – usually ultraviolet – through a photographic negative of the grid form. This induces local hardening of the resist, after which the soluble material can be etched away to leave the required pattern.

Characteristics of metal foil gauges

1. The foil gauge is essentially flat, and lends itself well to adhesive bonding.
2. The foil elements are thin: typically in the range 0.003–0.005 mm. It is, therefore, possible to produce grids of small area that nevertheless have electrical resistances in the range 100–5000 Ω. This is a convenient range for measurement.
3. It is easy to incorporate alignment marks into the gauge pattern, so that the gauges can be fitted into precisely known positions.
4. The flat construction of the gauge ensures that thermal gradients between the gauges and the surface to which it is attached are kept to a minimum.
5. The method of production provides for a very high degree of reproducibility between gauges, both in terms of resistance and in physical dimensions.
6. Most strain gauges are intended to measure strain in one direction, while remaining insensitive to transverse strains. The etched foil gauge can be formed with narrow, high-resistance, elements along its principal axis; the transverse elements can be made much wider so that they are of relatively low resistance.
7. Chemical etching introduces little or no strain in the foil from which the gauges are made. Etched foil elements, therefore, show good stability and minimal drift.

Since etched foil gauges have a large surface area, they may be susceptible to oxidation and insulation leakage at elevated temperatures: but these effects will not be apparent below 200°C, and for applications in tablet and capsule studies they can reasonably be ignored.

Gauge construction

As has been noted, the gauge element is in the form of a metal grid. The metal itself is usually an alloy, and it will be chosen for some specific properties since no single material is universally applicable. However, the two most widely used materials are copper–nickel alloy, or 'Constantan', and nickel–chromium alloy, Karma or 'K' alloy. Copper–nickel alloy grids can be used over the temperature range −75°C to +175°C, although they may exhibit slow changes in resistance when held for long periods over 70°C. Nickel–chromium has a somewhat greater range, extending up to at least 300°C, and also has slightly higher strain sensitivity.

Figure 2.2 shows a typical grid form. It can be seen that the gauge consists of many narrow parallel elements joined by end loops to produce an electrically continuous circuit. The end loops are relatively wide, and not only give reduced transverse sensitivity but also help to provide good mechanical attachment between the gauge and the substrate. Large pads are provided for the connection of electrical leads, and there are alignment marks that assist with gauge positioning. The 'effective length' of the element is defined as the distance between the end loops of the grid.

The fine grid makes it possible to achieve a suitably high resistance in the gauge element. This configuration provides maximum strain sensitivity along the measurement axis while remaining relatively insensitive to strains on the transverse axis.

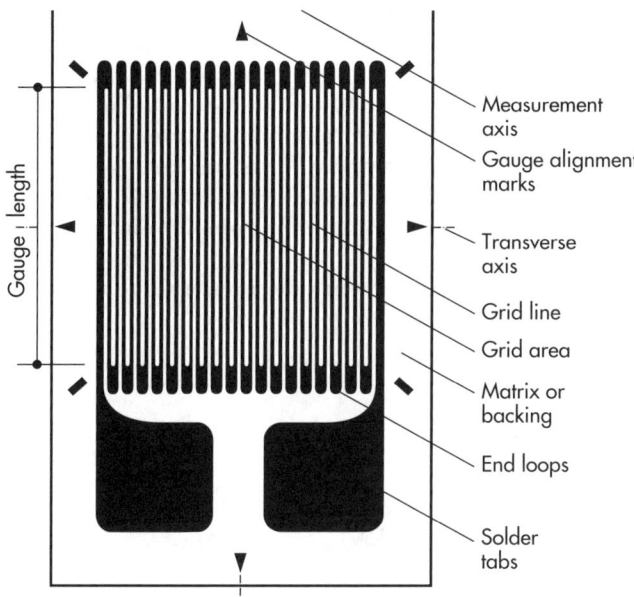

Figure 2.2 Foil strain gauge terminology.

There are some specialised gauge alloys that are not readily soluble in common etching fluids, and these materials may be punched out by the technique known as 'fine blanking', with an accurately ground punch and die assembly. However, for all practical purposes they can be ignored in our particular areas of interest, which in general involve the measurement of strain levels usually below 300 µε in materials at or near room temperature.

Gauge backing

Various insulating materials, such as fibreglass epoxy and resin-bonded papers, have been used to support strain gauge elements, though nowadays cast polyimide films are perhaps the most popular for general work, particularly when the gauges are to be used at moderate temperatures. Clearly any backing material must retain high insulation resistance at the working temperature, be dimensionally stable and transmit strain efficiently from the substrate to the gauge. For maximum efficiency in this last respect, the backing must be as thin as possible, typical values for thickness being around 25 µm. The sensing grid may be open faced, while some gauges are offered with a protective upper encapsulating layer of approximately 13 µm.

It is evident that gauges of this light construction will not affect the mechanical performance of any large structure to which they are bonded; however, they may have more effect on small components of about their own size. In designing instrumentation for the estimation of very weak forces, therefore, it is useful to bear in mind that the gauge and its adhesive can add materially to the stiffness of a very thin component. A final requirement is the ability for the gauge backing to form a strong adhesive bond with an appropriate cement.

Gauge configuration

The gauge grid is designed, as we have seen, to respond mainly to strain along one direction: normally the axis of principal strain. However, it is often very useful to be able to measure the pattern of surface strains on an object, and to facilitate this it is convenient to use composite or 'rosette' gauges, with two or more grids on a common backing. Figure 2.3 shows a selection of different forms that are available.

In the simplest example, two similar grids are mounted with their axes at right angles to each other. Double gauges of this form are often used to provide temperature compensation for one active gauge, the second gauge giving a smaller signal along the other axis. Since this smaller signal is derived from the transverse movement of the gauged surface, it is a function of Poisson ratio for the material involved, and the gauge is, therefore, often described as the 'Poisson gauge'.

There are many other common configurations, apart from the double gauge. For example, there is the three-axis rosette with symmetrical arms either at 0–60–120° to each other (delta) or 0–45–90° (rectangular), the latter being the more common configuration. These can be used in an investigative mode to determine the exact direc-

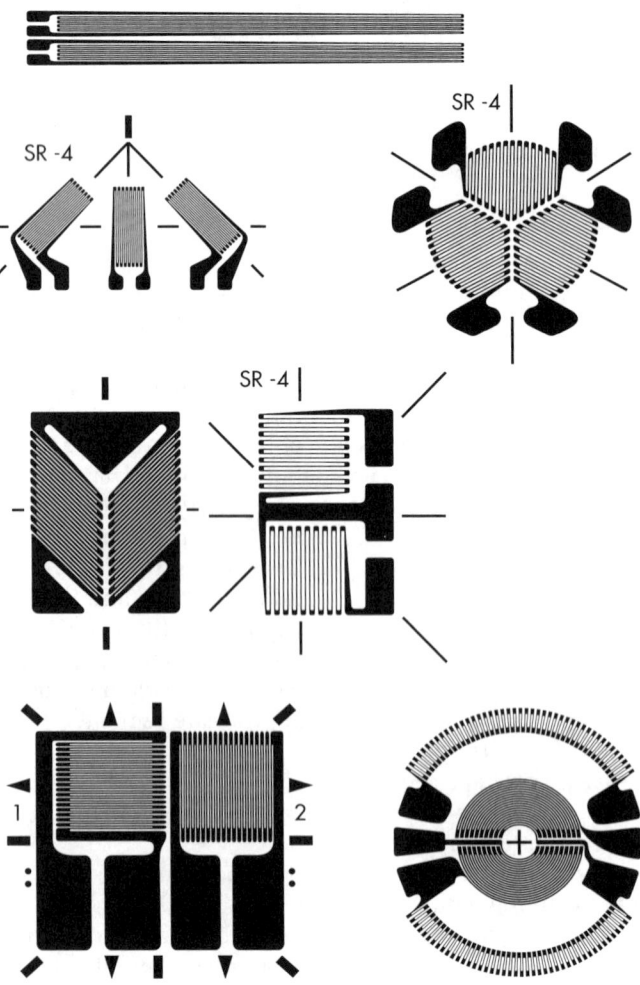

Figure 2.3 Resistive strain gauges are available in a variety of configurations for specific applications.

tion of the principal strain on a surface; the method of manufacture ensures that the grids are not only well matched for their electrical properties but also have accurate angular positions. There are also linear arrays of grid elements, which again can be used on a temporary basis to locate the best site for subsequent attachment of a permanent gauge, and circular gauges intended to fit the diaphragms of pressure gauges.

Electrical characteristics

We have noted that the strain sensitivity of a resistive gauge is indicated by its gauge factor, K, and that foil gauges in general have a K value that is close to 2. This apparently simple picture becomes less simple on closer examination, since the value of K is not invariant over the range of working conditions. In practice, gauge factors vary with temperature, and to some extent with the strain level itself.

Temperature variation of the gauge factor

The two most common gauge alloys, namely Constantan and 'K' alloy, have gauge factors that are almost linearly dependent on their temperatures. That of Constantan increases at approximately 1% per 100°C, while that of 'K' alloy decreases at a similar rate, as shown in Figure 2.4. As with many aspects of instrumentation, the picture becomes even more complicated on further study: a rise in temperature, for example, not only affects the gauge material but it also changes the elastic modulus of the structure being gauged, so that a given force is likely to produce a greater strain.

Modulus compensation

Gauges can be heat treated to give controlled rates of temperature variation in order to balance the modulus changes of a given substrate. The so-called 'modulus compensated' gauges then have an output that is almost independent of temperature when bonded to that particular substrate.

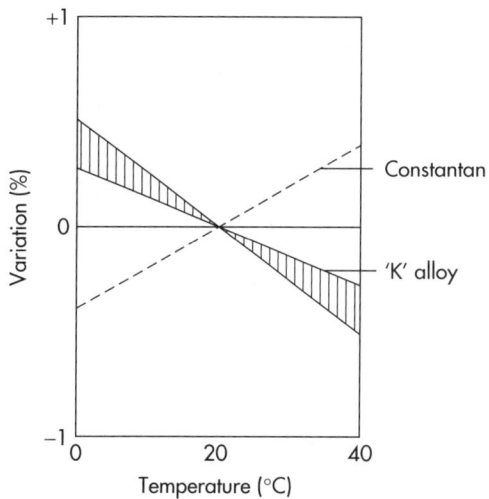

Figure 2.4 The variation of the gauge factors of Constantan and 'K' alloy with temperature.

Non-linearity of gauge resistance with strain level

Linearity of response is one of the desirable characteristics of a gauge alloy. Nevertheless, some degree of non-linearity does exist in most. A review by Chalmers (1982) quoted a deviation of between 0.05 and 0.10% of the maximum strain for a properly applied gauge. Larger deviations are only likely to be found if the gauge is taken up to strain levels of such magnitude that they begin to approach the elastic limit of its alloy. For much of the work involving strain gauge measurement, repeatability may be considered more important than linearity, since calibration can provide a reasonably accurate final figure. Linearisation techniques, involving computer manipulation of the original measurements, are normal industrial practice.

Temperature coefficient of gauge response

When a gauge and its substrate are bonded together, the gauge assumes the same temperature as the substrate for most practical purposes. If the gauge and the substrate have different coefficients of expansion, then any changes in temperature will set up strains in the gauge and will produce misleading signals. These will be read as 'apparent strain' (more recently termed 'thermal output') in the system and can be very large in terms of the expected signal. For example, a nickel–chrome gauge bonded to a mild steel bar may exhibit an apparent strain of as much as $1000\,\mu\varepsilon$ for a temperature rise of 40°C above ambient. For a measuring system capable of resolving as little as $0.10\,\mu\varepsilon$ under the best conditions, this would introduce a substantial error indeed.

Gauge compensation

Interestingly enough, it is possible to adjust the coefficients of expansion of certain strain gauge alloys in order to match those of common engineering metals. When the alloy is rolled into the thin foil that is needed for gauge construction, it becomes work-hardened. It can be taken from the fully hardened state to a fully annealed – or softened – state by controlled heat treatment, and during this process its expansion characteristics progressively change. With appropriate attention to the time and temperature of annealing, the alloy can be matched to a variety of different substrates over a temperature range of up to 25°C on either side of ambient. Gauges made from alloys that have been heat treated in this way are known as 'self temperature compensating' (STC) gauges. Commercial gauge manufacturers offer STC gauges to suit a range of different substrates, including aluminium, mild steel, stainless steel, concrete, titanium, and some plastics: in effect, most common structural materials. The gauges may be ordered as needed to match a given named substrate, or alternatively to match a given coefficient of thermal expansion. For example, a material with a measured linear coefficient of $16 \times 10^{-6}\,°C^{-1}$ could be matched well with a gauge such as the Showa N11MA 5 120 $\underline{16}$ 3 LW, where the underlined numeral indicates the average expansion coefficient in parts per million per degree centigrade. Other makers provide a similar facility, though it is worth noting that, as many of the gauge suppliers are either from the USA or have strong affiliations there, the temperatures quoted in their documentation may be in the Fahrenheit scale. The related code numbers may be misleading if this is not borne in mind.

Figure 2.5 shows how the 'thermal output' of a typical foil gauge can vary as a result of its heat treatment. The compensation achieved with an off-the-shelf STC gauge should be within 0.3–$0.4\,\mu\varepsilon\,°C^{-1}$. Where there are accurate and independent means of checking the gauge temperature, the data-logging system may be programmed with a compensating algorithm.

Dimensions of foil gauges

The photographic process that is used to form the etch resist pattern during manufacture of the gauges is capable of high resolution, and in principle could be used to form grid lines down to a few micrometres in width, or up to any larger size as required. For convenience in measurement, it is useful to be able to produce gauge elements whose electrical resistances fall in the range 100–$1000\,\Omega$, or occasionally up to $5000\,\Omega$ in some applications, and this requirement means that gauges are usually no smaller than $0.20\,mm$ in effective length. Even at this size there may be problems in ensuring that the strain to be measured is transmitted completely from the substrate to the element, and unless there is any over-riding need to have an extremely small gauge it is more usual to work with, for example, a 1.5–3 mm version.

Measuring range

It has already been noted that the levels of strain likely to be met in tablet presses or capsule fillers are relatively low, probably not exceeding 300–$500\,\mu\varepsilon$ at most. This is well below the rated capability of modern gauges. Some Constantan foil gauges, for example, are advertised as being safe for operation at strain levels of up to $50\,000\,\mu\varepsilon$, provided that the gauge length is greater than 3.0 mm. The limitation is not that of the backing, since polyimide is very flexible, but of the foil itself. There are special-purpose gauges with even greater capabilities, but they are hardly likely to be needed for applications in

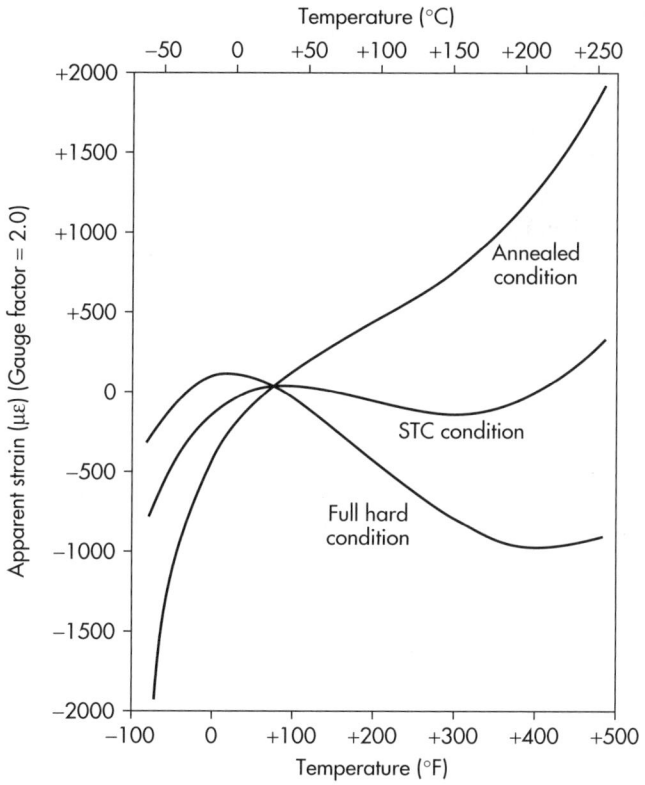

Figure 2.5 Apparent strain (or thermal output) characteristics as affected by temperature. Controlled heat treatment gives a gauge whose expansion characteristics match those of a given substrate (self temperature compensated (STC) gauges). (Courtesy of Vishay Measurements Group UK Ltd.)

tablet press instrumentation, and they are not discussed here.

Gauge life

It is important to consider what is meant by the term 'gauge life', since there are several components involved: the fatigue life of the element is one factor, while the adhesive bonding will be another. Moreover, the electrical connections themselves represent yet another region for possible failure in a dynamic system.

At a level of less than 500 με, most present-day foil gauges should be able to withstand many millions of operating cycles without measurable change. There is some falling off in performance above 1000 με, but such high levels may be thought unrealistic for the field covered by this book. Vishay Measurements Group, for example, quote an estimated 10^8 cycles of operation for a Constantan gauge under dynamic stress with an amplitude of ±1200 με: the same gauge at the increased level of ±1500 με gave 10^6 cycles. It should be noted that these figures refer to the number of operating cycles during which the gauge performance remains within stated limits – in this example, a 100 με zero offset – rather than its lifetime to catastrophic failure. A gauge will normally remain within limits for a longer time when it is used dynamically than when it remains under quasi-static load. Under good conditions, the electrical leads should last throughout the life of the gauge element, but this does depend on careful attention to detail during the wiring procedure.

Gauge resistance

We have noted that the early gauges, which were based on fine wire elements, usually had resistances in the region of 120 Ω, and a certain amount of instrumentation developed around that particular value. However, the photolithography method used to make gauges nowadays is capable of producing grids with a considerably extended range of resistances, and higher values, from 350 to 5000 Ω are in common use.

The value selected for a given application will generally represent a compromise between opposing considerations. If the gauge resistance is high, on the one hand, then it is possible to apply a higher voltage to the measuring bridge, and hence to extract larger signals for a given strain level. On the other hand, low ohmic values are less affected by electrical noise and insulation leakage, although leakage should be minimal at near-ambient temperatures. The 120 Ω gauges are still manufactured, but the 350 Ω variety are widely used, and those of 1000 Ω and above often appear in transducers of various types. The use of high-resistance gauges in portable equipment, which may be battery driven, can help to minimise the current drain.

Semiconductor gauges

The electrical resistance gauges discussed above have alloy grids, and it is mainly the changes in length and cross-section that provide corresponding changes in resistance when these gauges are strained. In other words, there is a largely volumetric effect. However, there is a further class of gauge that uses semiconductor technology and has much greater gauge factors, extending from 50 up to 200. They are usually made from doped silicon wafers, and their resistance change results from a heavily stress-dependent change in specific resistivity. It is possible to dope the silicon in modes that give either positive (P-type) or negative (N-type) gauge elements. Of these, the N-type can be made in temperature-compensated forms for a variety of substrates. It is also possible to make a complete four-arm bridge with N- and P-type gauges.

Since the specific resistivity of the semiconductor materials is much higher than that of the normal gauge alloys, it is not necessary to fabricate semiconductor gauges as fine grid structures. They are usually made as narrow strips, sliced from the silicon crystal, with a thickness of perhaps 0.01–0.05 mm and lengths varying from 0.75 to around 6.0 mm. Figure 2.6 shows a typical unbacked semiconductor gauge, with a scale in millimetres indicating its size. The connecting wires are made of small-diameter (30–40 μm) gold wire and are very soft and easily damaged. Unbacked gauges are extremely fragile and require great care in their application. Backed gauges are easier to apply.

Semiconductor gauges are available in a wide range of nominal resistances from 75 Ω up to 1000 Ω. Their fatigue life is not substantially different from that of foil gauges. At a dynamic load of ±1000 με, the expected life is over 10^8 cycles.

At one time, these gauges were considered to have poor stability, linearity and reproducibility,

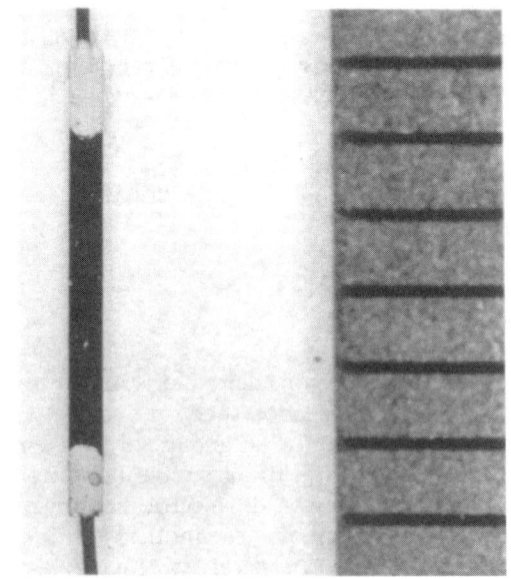

Figure 2.6 The semiconductor strain gauge. Because the electrical resistance of these materials is much higher than that of metal foil, these gauges are made up of single elements rather than grids. The scale on the right of the illustration is in millimetres.

though all these have been significantly improved, partly by better doping technology and partly by the use of automated sorting systems that can find matched sets of gauges from a production batch. It is claimed that a complete four-arm bridge of this type can have good temperature compensation, in the region of $\pm 0.015\%\,°C^{-1}$, but in general it has to be said that conventional alloy gauges still appear to be superior in terms of sensitivity to temperature variations and drift. However, our particular areas of interest in instrumentation do not usually involve large temperature excursions, so for many applications these points may not be critical. The major disadvantage of the semiconductor gauge appears to be the fact that its response to strain is quite significantly non-linear; consequently, the gauge factor varies appreciably with the strain level. It is suggested by the National Instruments Company that a semiconductor gauge may have a factor of -150 with zero strain, dropping non-linearly to -50 at $5000\,\mu\varepsilon$. It is, therefore, necessary to apply a correction curve to the raw data.

Semiconductor gauges have been used by, among others, Britten *et al.* (1995) in the construction of a capsule-filling machine simulator.

Sputtered thin-film gauges

There is an additional method of production that has been applied to the manufacture of resistive gauge elements; this uses sputtered thin-film technology that was developed for industrial electronics. An insulating layer is first evaporated on to the surface that is to be gauged. Then the resistive grid is directly deposited by sputtering under high vacuum. Thin-film elements are used in transducer assemblies such as pressure gauges and load cells, and the method of manufacture makes it possible to generate a complete four-arm bridge in one operation. These elements have also been applied to small cantilever beams, which are commercially available with a range of load capacities from 0.5 to 50 N. Since they do not use conventional adhesives to bond the gauge elements to their substrates, problems of adhesive creep and drift are minimised. Their long-term stability is said to be extremely good, though their application is naturally limited to use on relatively small components that can be conveniently loaded into a vacuum chamber.

Gauge selection

It will be apparent that it has only been possible to give a simplified picture of current strain gauge technology. However, the requirements of tablet press or capsule work are not particularly stringent, most operations taking place at or near room temperature, and the levels of strain to be measured do not usually exceed $300\,\mu\varepsilon$, so for most of our applications the standard Constantan or 'K' alloy gauges should be quite sufficient. In this context, we would strongly recommend that the gauge suppliers are best placed to give advice on the choice of gauge for a given application. The theory and practice of electrical resistance strain gauges were outlined in a series of articles by Mansfield (1985) and a reference work by Pople (1979) appeared in the *BSSM Strain Measurement Reference Book*, published by the British Society for Strain Measurement.

Gauge attachment

Years ago, any research group embarking on the instrumentation of a press or some allied equipment would have been obliged to carry out the gauging themselves. This is no longer to be recommended for any permanent instrumentation, although it may well be useful to carry out investigative gauging in-house, just in order to locate a suitable gauge position.

In Chapter 3, we shall see the detailed work that is necessary to produce an installation of professional quality. There are now quite a few organisations that offer gauge installation services of this standard, and some are listed in the Appendix at the end of the book. Once a suitable site for the gauges has been identified, our advice to the reader would be to leave permanent installation to the experts.

Siting strain gauges

If an unfamiliar piece of equipment is to be provided with strain gauges on a permanent basis, then it will be important to know where, and in what orientation, the gauges should be sited. We have already seen that tablet machines are normally designed for rigidity in use, so that little dimensional change occurs during a compression cycle. If a machine is to be gauged, it is, therefore, essential to site the gauges to the best advantage. It may be argued that the optimum positions on familiar machines are already well known, but to some extent we have attempted to start from first principles, assuming no prior knowledge, and illustrating the means that are available to identify gauging sites on any structure.

Stress analysis

There are two general methods of approach, namely modelling and direct measurement, each with a number of variations.

Modelling methods of stress analysis

Finite element analysis
Finite element analysis is a purely mathematical approach in which the surface of the object to be analysed is treated as if it were formed from a number of simple geometrical figures such as triangles, rectangles and so on, each of which is capable of individual stress analysis. Such elements as the thin plate, the cylinder and various types of loaded beam were analysed in great detail during the nineteenth century, and their behaviour under load can be predicted with considerable accuracy. The inter-relations of these elements are calculated, and the response of the object as a whole can then be deduced. The method becomes progressively more accurate as smaller and smaller elements are taken, though the calculation of the interactions naturally becomes more involved.

One important aspect of the finite element method is its ability to handle both real and imaginary structures. Provided that a surface can be described in sufficient detail, it can be analysed. It is, therefore, possible to estimate the likely effects of design changes in a structure or in a component without the expense of its manufacture and modification. In the study of specific machines, such as tablet presses, design changes may clearly not be permissible, but it is certainly possible to investigate, by this technique, the effect of siting gauges at different positions on a given machine. Modern finite element analysis packages can be run successfully on desktop computers and can be configured to show lines of equal strain or to display areas in false colour to give enhanced visualisation of a strain distribution pattern. Finite element analysis was used by Yeh *et al.* (1997) to determine the optimum positions of strain gauges in an instrumented die used for the measurement of die wall stress.

The over-riding consideration in any modelling system is, of course, that the model must be adequately representative of the real structure, taking into account all dimensions and materials, and any internal cavities, blowholes, etc., that might affect the mechanical behaviour of the structure. In this context, it is worth noting that reports on the Tay bridge disaster of 1879 found that some of the castings used in the construction of the ill-fated bridge had blowholes up to 2 inches deep (Lee, 1981). These holes had been disguised by the application of a black paste, which carried the French name of 'Beau Montage' but unsurprisingly was known in the UK as 'Beaumont Egg'. It had no mechanical strength and was applied as a purely cosmetic measure. Clearly here was an example of a structure in which finite element analysis might easily have given a dangerously misleading result.

Photoelastic models
There are at least two ways in which modelling can be used in this context. Firstly, some region of particular interest can be reproduced at full scale and tested in isolation. This may be quite useful if the original machine is complex, large and heavy. Secondly, and perhaps more usually, the whole structure can be reproduced as a small-scale model, with sufficient detail to ensure that its behaviour simulates that of the real machine for the purposes of the test. In each

case, the model will then be subjected to applied forces that have themselves been scaled to correspond to those in the original machine and will produce appropriate dimensional changes or strains in the model. If the model is made of transparent plastics material, it can be examined by transmitted light. An unstressed plastics model will not modify the light passing through it to any marked extent, but physical stress may cause the material to exhibit some degree of birefringence. If such a model is set up in front of a source of polarised light, and is examined through a suitable optical analyser, any stressed areas will be seen to show visible fringe patterns. White light produces coloured fringes, while monochromatic light can give clear, sharply defined fringes that are more useful for measurement. The effect can conveniently be seen by positioning the model between two sheets of 'Polaroid' with their optical axes crossed at right angles (Figure 2.7). The addition of a quarter-wave plate generates colours in the fringe system so that variations across the model are more easily seen. Suitable equipment for this operation is usually to be found in any glass-blowing laboratory, where it is used to detect residual stress in glass components before and after annealing.

The production of coloured fringes can be demonstrated quite easily in models made from 'Perspex' or similar types of acrylic plastic, but for careful analytical work it is customary to use specially developed polycarbonate or epoxy resins. One feature of these specialised materials is their ability to retain stress patterns after the stress itself has been removed. This retention can be facilitated by raising and lowering the temperature in a prescribed controlled way so that the pattern is 'frozen' into place for subsequent study. In this technique, the model is heated in a sufficiently large oven and is stressed while at an elevated temperature. It is then allowed to cool, still under stress, to room temperature. It is then stable enough for further analysis.

It must, of course, be remembered that stress distributions are usually three dimensional and are not always easy to determine accurately within a complex scale model. For the purposes of measurement, therefore, the model is usually cut into thin laminar sections that can be treated as two-dimensional forms. The cutting process must, naturally, be carried out without alteration of the frozen stress pattern. After sectioning has been accomplished, any further studies are likely to require a completely new model, so detailed analyses of complex structures by this procedure can be fairly time consuming.

In spite of its disadvantages, the photoelastic model does provide a simultaneous picture of

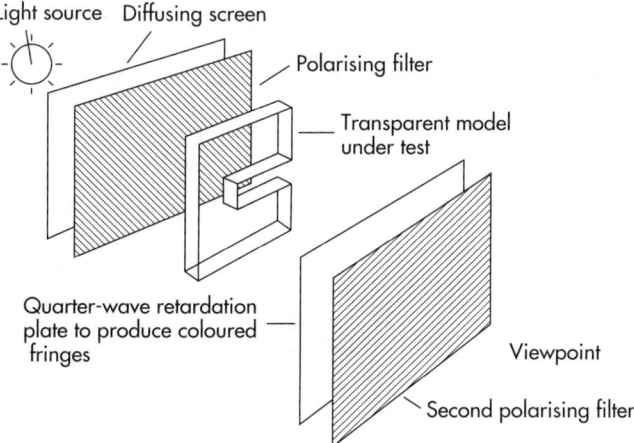

Figure 2.7 Stress analysis in polarised light. A transparent plastics model is placed between two crossed polarising screens. Stresses appear as variations in the transmitted light intensity.

the stresses, both internal and external, associated with a whole structure. If the model is subjected to progressive loading, changes in the fringe pattern can be observed as they happen. An increase in the stress level generally produces a corresponding increase in the number of fringes seen, but the particular spacing between fringes will depend on the composition of the polymer used for the model, and on its temperature.

Resins for modelling

The specially formulated casting polymers have higher strain sensitivity than that of the commercial sheet plastics and will, therefore, exhibit more fringes for a given load (Figure 2.8). Models cut from sheet 'Perspex' can, nevertheless, provide useful information on a rather more qualitative basis. They are best used when the component to be represented is itself relatively flat, for example a plate cut from sheet metal. In such cases, the stress pattern is likely to be nearly two dimensional and will be less confusing than that from a complex, solid, object.

It is, however, important to be aware of the possibilities of misinterpretation. Commercial sheet plastics often have internal stresses resulting from the method of manufacture, and additional stresses may be introduced by the subsequent cutting procedure. An example of this was seen in the writer's laboratory during the production of a simplified model of the upper arm for an F3 single-punch tablet machine. The outline of the arm was drawn on the protective paper covering a sheet of 5.0 mm 'Perspex', and the model was then cut from the sheet by a powered bandsaw. After cutting, the model was examined in a glassblower's polarising 'strain viewer' and was found to show a repeated strain pattern around the cut edges, with a highly stressed region occurring every few centimetres. This appearance might have led to wrong conclusions in a photoelastic analysis, but in fact it resulted solely from the method of cutting. Bandsaw blades are made from a narrow-toothed strip of metal welded into a continuous loop. The physical properties of the bandsaw blade are modified by the presence of the weld, and small projections may be introduced at the junction itself. In this particular instance, the welded joint was striking the 'Perspex' sheet at each revolution of the bandsaw blade, each impact causing localised changes in the material.

In addition to the strains produced by impacts, plastics may exhibit more generalised bulk effects. The F3 arm model mentioned above appeared substantially colourless when viewed directly through its front face. When rotated so that it could be seen edgewise, it appeared to show a range of coloured fringes. In fact, these

Figure 2.8 Fringes in a stressed model made from a strain-sensitive epoxy resin. (Courtesy of Sharples Photomechanics Ltd.)

colours resulted from a general internal stress, coupled with the wedge-shaped configuration of the model: different thicknesses showed different colours.

Annealing

If commercial sheet plastics are to be used for simple stress models, they should always be annealed after machining. Controlled heat treatment releases the internal stresses from the material and so they will not confuse the subsequent observation of load-related stress patterns. Information on suitable annealing programmes is available from the suppliers of plastics, but, as an example, it is usually possible to anneal polymethylmethacrylates such as 'Perspex' or 'Plexiglass' by raising their temperature slowly to approximately 85°C for 30 min or so, then cooling them down to room temperature over 1 or 2 h. Alternatively specialist suppliers of photoelastic materials can supply the plastic in a form that is stress free and requires only careful machining. The material is in the form of flat slabs and pre-cast blocks and rounds in various sizes. Sheets are available from 0.25 to 12.7 mm thick and up to 500 mm square.

The photoelastic technique outlined here has been discussed as being one means of determining the most satisfactory site for the installation of strain gauges. However, it has sometimes been applied as a measurement system in its own right. For example, Ridgway (1966) constructed a model die from 'Perspex', used the model to prepare tablets and estimated the die-wall stresses involved by counting interference fringes as the compression proceeded. The changing fringe patterns were recorded on cine film for later analysis. Accurate, quantitative interpretation of photoelastic patterns in solid, three-dimensional objects is not particularly easy to achieve, though the monograph by Frocht, published as early as 1948, is still a useful guide to this field.

Direct methods of stress analysis

It is perhaps worth repeating that both the methods outlined above have the disadvantage of dealing with an idealised concept of the structure being analysed rather than with the actual structure itself. In reality, as we have noted, materials may contain internal inhomogeneities. Methods designed to measure the mechanical performance of a real structure, by comparison, have the merit of including any of its weaknesses in the general picture that is obtained. Some of the available procedures are indicated below.

Photoelastic film measurement

We have already seen that mechanical stress modifies the optical properties of some transparent plastics, as in the analysis of resin models. The physical behaviour of these models is, naturally, that of the plastics from which they are made. However, it is also possible to use similar plastics materials as thin coatings on the real structures that are to be studied.

If the transparent plastic layer is fitted very closely, then its movement is controlled entirely by that of the more massive underlying structure. Interference fringes develop, as before, in the plastics material and may then be observed through a reflection polariscope. Measurements made by this technique can be of considerable value in any study of machine deformation under load but would not usually be attempted by a formulation laboratory: they are normally carried out by specialist companies who have the experience and the equipment to provide a reliable service, such as Vishay Measurements Group.

Investigative strain gauges

One commonly used procedure for stress – or, more accurately, strain – analysis is that of applying relatively large numbers of resistive strain gauges to the accessible surface of the test structure. Since neither the magnitude nor the direction of the surface strains are likely to be known at the outset, it is necessary to use patterns of gauges that can deal with a range of possibilities. Rosette gauges, with three elements at 120° are most useful in determining the directions of principal strains (see above). Another composite form of gauge, sometimes used for investigative work, carries many small-gauge grids arranged parallel to each other in a linear array. In some versions of this gauge, alternate grids are set at right angles to the axis of the assembly so that they provide a long line of compensated half-bridges. Their signals are compared for magnitude, linearity and noise. An example of an investigative strain gauge is shown in Figure 2.9.

Since these gauges are generally used on a temporary basis, it is not normally essential to provide the complete armoury of gauge-protection methods, or to provide strain-relief loops in all the wire connections. Gauges with integral leads, or with large pads for soldering, may be used, and they may be conveniently attached with cyanoacrylate 'instant' adhesives.

If gauges are applied to areas that seem likely to exhibit high strain levels, then the first experimental measurements should make it possible to deduce new gauge positions for a second test, and so on. The optimum positions can thus be approached through a series of iterative steps.

One advantage of investigative gauging over methods such as brittle lacquer testing is that it does provide much greater accuracy and sensitivity at the points measured. Though its main purpose is to identify a site for permanent strain gauging at a later time, it also puts an approximate value on the signal that may be expected from the permanent gauges. It does not give a general contour map of strain distribution, however, without a considerable amount of fairly tedious work and the fitting of many gauges.

Holographic interferometry

When a component is strained, then by definition its dimensions change. Points on its surface will move by small amounts from their rest positions, and that movement can be measured by various optical means. One sensitive optical method for the evaluation of small movements is holographic interferometry.

In one version of this method, a holographic exposure is first made of the component in its unloaded state. The component is then stressed, and a second exposure is made while the holographic film or plate is held in a fixed position. If the component has remained unchanged between the two exposures, a single holographic image results. However, where there has been surface strain, the composite hologram shows a pattern of light and dark fringes. These constitute the interferogram, from which the extent of surface movement can be calculated.

Holographic interferometry has become an accepted method for the study of small deformations in engineering components. It was employed by Ennos and Thomas (1978), for example, to evaluate the distortion of plastics after injection moulding, and it has been used by Ridgway Watt (1981) to observe the recovery of compressed tablets after their ejection from the die.

Various methods can be used for the production of the hologram. The classical procedure normally involves the use of a large optical table, together with a double-beam system for laser illumination. There are useful review articles on this subject from Robertson and King (1983) and Weingartner (1983). There is also a comprehensive monograph by Abramson (1981). The necessity for absolute rigidity of the optical table is such that only a few laboratories are likely to be equipped for the classical methods of holography. It is, nevertheless, quite possible to prepare useful holographic interferograms on

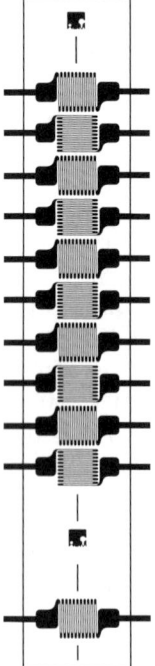

Figure 2.9 Investigative strain gauges. Groups of gauges are applied temporarily to the test surface and their signals compared for magnitude, linearity and noise.

inexpensive, simple, equipment by using the 'Denisyuk' white light method.

White light holography

The Denisyuk method avoids the problems of double-beam operation by using a single laser beam, which acts both as the signal and as the reference. With this arrangement, the mechanical properties of the equipment used are not particularly critical, and it is possible to produce useful holograms from small objects without undue complication. The object to be recorded is supported close to the surface of an unbacked transparent plate or film, such as Agfa Holotest 8E-75-HD. It is illuminated by laser light passing through the emulsion from the far side of the plate. Some of the light is reflected by the object and is returned back to the plane of the emulsion, where it interacts with the original or reference beam, giving rise to a three-dimensional interference pattern. The holographic emulsion records a thin section through this pattern, representing the object as seen from the plane of the film provided that the film is capable of resolving the fine structure of the interference fringes. In practice, if the light source is a helium–neon laser operating at a wavelength of 633 nm, the film must be able to resolve about 3000 lines mm^{-1}. The exposure time for a 2–3 mW laser will be in the range 0.2–10 s, depending on the beam width and the reflectivity of the test object. After development, the resultant hologram can be seen clearly in white light from a point source, or in direct sunlight. Surface movements in and out of plane will result in the appearance of fringes, each fringe representing a movement of half a wavelength of the laser light used.

The holographic camera

The Denisyuk method by itself is limited to objects of a few centimetres in size, but with the addition of a lens system it can be extended somewhat. Rowley (1979) modified a standard 35-mm camera by fitting a mirror immediately behind the normal plane of the film, and mounted a second mirror next to the subject. Laser light, falling on the subject and the adjacent mirror, enters the camera lens and is imaged on to the transparent Holotest 10E-75 film. Some light passes through the film and is reflected by the internal mirror, so producing interference fringes within the thickness of the emulsion. Rowley gives useful practical details, both of the modification to the camera and of the processing methods.

Real-time holographic interferometry

White light holograms can provide a sensitive measure of surface movement, but they are not normally able to display the movement in real time. For that it is necessary to be able to view the subject through a hologram of itself, so that differences between the subject and its image can be observed continuously. In practice this requires that the holographic plate must be developed *in situ*, since any movement would destroy the image. Dudderar and Gilbert (1985) used a dry-process holographic film in a Newport Corporation HC300 'instant holocamera' to achieve continuous observation of developing fringes.

The moiré grid technique

One of the practical limitations of the holographic method noted above is that of the power available from the laser. A subject area of, say, 100 cm^2, lit by a laser of 2 mW, needs an exposure time of a few seconds. Larger areas lit by the same laser would need correspondingly longer times, but these inevitably lead to movement during the exposure, and hence to holograms that are either degraded or non-existent. Higher-powered lasers can be used to illuminate structures that may extend to several metres both in height and in width, but they are expensive and potentially dangerous pieces of equipment. It is, therefore, convenient to have methods of strain analysis that can be carried out in daylight. One such method was developed by Burch and Forno (1975) at the National Physical Laboratory, and is based on the generation of moiré fringes.

The structure to be examined is first provided with a regular surface pattern. Usually, this is achieved by the application of paper that has been printed with a fine half-tone dot screen. The object is, in effect, 'wallpapered' with this material, and it is then photographed (Figure 2.10b). While the camera remains in a fixed position, the structure is mechanically loaded so that the resultant strain is transmitted to the dot screen paper. A second exposure is then made (Figure 2.10a).

It is clear that the small surface deformations of a large structure under moderate load would not be visible in a single photograph under normal conditions. A component stressed to 1000 με, for example, would appear completely unchanged to the unaided eye. Moiré fringes are generated if the two images are superimposed and these emphasise differences between the two (Figure 2.10c). The overall sensitivity of the system depends on the closeness of the dot pattern, and on the capability of the photographic equipment to resolve that pattern. However, since the camera is only required to respond to spatial frequencies corresponding to those near that of the dot pattern, it can be given enhanced resolution at those frequencies.

Burch and Forno (1975) used a slotted mask to optimise the imaging performance of a Pentax 35 mm film camera with an f2 57 mm Takumar lens over a limited range of spatial frequencies in both vertical and horizontal directions. Whereas a conventional iris diaphragm would have a single central hole, and would give a reasonably uniform response to a wide range of spatial frequencies, the mask used in this work had four linear slots, set along the sides of a square. When the camera was positioned at such a distance from the object that the observed dot resolution could be obtained, a value of 600 line-pairs mm^{-1} was reported.

With the modified camera and a finely printed dot screen, it should be possible to observe movements representing approximately 20 με, and to do so in normal white light. At the other end of the scale, the same method could be used for much larger displacements up to 10 000 με, without loss of contrast. It will be apparent that this particular system is only sensitive to in-plane, or 'sideways' movements, whereas the white light holographic system is sensitive to out-of-plane, or 'in- and -out' movements.

Speckle pattern analysis

The masked aperture camera, used for high-resolution moiré fringe measurements, was applied by Forno (1975) to a variation of that technique, in which random marks on the object surface were used instead of the printed grid. Clearly, when the surface to be covered has multiple curvatures, 'wallpapering' becomes more difficult.

In order to generate random marks with a suitably high spatial frequency, Forno (1975) applied retro-reflective paint to the surface of the object. This paint contains fine glass beads, each of which appears as a bright point when illuminated from a source near the camera, and, as before, the displacement of the points between two exposures gives the required information. The application of paint is relatively simple and overcomes the problem of dealing with complex, multi-curved surfaces.

(a)

The dot pattern shows an indistinct distortion

(b)

(c)
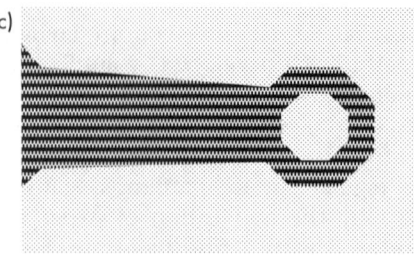

Figure 2.10 The development of moiré fringes. (a) The dot patterns show an indistinct image. (b) The dots on the reference pattern are regularly spaced. (c) Superimposition of (a) and (b) show a more visible image.

The method does require care, but it can give information not only on deformation of structures but also about their absolute shape. It is capable of practical application within the range 10–10 000 $\mu\varepsilon$.

Much work has been done with speckle patterns produced by laser illumination. These, unlike the patterns from dot screens or reflective beads, are not on the surface of the object but are three-dimensional interference patterns resulting from the use of coherent light. They have been studied as means of measuring the magnitude and direction of surface strains, and have produced valuable results. However, like holograms, they are dependent on the availability of laser illumination at an adequate level of brightness. As before, large structures, therefore, call for a high level of laser power. White light methods, where applicable, are much cheaper. Apart from their cost, laser speckle methods are difficult to use on surfaces that are tilted, whereas the white light method is tolerant in that respect.

Thermoelastic stress analysis
An interesting development in stress analysis has been that of stress pattern analysis by thermal emission (SPATE), originally proposed by the Admiralty Surface Weapons Establishment and subsequently developed by Sira Institute. The work was originally described in 1978 by Mountain and Webber.

The concept of adiabatic expansion or compression is well known as applied to gases, though it is less generally appreciated that similar effects occur in solids. When compressed, a steel bar shows a slight but measurable rise in temperature. Under tension, the same bar is slightly cooled. Similar changes can be observed in other solids, to an extent that is related to their coefficients of thermal expansion.

When a structure is subjected to regular cyclic stress, then a given point on its surface exhibits fluctuations of temperature above and below the ambient level. A stress change of $2 \, N \, mm^{-2}$ in a steel component, for example, induces a temperature change of approximately 0.002°C. If the small changes of surface temperature can be measured, they can provide information as to the distribution of stresses within the component or structure. The SPATE system achieves this by examining the infrared emission of each surface point in turn with a sensitive optical scanner.

When an analysis is to be carried out with the SPATE apparatus, the component under test is arranged to receive a cyclic load. The area of interest on its observable face is then sub-divided into as many smaller picture elements or 'pixels' as may be necessary to provide the required resolution. The field of view of the scanner can, in fact, be set up for a maximum of 256 elements along both the x and the y axes, although at the upper end of this range, the analysis time may be inconveniently long. Electrically controlled mirrors are used to direct the scanning beam from each pixel in turn on to an infrared detector for a predetermined number of stress cycles (Figure 2.11).

The alternating current (AC) output from the detector provides a measure of the excursion above and below ambient temperature and, when suitably filtered, gives a signal that is linearly related to the stress in each pixel. The signals pass to an electronic memory store as they are produced until the whole field of view has been scanned in a similar manner. Finally, the stored pattern of stress levels can be displayed on a colour graphics monitor, using a false colour scale to simplify the interpretation of small differences.

Since the whole system is under microprocessor control, it is comparatively easy to change the operating conditions as required. Thus, it is often useful to start an analysis with relatively few picture elements in order to obtain a rapid assessment of the stress pattern, and then to 'zoom in' on some area of particular interest at a much higher definition.

The resolution and stress sensitivity of this method are claimed to be in the same general order as that of strain gauges, but since no surface preparation and gauging are involved, SPATE can provide a very rapid survey of stress distributions. The longest time for a complete analysis would be that involving a scan over 256 elements on each axis. The present range of the equipment, in terms of distance, is from 0.5 m to approximately 10.0 m, with a field of view extending over 25°. Up to 16 levels of arbitrary colour may be displayed on the monitor; the discrimination from level to level can be

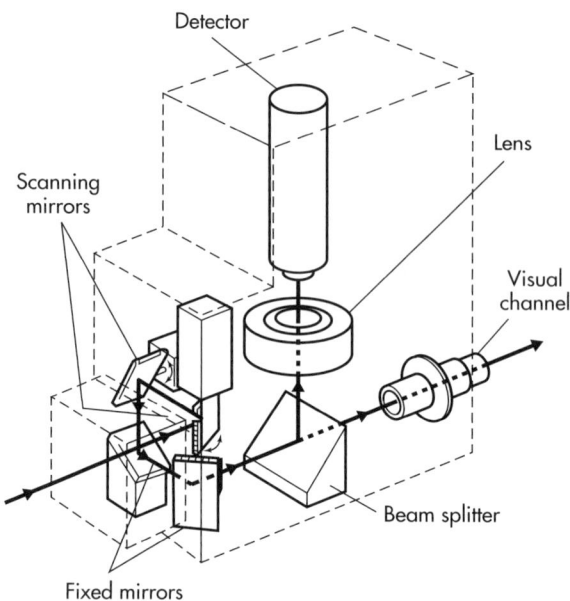

Figure 2.11 The stress pattern analysis by thermal emission (SPATE) system. Electronically controlled mirrors scan the surface reflecting successive point images to the thermal detector. Temperature changes are related to the stress levels involved (Courtesy of Ometron Ltd).

sufficient to distinguish stress changes of less than $2.0\,\mathrm{N\,mm^{-2}}$. The method appears to be both important and innovative, and in 1984 it received that year's Achievement Award from the Worshipful Company of Scientific Instrument Makers.

Brittle lacquer

A method of stress analysis that was popular in earlier years was to use a brittle lacquer. The principle of the method is simple. It involves the application of a specially formulated lacquer to the surface of the test structure. The lacquer is designed to be brittle at a specified temperature, and it cracks if it is subjected to tensile stresses above a certain threshold value. The individual cracks run along directions perpendicular to that of the principal strain on the underlying surface; consequently, observation of the cracking pattern makes it possible to construct maps of the surface strain. One proprietary range of brittle lacquers is marketed under the name Stresscoat'.

Experience has shown that the lacquer is not easy to work with, and tightly controlled conditions of humidity and temperature are required. Furthermore developments in health and safety regulations necessitate the use of contained breathing apparatus and protective equipment.

General considerations

The preceding notes have shown a variety of means for assessing the distribution of strains or stresses over the surface of a test object. It may often be convenient to combine several approaches to stress or strain analysis if the purpose of the analysis is to optimise the siting of gauges for long-term measurement. For example, either SPATE or white light speckle methods can identify highly strained areas on a machine. Then, in a second study stage, rosette strain gauges can be fitted on a temporary basis to provide more detailed information as to the magnitude and direction of principal strain. Finally, permanent gauge installations can be arranged in the locations found to be most suitable.

It must be remembered that the ultimate function of the installed gauges is that of measuring some applied force, rather than the dimensional changes of the gauged area itself. It is, therefore, of critical importance that there should be a constant relationship between these two parameters. Ideally, that relationship would be linear, but in the end it is reproducibility that is of the most value. The stress–strain characteristics of a simple element, such as a cantilever beam loaded within its elastic limit, are likely to be both reproducible and linear. However, a machine such as a tablet press is complex and has numerous interacting components.

When a force is applied to one element in a complex structure, that element will deform to some extent, and may well do so in a linear manner. If the force is increased, the element will deform further and may consequently make contact with another part of the structure. Should this happen, the first element cannot continue to move without also moving the part with which it has made contact. At this point, the element becomes mechanically stiffer.

Taking a hypothetical example, a force is to be transmitted to a gauging system at one end of a roll-pin in a rotary tablet press, while the other end acts as a fulcrum. Moderate pressure on the roller is transmitted to the shaft end, giving a reasonably linear calibration curve; however, as the applied pressure is increased, the shaft lifts until it lies obliquely across the bearing at the far end (Figure 2.12). With all free play in the bearing now taken up, the shaft now requires more force for a given displacement and a 'knee' develops in the calibration curve. It should be emphasised that this example is entirely imaginary, but it represents a possible way in which a calibration curve might be found to depart from the expected straight line.

A problem of this type would readily be identified by stress analysis studies, provided that the machine or its representative model is taken over the whole range of possible working loads. Stress analysis should be seen as a worthwhile prelude to a strain-gauging operation on any unknown type of machine, whenever gauges are to be installed for permanent use. It not only provides information as to the location and direction of the strain but also shows whether a particular site would exhibit linear and reproducible characteristics. It may be of use in the choice of a gauge type – whether it should be a foil gauge, a semiconductor or some other variety – or, indeed, whether some ready-made sub-assembly could be applied. Gauged sub-assemblies may be of two basic forms. One group consists of packaged strain gauges that can conveniently be attached to a structure by glueing or welding; these may be useful for applications in extreme

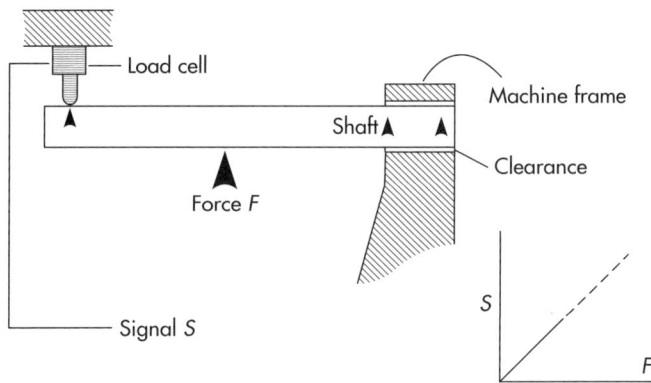

Figure 2.12 A non-linear response. If a shaft bends under pressure, it may take up any slack in its bearing, and it will require more force for a given displacement.

conditions of temperature or in corrosive atmospheres. The second group comprises more elaborate devices that are specifically designed to act as force transducers or 'load cells'. Load cells may contain resistive strain gauges or may operate on other principles.

The Wheatstone bridge circuit

Electrical resistance strain gauges, strained within their normal limits, only show small proportional changes in resistance. The strain level is not likely to exceed 300–400 ppm and even when multiplied by the gauge factor, the resultant alteration in resistance is still only a small percentage of the original value. In order to measure such changes conveniently, it is usual to connect one or more gauges together in the bridge circuit shown in Figure 2.13. This was devised by Samuel Christie in 1843. It was used extensively by Charles Wheatstone and has ever since been associated with his name.

In the original form of the bridge circuit, four resistors are connected to form the sides of a square. A voltage is applied across two opposite corners, while a sensitive voltmeter is connected across the two remaining corners. If the resistors are either all of equal value or are symmetrically disposed in two equal pairs, no current passes through the meter.

Wheatstone used this arrangement in a null mode to measure unknown resistances. The unknown device was connected in one arm of the bridge, while known resistances were switched into an adjacent arm until the meter showed no deflection. At that point, the two resistances were assumed to be exactly equal. The advantage of this procedure was that it did not need either an accurately stabilised power supply or a calibrated meter. But, it could not be used in that precise form for the measurement of rapidly changing values. To measure these, it is necessary to depart from the null mode to one in which a continuous output signal is generated from a dynamically unbalanced bridge.

If a resistive strain gauge is substituted for one bridge arm of exactly similar ohmic value, the bridge remains in balance, but if the gauge

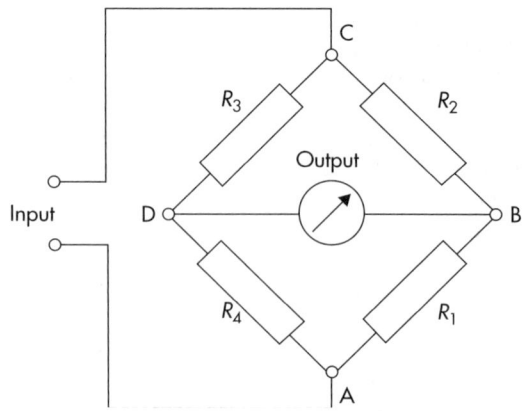

Figure 2.13 The Wheatstone bridge circuit. A voltage is applied to two opposite corners of the bridge, and a voltmeter is connected across the remaining corners. If the four resistors are of equal value, there will be no signal.

should then be subjected to strain, its resistance will be altered and a signal will appear across the output terminals. Provided that the power supply is stable, and that the meter does not draw an appreciable amount of current from the bridge, then the measured output signal will be a reliable, although not a linear, guide to the proportional resistance change of the strain gauge. It is possible to power a bridge either with AC or direct current (DC), but the DC version is most often used.

When the bridge circuit is used in this mode, the output signal is directly proportional to the applied voltage. It is, therefore, necessary to power the bridge from an accurately stabilised DC supply, which can either be set to operate at constant voltage or at constant current. A good unit of this kind can have a thermal stability in the region of $0.005\%\,°C^{-1}$ and should show less than 0.01% variation in output voltage for a 90% change in load. In practice, the bridge itself does not impose a significantly variable load on the supply, and the regulation can be extremely good. If all four arms of the bridge have an initial resistance R, and a voltage U is applied to terminals A and C of the circuit shown, then the voltage at the point D (V_D) will be given by Equation (2.1):

$$V_D = \frac{U \times R}{(R+R)} \quad \text{or} \quad \frac{U}{2} \qquad (2.1)$$

The voltage at point B will be the same as this, so there will be no potential difference between points B and D.

When the gauge connected between points A and B is now subjected to strain, changing its resistance to the new value of $(R + dR)$, the potential at point B is given by Equation (2.2)

$$U \times \frac{R + dR}{(2R + dR)} \qquad (2.2)$$

and the output signal is given by Equation (2.3)

$$U \times \left[R/2 - \frac{(R \times dR)}{(2R + dR)} \right] \qquad (2.3)$$

It will be apparent that this is not a linear function of dR/R, the proportional change in R, although the departure from linearity is small at low levels of strain.

Gauge resistance

As we have seen, the voltage level of the output signal is directly dependent on the input voltage, and on the proportional resistance change of the gauge. From this point of view, the nominal resistance of the four bridge arms is not critical. However, in practice, there are other considerations that must be taken into account when selecting a gauge resistance. For example, using a gauge of relatively high resistance reduces the effects of small resistance changes in the lead-wires and other circuit elements and also minimises the heat dissipation within the gauge for a given applied voltage. It is, therefore, possible to use a higher excitation voltage and so produce a larger signal. The disadvantage is that it does require more attention to the insulation resistance of the whole system, and it will be more susceptible to the possibility of errors from current leakage.

Historically, we have noted that the earliest industrial strain gauges were formed from thin wire elements, which tended to impose a certain limitation on the maximum resistance that could be produced. Accordingly, a value of $120\,\Omega$ became accepted as an industry standard. The development of photochemical etching allowed foil gauges to be produced at considerably higher resistance values, and $350\,\Omega$ is a popular value for currently produced gauges. In fact, some modern transducer-class gauges are offered at $5000\,\Omega$, the much higher resistance having the advantage of reducing current drain in battery-operated systems.

The proportional change in resistance, dR/R, will be a nearly linear function of the strain experienced by the gauge. However, we have already shown that the bridge output voltage does not itself change linearly with strain. It would be more accurately linear if the bridge arm was energised at constant current, but it is usually energised at constant voltage. So as the resistance of the gauge increases, the current through it correspondingly decreases, producing a slight reduction in the signal across the gauge.

Bridge configurations

Using one 'active' gauge, as in the previous example, provides what is generally described as a 'quarter-bridge' configuration. Two gauges give a half-bridge, and four gauges a whole bridge. The gauges may be active or passive, but in each case, the bridge output will be related to the number of active gauges.

Active and passive gauges

'Active' gauges are those that are bonded to a substrate in such a position that they respond to strain in that substrate. 'Passive' gauges are not intended to respond in that way, being primarily intended to provide temperature compensation for the active gauges. They may be attached to an adjacent unstressed surface.

Poisson gauges

Changes in length of a stressed component will always be accompanied by corresponding changes in width, as indicated by the Poisson ratio. If one gauge is bonded along the axis of principal strain, and the compensating gauge is bonded on to the same surface at right angles to it, the second gauge will not be completely passive but will contribute a signal of opposite sign from that of the first, or active, gauge at approximately one-third of its magnitude. Such a gauge is often referred to as a 'Poisson' gauge. The temperature compensation of the Poisson gauge is better than that of the passive gauge noted above, since it will be attached to the same substrate as the active gauge and will in general be close to it.

Lead resistance compensation

As the quarter-bridge configuration only involves one gauge, it has sometimes been used in civil engineering practice on such applications as the long-term monitoring of bridges and buildings. In this field, many gauging points must usually be employed, though absolute measuring accuracy is not necessarily a primary consideration, and the cost of the installation must be kept within limits. However, there are serious disadvantages in the quarter-bridge arrangement when particular accuracy of measurement is needed. For example, the single active gauge may have to be connected at a site that is comparatively remote from the remainder of the bridge circuit. Long leads introduce an additional resistance, shown as R_L in Figure 2.14, into the circuit; not only will this tend to unbalance the bridge but it will almost certainly be subject to thermal variations that differ from those of the other three bridge arms. The traditional solution to this problem has been to run an additional wire between the strain gauge and the bridge, as shown in Figure 2.14. This restores symmetry to the right-hand part of the bridge circuit and helps to balance the bridge again.

Zero drift

The quarter-bridge uses only one active strain gauge, and three bridge completion resistors. If one resistor is variable, and the other two have exactly equal resistances, then the bridge can be balanced precisely. With no strain affecting the gauge, the output meter can be set to register zero. However, problems arise if the temperature of the gauged structure varies, since temperature changes will not affect the resistances of the gauge and the completion resistors to the same extent. Moreover, there may be some degree of differential expansion between the gauge and its substrate, although the use of STC and modulus-compensated gauges (p. 18) should minimise

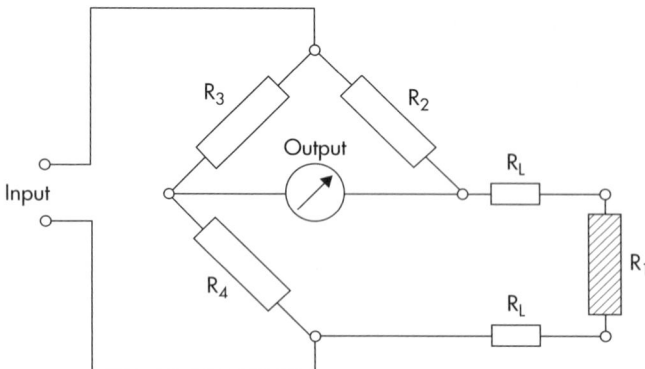

Figure 2.14 Using extended lead wires to a single gauge unbalances the quarter-bridge: it is usual to run two wires as shown to restore balance.

this particular problem. In any event, these two effects may result in a drift from the carefully balanced zero.

The most satisfactory solution to this problem is to replace the adjacent bridge completion resistor with a compensating gauge, which will have the same thermal characteristics as those of the active gauge. Provided that they are of a similar type, separate gauges can be used, one being bonded to the substrate along the axis of principal strain while the other is mounted to an unstressed area of metal at the same temperature. The bridge configuration now resembles that of a half-bridge, though under these conditions, the compensating gauge does not contribute to the signal. The output, therefore, is still that of a quarter-bridge.

A much better method of temperature compensation is that of using a composite or 'rosette' gauge, where two grids are made on the same backing film, one being at right angles to the other (Figure 2.15). The method of manufacture ensures that these double gauges are extremely well matched, and as they are physically close together, they will be as nearly as possible at the same temperature. Additionally, the composite gauge provides an improved signal through the Poisson effect. In most applications, rosette gauges give the better temperature compensation for zero drift, though whichever gauge type is chosen, one double or two singles, the balanced wiring arrangement largely overcomes the asymmetry of the quarter-bridge.

Gauge factor variation

We have previously noted that temperature changes not only affect the resistance of a strain gauge but also alter its strain sensitivity or gauge factor. One way of dealing with gauge factor variation is to make the bridge self-compensating by incorporating further variable components. For example, it is possible to vary the bridge excitation voltage by using a thermistor in series with the power supply, the thermistor being controlled by a resistor network with shunt and series elements that can be individually tuned to minimise these variations. In practice, tablet presses and capsule fillers are not usually exposed to large variations in temperature, and for many applications it may not be necessary to squeeze the last decimal point out of the measuring system.

Semiconductor bridge assemblies

We have already seen that it is possible to manufacture very sensitive gauges from doped silicon, and that they can be made to have positive or negative characteristics, although their gauge

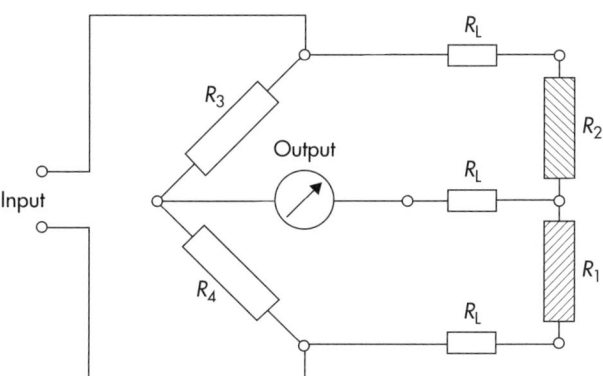

Figure 2.15 Compensating circuits are used to improve the thermal stability of the basic bridge. The half-bridge configuration gives good symmetry: R_1 and R_2 can be one composite gauge.

factors tend to be non-linearly variable with strain. Figure 2.16 shows that a complete four-arm bridge can be assembled in a small space by the use of two P-type and two N-type gauges, and this naturally gives the largest possible signal. Semiconductors can be tailored to have specific characteristics by selective doping, and this makes it possible to reduce temperature-dependent errors for a given set of circumstances. For example, the gauges have their own temperature coefficients of resistivity when in the unbonded state, but when they are attached to a substrate with a different expansion, the combination produces a new apparent or 'bonded' coefficient. If the bonded coefficient has the same value as the temperature coefficient of the gauge factor, and the bridge is excited at constant current, the output will be largely independent of temperature. This is because the bridge excitation voltage now increases at the same rate as the gauge factor decreases.

Linearisation of bridge output

Ideally all bridge circuits should be properly balanced, and compensated for every possible type of error, but for many applications this may not be strictly necessary. It is normal practice to linearise the bridge output by computer manipulation at the data-handling stage, provided that the characteristics of all the items involved are known. Commercial linearising units are offered by the manufacturers of data acquisition systems, though in the past some workers have claimed that it was necessary to build their own compensation circuitry. Welsh and Pyne (1980), for instance, fitted a resistor in series with their bridge and measured the voltage across it in order to determine the current drawn by the bridge. They then used changes in the bridge current to control the characteristics of a signal-conditioning amplifier, and reported a 10-fold reduction in temperature-associated errors with their semiconductor gauges, compared with those of a commercially compensated bridge. As usual, the requirements of the application must determine how much effort and expense go into the instrumentation system.

Bending and stretching

So far we have looked mainly at the two-arm or half-bridge. When the circuit is extended to include all four arms, there are two ways in which the gauges can be connected, as shown in Figure 2.17. If, for example, a simple rectangular beam is fitted with one active gauge on either side, parallel to the long axis of the beam, then those two gauges can either be wired into adjacent arms of the bridge or into opposing arms. In the first instance, the beam will be sensitive to bending. As the beam bends, one gauge resistance will increase, while the other decreases at a similar rate. The combined effect of these two changes will be to give an output related to the amount of bending. If the gauges are connected

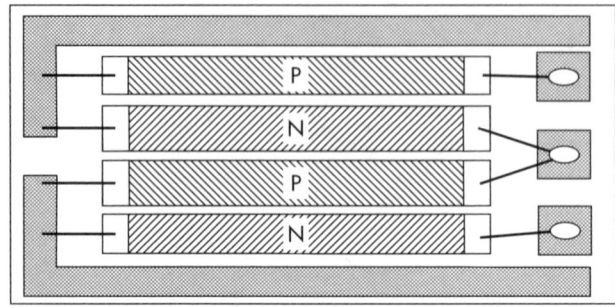

Figure 2.16 A complete four-arm bridge circuit constructed from P- and N-type semiconductor gauges. All arms are active, and so the signal from such a composite assembly represents the maximum possible.

into opposing arms, the system will now be more sensitive to tension or compression, and less sensitive to bending. Components such as load cells that are normally used in compression are often provided with a full set of gauges, connected so that they will balance out bending forces.

Effect of surface curvature

It is possible to fit thin modern gauges on to surfaces with appreciable curvature. They are, for example, routinely applied to the inner surfaces of relatively small holes in gauged bolts, punches and roll-pins. However, the differential expansion between gauge and substrate is affected by the amount of curvature, so care must be exercised when using gauges that are nominally of the STC type. The STC compensation is normally correct only for flat surfaces. A detailed treatment of this problem, as applied to semiconductor gauges, is given by Hoffmann (1990).

Bridge completion resistors

If a half-bridge circuit is to be used in order to provide temperature compensation, two fixed resistors will be needed to complete the remaining half of the bridge. Commercial standard resistors can be of many types, including metal film, carbon composition, wire-wound, metal oxide and thick film. Each type has different thermal characteristics. A typical carbon composition, for example, has a temperature coefficient of resistivity in the region of $\pm 1200\,\text{ppm}\,°C^{-1}$, while that of metal films may be as low as $\pm 50\,\text{ppm}\,°C^{-1}$. The lowest temperature coefficient of resistivity of any commercial offering appears to be that of the foil resistors from Vishay Measurements Group, with a claimed value of essentially zero (better than $1\,\text{ppm}\,°C^{-1}$) over the temperature range that would concern us in the tablet-making field. The two bridge completion resistors should ideally be of exactly the same value to maintain perfect balance. If they are not, then there are two possibilities: firstly, the

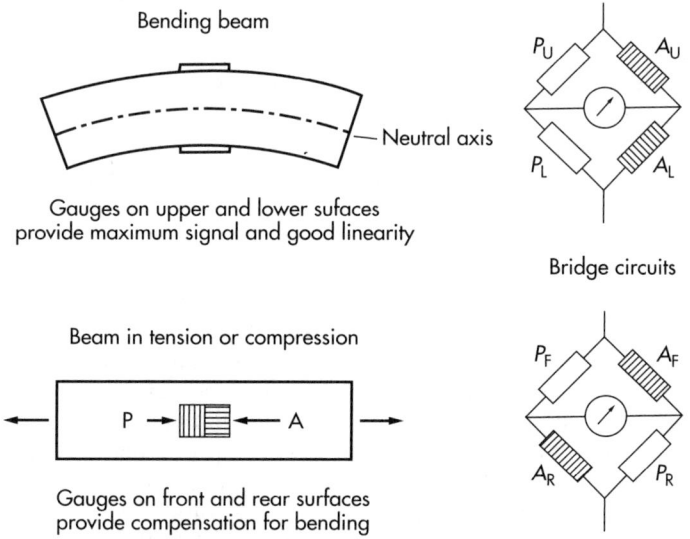

Figure 2.17 Bending and stretching: the gauges may be connected so as to measure tension or compression while being insensitive to bending, or vice versa. A, active; P, Poisson.

slight imbalance may be ignored, and offset in the data processing circuitry; secondly, additional resistive elements may be wired in series with the existing resistors and these adjusted by careful scraping. Scraping has the effect of reducing the cross-sectional area, and hence of increasing the resistance, so that accurate balance can be achieved.

Bridge excitation voltage

It is always convenient to have a transducer that provides a relatively large signal, since it is easier to detect that signal in the presence of noise. The output from a Wheatstone bridge is directly proportional to the input voltage, so there is a natural temptation to apply a relatively high voltage from the power supply. Inevitably, the power supplied to the bridge produces heat within the gauges and the bridge completion resistors, and there are, therefore, conflicting interests between the signal level and the errors introduced by heating.

The main source of heat loss in a typical strain gauge installation is the substrate to which the gauges are attached, and the power that can be absorbed varies according to the material of which that substrate is constructed. A gauge of high accuracy on a thick steel substrate under static conditions can be run continuously at a power density in the range $1.6–3.1\,kW\,m^{-2}$. About half this level would be acceptable on a stainless steel substrate, while a thermally conductive material such as heavy copper or aluminium could take twice that level. The experimental approach is, of course, to try a range of voltages and see what happens: raise the applied voltage in small steps until some thermal drift becomes apparent, then reduce the voltage below that level.

Taking the published figures as a guide, one can reasonably suggest that a gauge with an area of $10\,mm^2$ on a steel substrate should not be run for long periods at more than 3 V if the highest accuracy is to be maintained. Modern high-resistance gauges can tolerate higher voltages, but again we would recommend the practical method outlined above for any individual example.

Load cells

Load cells are self-contained assemblies for the measurement of applied forces. They are usually constructed in physical forms that are well adapted for convenient incorporation into engineering structures. Consequently, they may have the external characteristics of bolts, washers, beams, straight connectors or short cylinders. Internally, they contain some means of generating a signal that is related to an applied force, and this may be piezoresistive, piezoelectric or even magnetoelastic in character.

Load cells based on strain gauges

Load cells based on strain gauges comprise a deformable frame – the spring element – with strain gauges bonded in suitable positions to register movement under load. The rated loading of the frame is designed to remain well within the elastic limit for its material, and it can reasonably be assumed that the deformation is near to linear with reference to the applied load.

Early load cells were usually of a simple configuration, such as that of a short column with gauges fitted symmetrically around it in a full bridge circuit (Figure 2.18). It was intended that the symmetry of the system would compensate for off-axis loading, but the compensation is not necessarily perfect for these gauge elements in compressive duty. The load is, therefore, often applied through a coupling device that will allow a certain amount of misalignment between the column and the load. For example, a form of ball-and-socket joint may be used. Alignment problems are likely to be minimised when the cell is to be used in tension, rather than in compression, although this may not be a practical option.

One characteristic of the simple gauged column load cell is that it can be mechanically stiff. Typical cells of this form can have deflections of approximately 0.025 mm or less at full load so that they do not move through large distances for an average range of forces.

Later work in this field seems to have been primarily concerned with the development of configurations that are more tolerant of off-axis

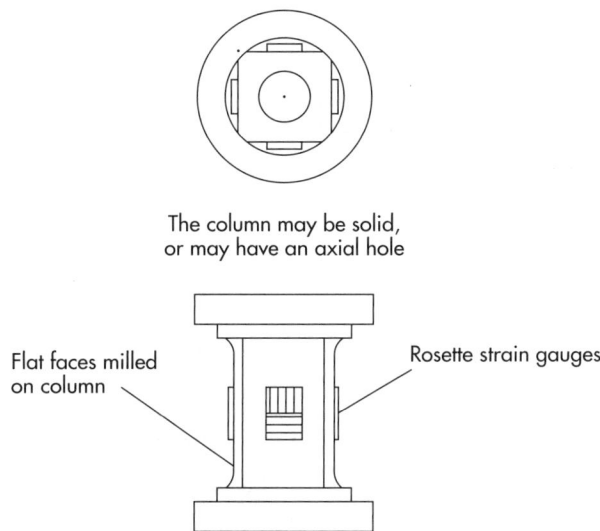

Figure 2.18 The gauged column load cell. This is simply a metal column with strain gauges in balanced positions on its surface. Two gauges are active and two are Poisson gauges. Such cells are sensitive to off-axis loading and are more satisfactorily used in tension than in compression.

loading. This has usually been achieved by departure from the simple column, in favour of double cantilevers, concentric rings, shear beams and other shapes. However, these alternative constructions may be less rigid than the plain cylinder, and it is important to check that the operation of a machine incorporating such devices is not changed by their incorporation.

Figure 2.19 shows a concentric element load cell. This has a short central column supported from an outer ring on gauged elements. This construction ensures that the central column is constrained solely to vertical motion, and this makes it insensitive to off-axis loading. The gauges may all be active, and the deflection under load is usually greater than that of a gauged column load cell. Since the position of the applied load does not affect the output signal, such cells are particularly suitable for tasks such as weighing.

Shear beams

If a simple cantilever beam is loaded at one end, its deflection can be calculated reasonably accurately provided that its mechanical properties, its dimensions and the applied force are all known. However, the deflection will be critically dependent on the distance between the supported end and the point where the load is placed. There are many applications in which it would be desirable to have a gauged beam where its output was largely independent of the load point, one example being in the support of the ejection ramp in a rotary press. In this instance, the load point moves continuously as the punch travels across it, and a gauged cantilever beam would give a correspondingly variable signal even though the load remains nearly constant. The shear beam overcomes this problem by using a construction that, in effect, forms the four sides of a parallelogram, as shown in Figure 2.20. When the far end is loaded, the geometry of the system ensures that it can only move vertically, rather than tilting. As a result, the movement itself does not depend on the point at which the load is applied. A strain-gauged element can be fitted between the two sections of the shear beam and will give an appropriate output when the beam is loaded.

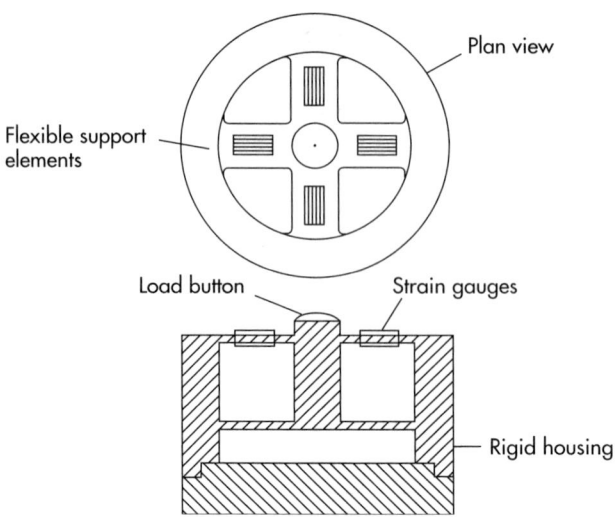

Figure 2.19 A concentric element load cell. The central column is constrained to vertical motion only, and is hence less sensitive to off-axis loading.

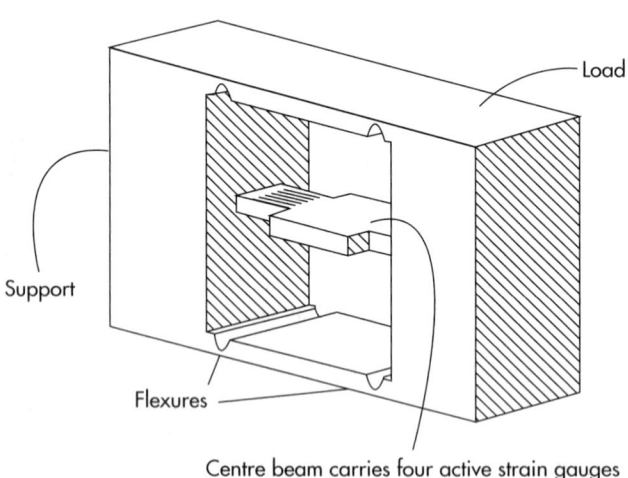

Figure 2.20 The shear beam. These are reasonably independent of the point at which the load is applied.

Strain-gauged roll-pins

Shafts with internal gauges have been used for many years in heavy industry as weighing accessories for crane hooks and associated lifting equipment, where the gauging elements can be protected against mechanical damage. They are, however, applicable in principle to the measurement of any radial forces acting on rollers, and they have been used in tablet presses as the basis for instrumented roll-pins. The usual arrangement is that a standard form of roll-pin is bored out to give a central hole extending along its axis for most of its length. When the roller or the pulley is in its normal position and is loaded radially, the central hole is subjected to shear forces at the points corresponding to the two sides of the roller (Figure 2.21). If shear-sensitive strain gauges are fitted inside the hole at these two points, the radial force can be measured as with any strain-gauged load cell. Of course, fitting gauges in exactly the right positions down the inside surface of a narrow hole is not something to be tackled lightly, and as in other examples, we strongly recommend that this should be left to the professionals. However, this particular type of load cell would appear to be a very useful accessory for a rotary press, since it requires little or no modification to the original machine and can simply replace the plain roll-pin supplied with the press.

Gauged bolts

The gauged roll-pin usually uses strain gauges in shear mode, but it is possible to fit internal gauges to other components in such a way as to respond to tension or compression. This approach allows the production of strain-gauged bolts. Fitting the gauges along the inside surface of a narrow bore can reduce the sensitivity of the device to bending, and at the same time affords considerable protection against mechanical damage. A similar arrangement has been applied to the internal gauging of punches.

Piezoelectric load cells

Certain non-conducting materials, usually crystalline in nature, have the property of converting mechanical energy directly into electrical energy. If a crystal of quartz, for example, is subjected to physical pressure, electrical charges develop across opposing faces of the crystal (Figure 2.22). The magnitude of the charges is accurately related to that of the applied force. Barium titanate behaves similarly, and there are other piezoceramic materials such as lead zirconate titanate (PZT), in which this effect is even more pronounced. Examples may be seen in domestic gas lighters, where a light impact generates a considerable spark.

The 'piezoelectric' effect is used as the basis for a range of force-measuring devices, which have themselves been applied quite extensively in the field of tablet machine instrumentation. Piezoelectric load cells are both compact and mechanically rigid; consequently, they have very little effect on the mechanical properties of any machine to which they attached. A typical value

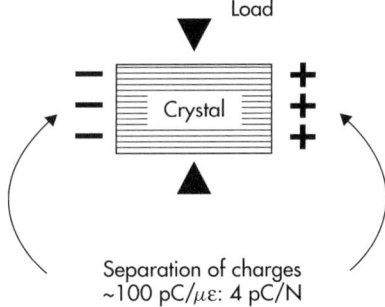

Figure 2.22 The piezoelectric effect. Force applied to the opposing faces of the piezoelectric material produces a separation of electrical charges across the other faces. The quantity of electricity involved is very small.

Figure 2.21 An internally gauged roll-pin with strain gauges fitted at two regions in order to sense the shear forces applied by the pressure roll. The roll-pin has an axial cavity; the shear-sensitive gauges are mounted inside the grooved regions in the cavity.

for deflection under load would be 1.0 μm per 20 kN, while the corresponding sensitivity would be approximately 4.2 pC/N. The thickness of a load cell may be as little as a few millimetres.

These cells are quite sensitive to side loads, and for that reason the surfaces between which they are compressed must be very accurately parallel. This requirement would represent a problem in machines that may only be fitted to moderate degrees of flatness. It is, therefore, normal practice to bolt the transducers against a nested pair of part-spherical washers, which are able to take up errors in alignment (Figure 2.23).

Piezoelectric load cells usually have a resonant frequency in the range 10–100 kHz, so they respond very rapidly to changes in applied force.

Advantages of piezoelectric transducers

The combination of high sensitivity, rapid response, small size and general rigidity seems particularly attractive; indeed, many workers have chosen piezoelectric transducers for use with tablet presses. Sixsmith, in his 1977 review paper, suggested that they had the following advantages:

- greater sensitivity
- no bonding to machines
- less temperature sensitivity: better than $0.01\% \,°C^{-1}$
- the signal caused by initial loading can be cancelled out, simply by earthing.

Piezoelectric transducers, produced by Kistler, were used by Manesty Machines Ltd, in the first version of their 'Tablet Sentinel' weight-control system. They have been recommended by various research workers since then, for example, Williams and Stiel (1984) and Hoblitzell and Rhodes (1985). Schmidt and Koch (1991) compared piezoelectric force transducers with resistive strain gauges in a rotary tablet press, and Cocolas and Lordi (1993) measured die wall pressure with piezoelectric force transducers.

Disadvantages of piezoelectric transducers

There is, inevitably, a practical disadvantage to their use, which is that they are not well suited to the measurement of static loads. Although the signal output may represent many volts, it is microscopically small in terms of power, and it is only generated by changes in the applied load. If

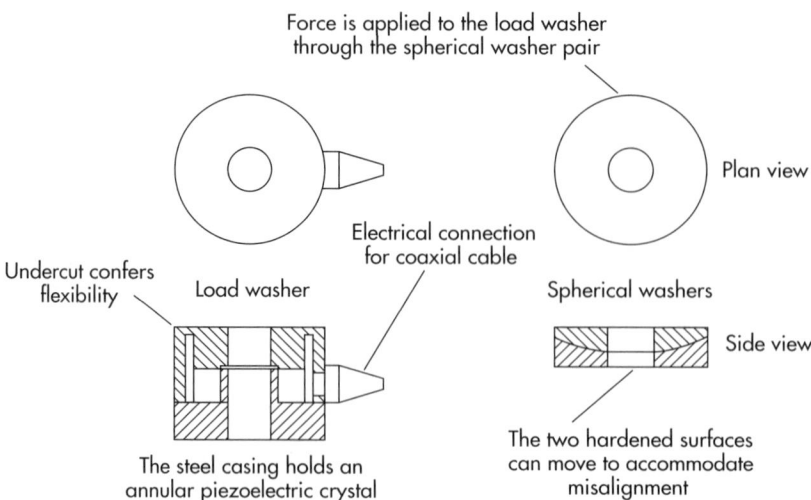

Figure 2.23 The piezoelectric load cell. They are very compact but are sensitive to uneven loading. It is, therefore, normal practice to load them through hardened steel washers that can take up small amounts of angular misalignment.

the loading on the cell is maintained at a constant level, the signal tends to leak away, so that it appears as if the applied load is decreasing.

The insulation resistance of the cell itself must be very high, a typical value being $10^{12}\,\Omega$. Similar considerations apply to any cables and connections. In order to minimise the loss of signal, it is necessary to use an amplifier with a correspondingly high insulation resistance at the input: a so-called 'charge amplifier'. The input section of the charge amplifier usually contains a field effect transistor, or FET, which has suitable characteristics, and there is an internal capacitive feedback loop that can be used to alter the measuring range of the amplifier.

A further capacitor is connected across the input terminals of the charge amplifier. The choice of this particular component affects the frequency response of the system, and it must, therefore, be chosen carefully to suit the requirements of the work. An incorrect choice of capacitor at this point may produce misleading calibration curves. Compared with the strain gauge amplifier circuitry, this system is appreciably less tolerant.

Integral charge amplifiers

The problems associated with the use of individual cells, cables, connectors and charge amplifiers are generally recognised as serious disadvantages of the piezoelectric devices in certain circumstances. However, developments in microelectronics have made it possible to reduce the physical size of the amplifier to such an extent that it can now be incorporated within the same housing as the load cell. Special interconnecting cables become unnecessary with such an arrangement.

When forces are changing continuously, as they do in production runs at normal machine speeds, piezoelectric transducers are at their most successful. It is always important, however, to ensure that some means is found to check the baseline of the output signal from time to time, in order to avoid the occurrence of a steady drift. Early versions of the Manesty Tablet Sentinel encountered the problem of zero drift and were known to register an apparent reduction in compression forces when those forces were in fact increasing. Since then, the whole system has been redesigned with conventional strain gauges.

In recent years, there has been considerable research interest in the study of compression forces under static conditions, for example the time-dependent decay of the stresses in a tablet within the die. For studies of this type, strain gauges probably still represent the simplest and most unambiguous means of measurement, particularly since strain-gauged load washers have become available in sizes that approach those of piezoelectric washers. It must also be added that most of the major tablet press manufacturers seem to have adopted strain gauges for their own commercial product lines.

Another consideration if an instrumented unit is to be set up is that the piezoelectric washers require a different signal amplifier from that used for resistive strain gauges. There is much to be said for having spare amplifiers available, and having gauges that all work in the same way reduces the number of different units that must be kept in hand.

Magnetoelastic cells

Mechanical stresses can affect materials in a variety of ways. If a ferromagnetic material is stressed, for example, its magnetic properties may be changed to a measurable extent. This principle has been used as the basis for a third type of load cell, described by Erdem (1982). In this construction, the body of the transducer is a cube of a ferromagnetic material. It has four holes drilled through it, one near each corner of one face. Then two windings are fitted diagonally through these holes, at right angles to each other (Figure 2.24). When one winding is excited with an AC input, the coupling between that and the second winding is dependent on the direction of the magnetic flux within the block. This, in turn, is a function of any applied force across opposing faces. The cell is excited at mains frequency. Erdem quoted a repeatability of 0.02%, a non-linearity after compensation of $\pm 0.05\%$ and a temperature coefficient of approximately $\pm 30\,\mathrm{ppm}\,^\circ\mathrm{C}^{-1}$.

The magnetoelastic cell is commercially available under the name Pressductor Millmate from

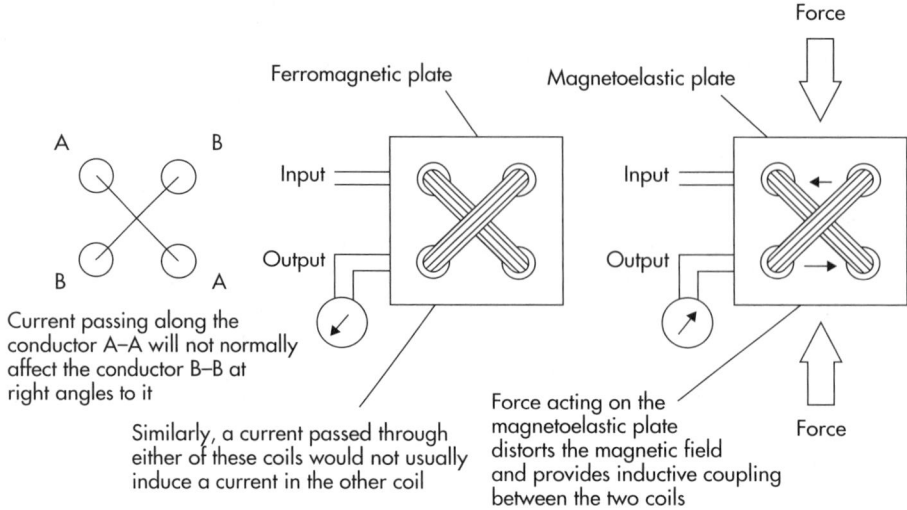

Figure 2.24 The magnetoelastic load cell. Forces applied to the opposite faces of the cell modify the coupling between two pairs of coils.

ABB, and has been described by Johannson (1981). It has been applied, *inter alia*, to measurement of the forces on the rollers in steel mills, but it does not appear to have advantages over the strain-gauged load cell for applications on tablet and capsule-filling equipment.

The linear variable-differential transformer load cell

Since forces are generally inferred from the strain that they produce, the distinction between force- and displacement-measuring devices often becomes blurred. Some 'weighing' cells are constructed as frameworks that can be deflected by the application of a load, while the deflection is measured by means of a linear variable-differential transformer (LVDT; Figure 2.25). Clearly cells designed to give a large deflection under load have the potential capability of providing greater accuracy, but their stiffness and frequency response become less suitable for tablet machine studies. Notes on this type of cell are given by Seippel (1983). The LVDT itself is described in Chapter 4.

Advantages of load cells

The value of load cells in tablet machine instrumentation lies in the fact that they are self-contained and portable. A load cell, unlike a bonded strain gauge, can be removed at any time and checked for accuracy. It can be calibrated in many instances by dead-weight testing, and in the unlikely event of failure, it can be replaced by a similarly calibrated unit.

Shear-beam load cells are now becoming widely available, and as we have seen, these have the useful property of providing an output that is substantially independent of the point at which the load is applied. Their function is often to act as weighing elements and to support platforms or scale-pans. Since the load cell is a factory-made device, it can be assembled and sealed under carefully controlled conditions and can be tested before delivery.

These considerations do not normally apply to the strain gauge bridges that are fitted by individual laboratories to existing machines. They can neither be calibrated before bonding nor can they be removed without destruction of the gauges. However, bonded gauge installations can

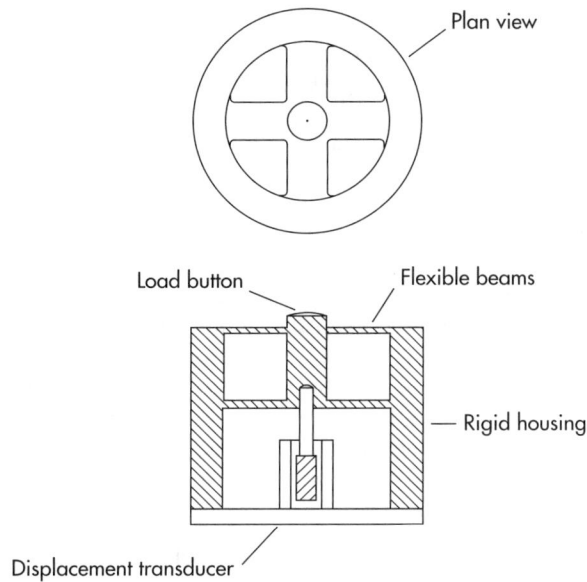

Figure 2.25 The linear variable-differential transformer (LVDT). This type of load cell can be seen as another version of the spring balance, in which deflection of an elastic component is measured by means of an LVDT. It may be fitted with a damping dashpot to reduce high-frequency oscillation.

be calibrated quite conveniently *in situ*, through the use of a suitable load cell that can be inserted as necessary between the punch faces of the machine.

The resolution available from a load cell is essentially limited only by its amplifier and electronics, but it is commonly taken to be approximately 1 in 10 000. Temperature compensation can be very good, of the order 0.001% °C^{-1}.

Disadvantages of load cells

There are a few possible disadvantages associated with the use of load cells compared with separately fitted strain gauges. Firstly, their physical size may preclude their use in a restricted space. Strain gauges, as we have seen, can be fitted in an area of a few square millimetres. Load cells are likely to occupy several centimetres both in height and in diameter. Admittedly, some very small load cells do exist. Those offered as bolt force sensors, for example, may be less than 4.0 mm in height provided that the load capacity is restricted to less than 1000 kg. For direct calibration of a tablet press, however, a working range of 5000 kg would be more practical, and a correspondingly larger washer would be required.

Secondly, there may be no convenient position for a load cell within the region suggested by stress analysis. It is, of course, possible to economise in space by using gauged bolts in place of ordinary bolts, load washers in place of ordinary washers, and so on. However, the measured strain levels at these points may be inadequate or non-linear, and in this event it may be necessary to modify the machine that is to be gauged. For example, if the force on a compression roller is to be monitored, it may be possible to remove the bearings that hold it, then to re-fit them in such a way that they are supported on shear beam load cells. Such a procedure would call for machining operations, in order to make room for the additional components, though otherwise it could be seen as sound enough practice. Nevertheless, the substitution of instrumented roll-pins would, in this example, be quicker and easier.

Finally, we have already noted that some load cells are mechanically stiff, while others have more of a springy characteristic. It would be important to choose the more rigid variety of cell for any situation in which the press performance might otherwise be altered. As we have noted above, cells designed for weighing often have a large displacement for a given load, though that is of no consequence in that application.

The point may perhaps be emphasised by looking again at the previous example. Load cells at each end of the roll-pin could certainly provide an accurate measure of the force on the compression roller. However, any transmission of force to the load cells must cause them to distort, so the roller and the roll-pin will move slightly more than usual. If the additional movement is appreciable, then the compression characteristics of the machine may be altered from those of the standard model. By contrast, it is clear that the attachment of bonded strain gauges to the frame or the punches of a tablet press should not alter its mechanical response in any measurable way.

The suppliers of load cells, load washers and the like provide figures to show the mechanical deflection of their devices at full-scale load. This may be as little as 0.03 mm but can also be as high as 0.30 mm. The larger deflections usually seem to be associated with those load cells that are designed for insensitivity to off-axis loading. The LVDT load cells also have inherently larger deflections, which may be as much as 1.5 mm at full load, since a greater deflection clearly gives a more open scale of measurement.

Miscellaneous methods

The text relating to surface preparation and strain gauge bonding in Chapter 3 emphasises the fact that a high degree of cleanliness is essential for this work, together with a careful attention to detail. In general, the standards demanded for gauge bonding should readily be attainable in the laboratory, or in that of the original machine manufacturer. However, if gauges are to be retro-fitted to existing industrial equipment, which may be continually exposed to surface contamination, then the task of achieving those standards may be very difficult.

An example of a problem in this category is that of fitting gauges to the support legs of North Sea oilrigs. Although this may seem far removed from tablet compression work, nevertheless both areas have a need for gauging methods that can be used in the presence of contaminating materials with a minimum of surface preparation.

One way in which surface preparation can be minimised is by the use of gauged sub-assemblies that can be attached quickly and easily to the main structure. These sub-assemblies may embody electrical resistance strain gauges, or they may operate on other principles.

Encapsulated gauges

Some commercial sub-assemblies are produced by the encapsulation of a normal gauge and its connecting leads. This approach is used, for example, by HBM for their 'Strain Transducers' DB1, DA2 and DA3. These are standard foil gauges in a weatherproof outer casing, which can be used under water or in other difficult conditions. Unless they are completely embedded, they can also be removed and re-used.

These encapsulated gauges are useful for investigative work, particularly on large structures. However, they themselves are also large, and they would be unsuitable for use in restricted spaces. In general, any form of encapsulation leads to a reduction in the efficiency of strain transfer from the substrate to the gauge. Using long gauges helps to minimise the loss of signal. For example, some versions designed for embedment into concrete have an active length of as much as 150 mm.

Weldable gauges

A standard foil gauge is bonded, under clean room conditions, to a stainless steel foil ('shim') and is provided with suitable environmental protection to form a sealed package (Figure 2.26). The shim is then attached to the structure under test by electrical spot-welding. Surface preparation for such a gauge need only involve the removal of paint, grease and loose scales of metal oxides. The substrate must, of course, be of a ferrous or similar material that will accept spot-welding. The weldable gauge is bulkier and less sensitive than a conventional strain gauge.

Vibrating element gauges: the S-sensor

Not all strain gauges use the principle of electrical resistance measurement, even though resistive gauges are still the most common. An alternative principle is that of using variations in surface strain to alter the tension in a vibrating element, so varying its resonant frequency. Hornby (1982) noted that large vibrating wire gauges have been used since 1928. They have very stable characteristics and are suitable for long-term gauging applications. They have been used, for example, to monitor strains within concrete beams in structures such as bridges. The same principle has been applied to the design of a relatively small sensor for use on machines and metal structures in general, which is marketed under the name S-sensor.

The S-sensor consists of a thin metal ring with two diametrically opposed brackets, some 40 mm apart, that can be spot-welded to a metal surface. Across a diameter of the ring is fitted a strip of spring steel, formed into a shallow curve of 'S' profile; this is the sensing element of the device. The output signal is in the form of a variable-frequency oscillation, and so it is largely unaffected by electrical noise. Like the weldable gauge, the S-sensor is too large for application to small components.

When the test surface is strained, the thin ring undergoes a corresponding deformation, thereby altering the geometry, and hence the resonant frequency, of the steel strip. On the standard model of the S-sensor, a stress change of $50\,N\,mm^{-2}$ causes a frequency shift of 160 Hz. The base frequency is 1600 Hz. External electronics continually monitor this frequency, using an integral magnet coil to maintain resonance.

We have seen that conventional strain gauge bridges are susceptible to errors introduced by the resistance of their electrical leads, and by inadequacies of their insulation. The vibrating element gauges do not suffer from these particular problems, because their signal is in the form of a variable-frequency oscillation. Within fairly wide limits, the signal frequency is independent of the resistance of the connecting cables, so this type of gauge shows great immunity to electrical interference in general. Nevertheless, the resonant frequency of their spring elements will be affected by changes in temperature, and this must be taken into account when the readings

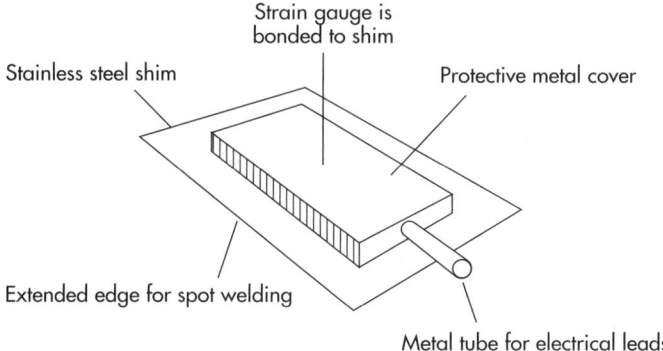

Figure 2.26 The weldable gauge. The gauge has been mounted on to a metal backing plate, which, in turn, can be welded to a test structure by spot welding should normal techniques of adhesive bonding be impracticable. A very robust assembly results.

from such a sensor are to be interpreted. Resonating sensors have been developed for quite a range of variables, including not only force but also fluid flow, level, density and viscosity. The subject is covered in a review by Langdon (1985).

The S-sensor itself is hermetically sealed and can withstand mechanical overloads of 10 times its rated range; it can operate in conditions that might be unsuitable for more conventional types of gauge. The rated range is equivalent to $250\,\mu\varepsilon$; the resolution is quoted as better that 0.01%, and the temperature stability better than $\pm 0.1\%\,°C^{-1}$. A device with these characteristics might not represent a first choice for research work and could not be fitted on to a small component such as a tablet punch. However, it could be applied to production monitoring, where its robustness and long-term stability would be valuable.

Capacitance gauges

It is possible to use changes in capacitance instead of changes in resistance as the basis for measuring quite small levels of strain. In principle, a capacitor is made up in such a way that one of its plates is attached to some part of the test structure, while the second plate is attached to another part. Strain in the structure then varies the gap between the plates, producing a change of capacity.

If the points of attachment are widely spaced, then relatively large movements can be produced at the gap between the plates. Various configurations have been used, including rhombic frame and concentric tube systems: these can have capacitances in the range 1–100 pF and can have sensitivities of approximately 0.01–0.1% of full-scale displacement.

Capacitance gauges have been reviewed in an article by Procter and Strong (1982). Their advantages are generally related to their ability to operate at very high temperatures, in some instances up to 1100°C. Like the other gauged sub-assemblies, they require very little surface preparation at the point of attachment. Their disadvantages include a rather high gauge-to-gauge scatter, a considerable mechanical complexity, and some large thermal drift effects. While it is useful to know that these gauges exist, it does not seem that they are likely to replace conventional strain gauges for tablet compression studies at or near room temperature.

Magnetostrictive gauges

Yet another physical principle has been applied in the magnetostrictive strain gauge described by Wun-Fogle et al. (1987). Here the sensing element is an amorphous ferromagnetic ribbon that has been annealed in a magnetic field so that the residual magnetic moment is at right angles to its long axis. The application of an external magnetic field produces dimensional changes in this ribbon; conversely compression or extension causes large changes in its magnetic permeability.

In the work described by these authors, transversely annealed amorphous ribbons (Metglas 2605SC) were bonded to the upper and lower surfaces of aluminium beams, while the permeability of the material was measured via a set of coils operating at a frequency of 1 kHz. The beam was subjected to cyclic loading at frequencies ranging from 0.01 to 1.0 Hz, and the resultant permeability changes were plotted against strain. The equivalent 'gauge factor' appeared to be around five orders of magnitude higher than would be expected for resistive strain gauges, and it was possible to obtain significant changes in permeability with strain levels below $10\,\mu\varepsilon$.

The fibre optic strain gauge

We have seen that forces are nearly always estimated indirectly by the strain that they produce in a component. On this basis, any device that is able to measure displacement with a sufficient degree of sensitivity and accuracy can also be employed as a force gauge.

Changes in optical properties can also be employed in the measurement of strain. The development of fibre optic transducers, in fact, has probably started one of the most significant trends in modern instrumentation. At one time, fibre optics merely represented a convenient means of conveying light to otherwise inaccessible places. It is now able to provide the basis for a wide variety of intrinsic sensors, in which the properties of the fibre are modified by its interaction with the parameter that is to be measured. Fibre optics used in this way has given rise to a range of sensitive transducers that are totally immune to the effects of electrical interference.

Several designs of optical displacement transducer have been reported. Probably those with the greatest sensitivity, and the most suitability to strain gauge work, use the principle of the interferometer. In such an arrangement, light from a common source is split into two beams of roughly equal intensity. The two beams are directed along fibre optic elements, and are subsequently re-combined at the detector. If either of the optical paths is varied because of changes in the geometry of the fibre optic, phase changes between the two beams can be measured at the detector. It is possible for one of the two fibres to be attached between two points on a surface so that strain in that surface is communicated to the fibre. Then the fibre becomes an intrinsic strain sensor. Uttam *et al.* (1985) have described a system that operates over the range $0.1–1000\,\mu\varepsilon$, using a frequency modulation detector arrangement run through a standard FM radio receiver that acts as a demodulator. When a helium–neon laser operating at 633 mm was used as the light source, phase changes were produced at 11 radians $m^{-1}\mu\varepsilon^{-1}$. The applications of fibre optic strain gauges have been described by Culshaw (1983).

Laser speckle strain gauging

When an object is illuminated by laser light, it exhibits a characteristic 'speckle' pattern over its surface. The speckle pattern is produced by interference between reflected rays of the coherent light and is dependent on the fine structure of the object surface. If the surface is deformed by strain, the speckle pattern is displaced accordingly, and so the speckle displacement can be used as a measure of surface strain. An automatic method of evaluating the speckle displacement by a cross-correlation technique has been proposed by Yamaguchi (1981). The suggested advantages of such a method were that it involved no contact with the test object and that, in principle, it could be used to measure punch strains without the usual problems of telemetry. Its perceived disadvantage was that its sensitivity was believed to be only approximately $20\,\mu\varepsilon$.

References

Abramson N (1981). *The Making and Evaluation of Holograms*. London: Academic Press.

Britten JR, Barnett MI, Armstrong NA (1995). Construction of an intermittent-motion capsule filling machine simulator. *Pharm Res* 12: 196–200.

Burch JM, Forno C (1975). A high sensitivity moiré grid technique for studying deformation in large objects. *Opt Eng* 14: 178–185.

Chalmers GF (1982), Materials, construction, performance and characteristics. In Window AL, Holister GS (eds), *Strain Gauge Technology*. London: Elsevier Applied Science, pp 1–38.

Cocolas HG, Lordi NG (1993). Axial to radial pressure transmission of tablet excipients using a novel instrumented die. *Drug Dev Ind Pharm* 19: 2476–2497.

Culshaw B (1983). Optical systems and sensors for measurement and control. *J Phys E Sci Instrum* 16: 978–986.

Dudderar TD, Gilbert JA (1985). Real-time holographic interferometry through fibre optics. *J Phys E Sci Instrum* 18: 39–43.

Ennos AE, Thomas K (1978). Holographic measurement of distortion of plastics after injection moulding. *Plastics Rubber Processing*, December: 165–171.

Erdem U (1982). Force and weight measurement. *J Phys E Sci Instrum* 15: 857–872.

Forno C (1975). White-light speckle photography for measuring deformation, strain and shape. *Optics Laser Technol* 7: 217–221.

Frocht MM (1948). *Photoelasticity*. New York: John Wiley.

Hoblitzell JR, Rhodes CT (1985). Recent developments in the use of instrumented tablet presses. *Pharm Int* 6: 45–49.

Hoffman K (1990). *How to Avoid or Minimize Errors in Strain Gauge Measurement*. Darmstadt: Hottinger Baldwin Messtechnik GmbH.

Hornby IW (1982). The vibrating wire strain gauge. In Window AL, Holister GS (eds), *Strain Gauge Technology*. London: Elsevier Applied Science, pp 327–348.

Johannson E (1981). *Pressductor Force Transducers: Principle of Operation.* [Pamphlet YM21–10E] Manchester, UK: ASEA.

Langdon RM (1985). Resonator sensors: a review. *J Phys E Sci Instrum* 18: 103–115.

Lee A (1981). The Tay Bridge. In *The Innovative Engineer.* London: Morgan-Grampian, pp 42–47.

Mansfield PH (1985). Electrical resistance strain gauges. *Transducer Technol* 8: No 1, 17–19; No 6, 9–11.

Mountain DS, Webber JMR (1978). Stress pattern analysis by thermal emission (SPATE). *Proc Soc Photo-Opt Inst Eng* 164: 189–196.

Neubert HKP (1975). *Instrument Transducers, an Introduction to their Performance and Design,* 2nd edn. Oxford: Clarendon Press.

Pople J (1979). *BSSM Strain Measurement Reference Book,* Newcastle upon Tyne, UK: British Society for Strain Measurement.

Procter E, Strong JT (1982). Capacitance strain gauges. In Window AL, Holister GS (eds), *Strain Gauge Technology.* London: Elsevier Applied Science, pp 291–326.

Ridgway K (1966). The use of photoelastic techniques in the measurement of die wall stress in tabletting. *J Pharm Pharmacol* 18: 176S–181S.

Ridgway Watt P (1981). Holographic interferometry in tablet relaxation measurement. *J Pharm Pharmacol* 33: 114P.

Robertson ER, King W (1983). Holography and its applications. In Luxmore AR (ed.), *Optical Transducers and Techniques in Engineering Measurement.* London: Elsevier Applied Science, pp 161–201.

Rowley DM (1979). A holographic interference camera. *J Phys E Sci Instrum* 12: 971–975.

Schmidt PC, Koch H (1991). Single punch instrumentation with piezoelectric transducers compared with a strain gauge on the level arm used for compression force–time studies. *Pharm Ind* 53: 508–511.

Seippel RG (1983). *Transducers, Sensors, and Detectors.* Reston, VA: Reston Publishing.

Sixsmith D (1977). Instruments for tablet technology. *Mfg Chem Aerosol News* 1: 17–21.

Uttam D, Culshaw B, Ward JD *et al.* (1985). Interferometric optical fibre strain measurement. *J Phys E Sci Instrum* 18: 290–293.

Weingartner I (1983). Holography: techniques and applications. *J Phys E Sci Instrum* 16: 17–23.

Welsh BL, Pyne CR (1980). A method to improve the temperature stability of semiconductor strain gauge transducers. *J Phys E Sci Instrum* 13: 816–818.

Williams JJ, Stiel DM (1984). An intelligent tablet press monitor for formulation development. *Pharm Technol* 8: 26–38.

Wun-Fogle M, Savage HT, Clark AE (1987). New magnetostrictive strain gauge demonstrates high sensing potentials. *InTech* 34: 51–54.

Yamaguchi I (1981). A laser-speckle strain gauge. *J Phys E Sci Instrum* 14: 1270–1273.

Yeh C, Altaf SA, Hoag SW (1997). Theory of force transducer design optimisation for die wall stress measurement during tablet compaction. Optimisation and validation of split web die using finite element analysis. *Pharm Res* 14: 1161–1170.

3

The installation of strain gauges

Anton Chittey

Introduction

Installing strain gauges is generally not difficult, but there are many possible pitfalls to consider. Meticulous cleanliness and attention to detail are essential to get accurate reliable data. A well-specified and executed installation should last many years of service, whereas a sloppy or haphazard approach could render any data invalid.

The installation process will be covered in some detail but it cannot cover every eventuality. It may require more research or advice from your strain gauge supplier for specialist or unusual applications. Many instructions are available from manufacturers, some of which are listed in the bibliography.

This section starts with the assumption that the strain gauges, location and arrangement have already been specified. The processes that must be followed are:

1. Surface preparation
2. Adhesive bonding
3. Leadwire attachment
4. Protection.

However, before any of the processes are attempted, thought must be given to health and safety considerations.

Health and safety considerations

The processes and especially the chemicals and materials used, are considered to be hazardous to health. In many cases, common sense will enable the installer to avoid obvious hazards from exposure to toxic chemicals. This includes physical protection for eyes and hands and good ventilation to avoid inhalation of fumes. Less obvious are hazards by ingestion and from flammable compounds. Some materials, such as beryllium copper, are toxic if their particulates are inhaled.

Good hygiene and cleanliness are essential, not only to protect the installation but also to avoid ingesting compounds (for example the lead in lead-based solder). It is obvious to avoid eating, drinking and smoking in the working area as well as to provide ventilation and extraction.

Despite your best efforts, situations may occur such as spillages that may require immediate action. First aid facilities and eyewash stations must be provided. The chemicals and compounds used will be supported with Material Safety Data Sheets specifying the precautions and actions in the case of accidental or excessive exposure to the product. Additionally a COSHH (control of substances hazardous to health) assessment should be carried out before work commences to ensure the risks are minimised and that a suitable procedure is in place in case of emergency.

Surface preparation

The strain gauges must be attached to the surface of the component as intimately as possible to accurately reflect the movement of the substrate. A common problem with strain gauge installations, and the obvious first problem when poor overall performance is experienced, is insufficient understanding of what is meant by surface preparation. The three aims of preparing a component for strain-gauge installation are to achieve a surface that is

- mechanically sound
- chemically clean
- chemically inert.

The first of these, mechanical soundness, is obvious. The surface must be free from corrosion and visible surface finishes, including paint and plating. Removing surface plating is especially important, since plating may be considered a good surface for bonding as it can appear to be sound. However, many plated finishes are not an integral part of the base metal. Anyone who has seen chrome plating peeling or flaking off a spanner will be witness to this fact.

To achieve 'chemical cleanliness' requires the removal of all the invisible contamination present on all surfaces. Oil, grease, fingerprints and any other environmental contaminants must be completely removed. It is not generally considered sufficient to wipe the surface with a clean cloth or even to use detergent. This may sometimes apparently work but will be inconsistent and may compromise the overall accuracy and stability. Considering the time spent on the overall process, it is not prudent to take short cuts.

Finally the surface can be left with an acidic pH from the previous process, so a neutraliser stage is necessary to give a neutral pH. Many adhesives will simply not adhere to an acidic surface, so failure at this final stage can cause significant problems later on, from an obvious failure to bond to more subtle instability problems once the installation has been calibrated.

The process involved in achieving the above-mentioned surface criteria will now be described in some detail.

Surface degreasing

Before proceeding further, the surface must be thoroughly degreased. This is to prevent grease being driven into the surface by later processes, especially that of abrading. A range of degreasers can be used, providing they are compatible with the substrate material. Aerosol degreasers are especially effective and have the advantage of being in a one-way container, thus preventing contamination. Less effective but considered safe on many materials including plastics are solvents such as isopropanol, but this is highly flammable. Unsuitable degreasers include many aqueous or citrus-based products.

Try to degrease the whole of the component. Failing this, degrease a much larger area than the gauge installation actually needs to prevent accidental contamination in later stages of the installation process.

Surface abrading

The surface must be abraded to remove obvious signs of instability, such as plating, paint and corrosion. Surface imperfections must also be removed as many adhesive systems will not fill voids. A suitable finish is essential to ensure a good 'key' for the adhesive: too smooth is undesirable, too rough and the gauge and the adhesive may not perform adequately.

A variety of processes may be necessary to achieve the desired finish. Wire brushes can be useful for the removal of paint and corrosion. Rotary brushes, driven by an angle grinder or drill, are especially effective. For more serious defects, a flexible grinding disk or abrasive flap wheel can rapidly remove significant amounts or material and leave a good surface finish. However, they can also be quite aggressive if not used carefully. Hobbyist hand-held low-voltage drills are available with a variety of miniature attachments, making them useful in many situations.

Files may be used but with great care in order to avoid gouging the metal surface. So-called 'needle files' are useful for areas with limited access, such as the interior of holes.

Final surface abrading is normally manual using silicon carbide paper of varying grades. Abrading should progress from working from a coarse grade to using finer grades until the abrasion marks from the previous grade are completely covered. The 220 grit paper is quite coarse, whereas 320 grit (for steel) or 400 grit (for aluminium) are ideal for final finishing. Where the surface material is compatible, an acidic conditioning solution should be used to aid the final abrading process. This conditioner is normally a mild phosphoric acid and helps to etch the substrate while removing contaminants.

Air abrading (grit blasting) is used in situations where lots of items must be prepared. 100–400 grit alumina powder is ideal in combination with a clean air supply. However, a poor surface can be masked by this process, so you must ensure a flat, even surface is present before blasting.

Gauge location lines

The overall performance of the installation depends in part on the accurate alignment of the gauges to the component to be gauged and to each other, assuming a multiple-gauge layout. Clear, accurate, unambiguous alignment marks are essential and can easily be achieved with a few considerations.

Firstly, a line should not be scratched as this can be detrimental to the strain gauge, the adhesive and the component under test. Instead a burnished line enables accurate installation with no effect on the overall measurement. On aluminium and other soft metals, a 4H graphite pencil can give good results, though care must be taken on high-strength alloys as graphite can cause embrittlement. On steel, an ink-free ballpoint pen held normal to the surface is ideal: alternatively a brass rod with a rounded point can be used. Marking out will recontaminate the surface and so this must be removed in the next step.

Surface conditioning

To remove all of the remaining contamination, the surface must be scrubbed with conditioning compound (assuming compatibility with the substrate material) and a cotton-tipped applicator, until a clean tip is no longer discoloured. The conditioner must not be allowed to dry on the surface: it must be removed with a clean tissue or medical-grade gauze pad. The drying process starts by wiping from the centre of the prepared area out to one edge. Then a fresh gauze pad is used to wipe from the centre out towards the opposite edge. At no stage should the gauze pad be returned to the centre and used again as this will introduce contamination from the unprepared area. This conditioning stage is repeated until further cotton-tipped applicators remain clean.

Neutralising

Many adhesives will not bond to an acidic surface. The previous section advises, where compatibility can be ascertained, to use an acidic conditioning solution. Therefore, a further stage is necessary to bring the surface to an approximately neutral pH (pH 7.5). To achieve this, a neutralising solution (weak ammonia solution) must be applied to the surface and scrubbed using cotton-tipped applicators. Then the solution must be dried in a similar way to that described in the last section by using a clean gauze pad to wipe from the centre of the prepared area to the edge and then again from the centre to the opposite edge with a fresh gauze pad. The neutraliser must not be allowed to dry on the surface as this will recontaminate the prepared area.

The surface is now ready for bonding. To avoid further contamination, the adhesive bonding must be carried out as soon as possible. Recommendations are available for different types of material, but a period of more than an hour or so is undesirable. Apart from the obvious threat of airborne contamination, the metal will start to re-oxidise.

Bonding with adhesive

There are many types of adhesive available for various specialist applications, but the three main adhesives can be grouped as follows.

- instant: cyanoacrylate adhesive, sometimes called 'superglue'

- cold-cure: 100% solids epoxy resin
- heat cure: solvent thinned epoxy-phenolic.

Note that general-purpose adhesives are not normally suitable, so a certified strain gauge adhesive must be used. It is tempting to use a low-cost alternative, but the implications of saving money here are significantly offset by the overall costs of installing and calibrating the transducer.

Each adhesive requires different treatment during installation and will have varying life expectancy. More details will be given below as each procedure is described.

Correct storage and usage is essential to ensure a reliable and repeatable installation. Some adhesives should be stored in a refrigerator before opening to ensure life expectancy up to the manufacturer's 'use-by' date. However, it is not generally recommended to refrigerate opened containers. If the product is to be used only occasionally, refrigeration is permitted. However, after removal from the refrigerator, the container must be allowed to warm up to room temperature before opening to avoid condensation in the bottle.

As a reminder, before proceeding it is essential to ensure meticulous cleanliness throughout the bonding process. The strain gauges will be clean in their pack, and so further cleaning of them is not required. Avoid manipulating the gauges by hand and ensure that any surfaces that they come into contact with are degreased with a solvent, followed by neutralisation.

Gauge handling

Until this point, the strain gauges should be left in their pre-sealed pack as this excludes contaminants. The gauges can now be removed from the pack using a pair of pre-cleaned gauge-handling tweezers and placed on a pre-cleaned gauge-handling area. A degreased and neutralised glass block is ideal. The gauge can then be picked up with a suitable adhesive tape for transfer to the pre-cleaned and marked surface: the positioning of the tape on the gauge and the type of tape is somewhat dependent on the adhesive being used. The tape is merely for handling and enables accurate positioning of the gauge to the alignment marks. As overall performance is affected by accurate alignment, it is desirable to spend some time getting the gauge precisely on the marks: a good-quality tape will enable repositioning many times without leaving any residue.

While the gauge is being handled in this way, it is in a vulnerable position. Excessively rough handling, such as creasing the gauge, can adversely affect its performance later on. Try to hold the gauge as flat as possible, especially when lifting the gauge from the handling surface or when repositioning.

During the bonding process, it is desirable to position a bondable terminal in a convenient location during the same process. Surface preparation and adhesive application should be identical to that of the gauges. While strain transmission is not required, a good anchor for the main leadwire, which may be of heavy gauge, is necessary to prevent failure.

Once positioned, the installation is ready for the adhesive application.

Instant adhesive

Cyanoacrylates are sometimes considered to be an all-purpose solution for many applications. However, this not true for strain gauge installations. This adhesive is really only suitable for short-term evaluation (up to a year is a typical recommendation) as it is moisture, time and temperature sensitive. It can, therefore, be used for 'stress analysis' applications and for evaluating the suitability of a component for conversion into a transducer. In fact, the performance of a cyanoacrylate certified for strain gauges is excellent in the short term (a few days or weeks) and can, therefore, closely match the performance of an installation designed for long-term use. Do not be tempted to use general-purpose cyanoacrylates (superglue) as this can seriously compromise the installation.

The benefits of this adhesive are many: it is clean, convenient, high performance and quick (instant) as well as being easy to remove. Commercial cyanoacrylate removers are available. It does, however, also have its drawbacks. It has a relatively small temperature window: -25

to +65°C long term. Many manufacturers will claim a much wider temperature range but this is not in relation to its ability to transmit strain. This will degrade with time, although the gauge will not actually 'fall off' as some people expect. Further, the peel strength is low even though its shear strength is very high. Do not be tempted to test the bond strength by trying to peel a gauge off the surface, as this will prove very easy.

Instant adhesive application

The gauge must be picked up with cellophane tape with the entire gauge inside the edges of the tape. Approximately 100 mm of 20 mm wide tape is generally suitable and convenient. Once positioned, the tape is pulled gently up to expose the back of the gauge and a catalyst applied according to the manufacturer's recommendation to the gauge only. One drop of the adhesive is applied to the tape–surface interface and the tape is then held by the free end close to the surface to enable the adhesive to spread to the width of the tape. A folded gauze pad is used with firm, even pressure to wipe through from behind the adhesive all the way through and over the gauge. Finally the gauze is disposed of and pressure is applied with the thumb on top of the gauge for 1 min.

After a further 2 min in normal circumstances, the tape can be removed by pulling one end right back over itself. If the adhesive is still liquid, the thumb pressure should be applied for a little longer and a few more minutes 'rest' allowed before attempting removal again.

There can be many reasons why a gauge does not bond, from insufficient or inappropriate surface preparation to low temperatures (below $-20°C$) or low humidity, since cyanoacrylates require a certain amount of humidity to polymerise. If there should be a failure, a telephone call to the strain gauge adhesive suppliers can help diagnosis.

Cold cure adhesive

Also called 100% solids epoxy, these adhesives are not really 'cold' cure but are in fact room temperature curing. They have no solvents and are a little more viscous than other types of strain gauge adhesive, so glue lines can be a bit thicker. This makes them suitable for filling porous and uneven substrates. They are very stable and can be used for a service life of many years. Once again, general-purpose epoxys should not be used: these are entirely unsuitable for strain gauge installation, especially as they may be very viscous. Instead use a certified strain gauge adhesive.

These adhesives are not as quick or as convenient as the instant adhesives. They must be thoroughly mixed in accurately defined ratios. This often leads to wastage as a minimum mix quantity, typically 10 g, must be made since smaller mixes can result in large mixing ratio errors. To ensure accuracy, these adhesives are generally available in pre-weighed kits. Bulk quantities are available but require a weigh scale accurate to at least 0.1 g, calibrated pipettes, suitable clean glass containers and plastic stirring and application rods.

Once mixed, they have a limited 'pot life' or working time of around 15 min, depending on environment. They must be clamped for the duration of cure, typically 6 h at room temperature (overnight is ideal) or shorter at slightly elevated temperatures. They will not generally cure at temperatures lower than 20°C.

Cold cure adhesive application

The gauge must be picked up with cellophane tape with the entire gauge inside the edges of the tape. Approximately 100 mm of 20 mm wide tape is generally suitable and convenient. Once positioned, the tape is pulled gently up to expose the back of the gauge ready for adhesive application.

The adhesive is mixed thoroughly according to the manufacturer's recommendation; typically 5 min of gentle stirring is necessary to ensure even and essentially bubble-free mixing. Then the adhesive is smeared on both the gauge and the substrate using the stirring rod. The smearing action should move away from the tape–surface interface to remove any bubbles or other inclusions away from the gauge location.

Gently the tape and gauge are brought down to the surface using a gentle wiping action with a gauze pad. Then a small piece of silicone gum is placed over the entire gauge and a small metal backing plate contoured to the surface profile is placed over the silicone pad; the assembly is then clamped using a suitable spring clamp. The clamping pressure should be within the range recommended by the strain gauge adhesive supplier. The system is left for the desired curing time (overnight is preferable) and then the clamp assembly is removed and the tape peeled back.

Heat cure adhesive

There are many forms of adhesive that require curing at elevated temperatures. As mentioned briefly above, a cold cure adhesive can have its cure time shortened by raising the temperature. Heat cure adhesives, by comparison, simply will not cure at room temperature. While there are some 100% solids epoxy adhesives requiring elevated temperature curing, this section is concerned with the solvent-thinned epoxy phenolics.

Adhesives in this group offer extremely thin, high-performance gluelines for extended lives of 20–30 years or more. They are the first choice for professional transducer manufacturers and are used, for example, in commercial weigh scales in 'legal-to-trade' situations.

Most but not all, are two component systems, so mixing is required. They will generally be supplied in two half-full bottles, one to be poured into the other and then shaken to mix. After standing for 1 h, they are ready to be used. Owing to the elevated temperature cure requirements, the pot life can be from 3 to 12 weeks depending on specific formulation and storage considerations.

Cure times are dependent on temperature and can be from 80°C for 4 h to 175°C for 1 h. Curing must start in a cold oven and the temperature must be raised slowly to avoid solvent entrapment. Furthermore stringent clamping is required and a further post-cure may also be necessary after removal of the clamps.

Heat cure adhesive application

Owing to the high curing temperature, cellophane tape is not suitable. Instead a high-temperature alternative (Mylar) is required. This should be placed across the terminals of the strain gauge to avoid solvent entrapment. Once positioned, the tape is peeled back to expose the back of the gauge for adhesive application.

Once the adhesive is mixed, it is applied to both the back of the gauge and the surface by painting it on with a small brush. This is often supplied as a brush integral to the cap of the adhesive bottle. The surfaces are allowed to air-dry for the recommended time and then the gauge and the tape are folded back on to the substrate.

Clamping is a little different as the gauge and the adhesive are now exposed. A thin Teflon slip-sheet must be positioned to cover the area before a silicone pad is applied plus a suitably contoured metal backup plate. Clamping can be more stringent, so careful attention should be given to the manufacturer's recommendations. Spring clamps are now essential, as raising the temperature may cause a major change in clamping pressure if mechanical clamps are used, owing to differential expansion between the component and clamp materials.

The assembly is placed into a cold oven and the oven is set to rise slowly to the required cure temperature. The component must have reached the desired temperature before the clock is started, as a large component may take some time to get to temperature. Once a suitable time and temperature have been reached, the oven is switched off and allowed to cool down slowly. The clamps are removed and, if required, the assembly is post-cured according to the manufacturer's instructions. While not always necessary, post-curing is essential for best accuracy and repeatability.

Leadwire attachment

Soldering can be one of the most challenging parts of strain gauge installation, and yet, with a consistent technique and appropriate tools, it can be both quick and easy. Common mistakes may easily be avoided: use a good-quality soldering iron (cheap irons are seldom suitable), use fresh solder every time a joint is touched up (fresh flux is actually what is needed) and try not to be too quick. The 2 s rule will ensure that there is enough heat in the joint without danger of burning the gauges or the terminals.

Suitable solder and leadwires should be chosen for the job. If these are too heavy, they will be difficult to handle, and there will be a danger of damaging the gauges.

Finally, thorough cleaning and removal of the flux residue is of paramount importance to ensure a stable output over time.

Intra-bridge wires

Generally fine single-conductor enamel insulated wires are used. Some have insulation that can be burnt off with a soldering iron and some must be mechanically stripped. In some situations, wire insulated with polyvinyl chloride or Teflon can be used, but this can be more difficult to work with.

Main leadwire

A multi-core cable, both twisted and shielded, is ideal for the main leadwire. For full bridge installations, this can be either 4 or 6 core and can be of any suitable thickness. In general, a thicker cable should be used for longer lead runs, while a finer wire must be used if the component is moving in order to maximise flexibility and minimise chances of fatigue failure. Some 4-core cables are available with diameters of approximately 1 mm.

For quarter-bridge installations using a single gauge for stress analysis, 3-core flat ribbon cables are easy to work with.

Solder

Ideally, lead-based solder with a rosin (flux) core of a small diameter (0.5 mm or less) should be used. However, as lead and rosin flux are both toxic, there are many lead-free and rosin-free products available. General-purpose solders and fluxes should not be used as many of these can be detrimental to the long-term stability of the strain gauges.

Soldering station

A user-selectable temperature-controlled soldering iron with an iron-clad screwdriver tip is essential for achieving good connections to the strain gauges and terminals. The iron must be set to the correct working temperature, some 30 to 50°C above the melting point of the solder, and the tip kept meticulously clean. Using a wet sponge is not a good idea as this can cause the tip to oxidise rapidly.

Soldering technique

To avoid damage and excessive solder flow, paper drafting tape is used to mask all but the small area where the wire is to be attached. This is especially important on the strain gauge, as the terminal area can be much larger than needed. In general, the exposed area should be a third to a half that of the terminal.

All points must be tinned, which means applying a small amount of solder to each wiring point. This is achieved by placing solder over the terminal, pressing the soldering iron firmly over the terminal and holding for 2 s. Immediately the iron and solder are lifted simultaneously from the terminal. A small low-profile mound of solder should be present. If it is not, the procedure is repeated.

The leads are prepared appropriately according to the type. Only a small amount of wire should be free from insulation to avoid possible paths to the substrate. The end of the wire is tinned then positioned over the terminal and secured using drafting tape. It is then soldered using the same technique as above, remembering to use more solder, as the flux core is essential to achieve good joints. The joint is pressed firmly and held for 2 s. This is repeated if necessary.

The intra-bridge wires must be in close contact with the substrate as they are routed. If they are not, any small change in temperature will be shown as a rapid zero shift because of the low thermal mass of the fine wires.

Cleaning

All signs of flux residue are removed using a suitable solvent. A mixture of toluene and isopropanol is a good method that can quickly dissolve the flux and also effectively remove any drafting tape. As with any cleaning process, the suspension of the contaminants must be blotted off the surface using a gauze pad. The solvent should not be allowed to evaporate as this will redistribute the flux on to the installation.

Protection of the installation

Some form of protective coating must be applied to the installation. This is to provide both environmental and mechanical protection and should cover the gauges, leads and the adhesive.

Suitable compounds must be used as recommended by the strain gauge supplier. General-purpose compounds must not be used. For example, silicone rubber (RTV) is a good product but must be free from corrosive ingredients. The type used as a kitchen and bathroom sealant is based on acetic acid and will destroy the installation in minutes. Many other compounds are slightly conductive and can, therefore, compromise the installation.

There are many products available, and these are selected according to the required level of mechanical and environmental protection, as well as the length of time the installation must survive. Some examples are listed below.

Paint-on (air-drying) products. Acrylic and polyurethane varnishes are convenient single-component products. Painted on in thin layers, they generally air-dry in 20 min but must be allowed to fully cure for 24 h if further coatings are to be applied. If used on their own, they are only suitable for low-humidity laboratory conditions and provide no mechanical protection as they form only a thin layer. They are very useful for anchoring the intra-bridge wires.

Silicone rubber. This is a convenient coating that offers good moisture and mechanical protection. There are both self-levelling and non-drip varieties. Applied in 2–3 mm layers, it cures in approximately 6 h.

Microcrystalline wax. This is an excellent moisture barrier and is used by many professional transducer manufacturers. It has obvious temperature limit issues as it melts around 65°C and offers little mechanical protection.

Polysulphide rubber. This two-component coating offers excellent mechanical and moisture protection, including for immersion in water. It is conductive and is normally used in conjunction with an insulating layer such as bondable Teflon sheet or a paint-on coating as described above. If used with an air-drying coating, this must be allowed to fully cure before applying the polysulphide. Otherwise there can be an interaction between the two.

Inspection and testing

To ensure a good installation, a continuous inspection process should take place throughout the installation procedure. This starts with visual inspection and, in later stages, includes electrical testing.

The adhesive

The first inspection should take place immediately after bonding. The gauges should be accurately positioned to the alignment marks, within 0.4 to 0.5 mm is realistic. Furthermore, the glue line should not have any voids (seen as lighter areas) or inclusions (lumps) under the grid. The bond cannot be checked by attempting to peel the gauge as some adhesives have very low peel strength.

Solder and leadwires

The intra-bridge leads should be neat and undamaged. The tinned end should entirely overlap the edges of the gauge and terminals to avoid the possibility of shorting to the substrate. Main leads should be similarly positioned and the insulation must be free from burning from the soldering iron. This is especially common when polyvinyl chloride (vinyl) leads are used.

Solder joints are becoming more difficult to inspect. Lead-based solders are easy to work with and easily produce beautiful joints. An ideal joint should be bright and shiny, with a low profile (more like a small bump than a ball) and be free from spikes and holes. Joints made from lead-free solder must meet most of the above criteria, but there is one big difference. Generally they produce a slightly dull 'crystalline' finish. This makes it more difficult to differentiate a good joint from a 'dry' one.

Electrical tests

The installation may fail despite rigorous visual inspection and so must be tested electrically. These tests must be performed both before and after applying a protective coating to ensure that the coating or its application have not affected or damaged the installation and to permit early rectification of faults.

Basic continuity is the first check. Are all the wires connected? Is there a short circuit? Judgements can be made knowing the gauge resistance as to whether everything is in order. For single gauge (quarter-bridge) installations, it is obvious that the resistance should be close to the gauge resistance shown on the packet the gauges came from, typically 120 or $350\,\Omega$. For full-bridge installations, the resistance between the two power leads or the two signal leads should equal the resistance of one arm. Between any power and any signal lead, there will be three-quarters the resistance of one arm. Thus for a typical four-gauge bridge with $350\,\Omega$ gauges, the resistance across power leads should be $350\,\Omega$ and the resistance between any power lead and any signal lead should be approximately $262\,\Omega$. These values are approximate since they do not take lead resistances and tolerances in the gauge resistance into account.

Finally, the installation should be checked for isolation from the substrate. This cannot be effectively measured with a multimeter in the ohmic range as this instrument is simply not sensitive enough. A purpose-built insulation tester must be used. The test voltage must be adjustable to a low setting; 25 V is typical but test voltages of over 100 V should never be used as this can cause the insulation of a perfectly sound insulation to fail. Before protection, it should be possible to achieve a resistance of $2 \times 10^{10}\,\Omega$ ($20\,G\Omega$) and this should not reduce to below $1 \times 10^{10}\,\Omega$ after protection.

Load tests

After performing all the above testing and inspection, there is still no absolute way to guarantee the installation. The only way to do this is to connect it to a suitable instrument, apply and then remove a reasonable load and monitor the output. A good installation will be stable at the initial zero, will be stable at load and then return back to the original zero point. Instability can indicate voids under the grid, ineffective surface preparation or poor adhesive application and curing. Further tests can be carried out for a full bridge as it will normally be desirable to calibrate it to a known load. When checking calibration, the load is normally applied and removed incrementally to ascertain linearity, repeatability and lack of hysteresis.

Specialist applications

There are many applications that require additional information, skills, tooling and ingenuity. For example, there may be restricted access to the gauging area: perhaps in a narrow slot, an internal bore or a small hole. Higher temperatures or unusual substrate materials need special attention, and in some installation environments, there may be other suitable techniques and products such as weldable gauges.

Leadwire attachment to unbonded gauges

It is obvious that after gauges have been installed in holes or small gaps and channels, the gauges may not be accessible to solder the leads. On a large bore with gauges close to one end, there is not really a problem, but smaller bores and gaps will require the gauges to be pre-leaded. Gauges can either be purchased with leads already attached or fine wires can be pre-attached using normal solder procedures with some slight modification. The main terminal can either be installed in a more convenient and accessible location elsewhere on the component or immediately inside the bore for easy access.

The leadwire used should be perhaps 34AWG maximum to avoid damaging the gauge during handling; enamel-insulated solid copper wire is ideal. Firstly, a flat smooth-surfaced aluminium plate is cleaned using degreaser and neutraliser. The gauge is taped to the aluminium plate with paper drafting tape and the procedure described above is followed. The solder flux residue must be thoroughly cleaned away and the tape removed using rosin solvent. The top of the gauge must be clean before it is turned over, with the leads held, so that the back can be cleaned with more rosin solvent. The gauge should be gently blotted dry. Finally a cotton-tipped applicator dampened in neutraliser is gently used to wipe the back of the gauge. The gauge is placed back into its packaging to protect it until it is installed.

Gauge positioning

The problem is that gauges down a hole or bore must be installed 'blind', so accurate positioning can be a challenge. The leaded gauges must be temporarily attached to a flexible carrier such as Mylar film. Double-sided tape or rubber glue is sufficient as it must be removable. This carrier should be longer than the distance from the edge of the hole to the desired location so that the carrier can be marked with a line. It can then be rolled back, inserted to the line and secured with tape to the edge of the hole.

Clamping

This is the most difficult part of the process and may require the building of custom clamping rigs.

For larger holes, a rubber bung with a hole through the middle can be machined, the bung being slightly smaller than the bore. The central hole has a bolt inserted with washers at each end of this 'cotton reel' so that the rubber bung will barrel out when the bolt is tightened and so clamp the gauge.

For smaller holes, a narrow silicone tube, blocked at one end and then pressurised, can be adequate. Alternatively, a round bar can be inserted and then pushed in the opposite direction to the gauge location; this does not give even pressure over the gauge and so must be used with caution.

For other situations a custom rig can be built, with arms that can be moved by a cam connected to a threaded bar or bolt similar to an adjustable reamer. As the bolt is turned, the arms move outwards to apply pressure. The custom rig requires some ingenuity to design and build and it can be expensive. For obvious reasons, this solution should only be pursued if there are many installations of the same type or if there is simply no other way.

High-temperature installations

Conventionally bonded strain gauges can be used up to 260°C for extended periods and up to 400°C for special or short-term applications. Careful selection of adhesives, leadwires and solders is necessary to ensure survival. Beyond 400°C, the conventional gauge can no longer be used.

For temperatures up to perhaps 1000°C, free-filament gauges are available. These are basically a special strain gauge with a temporary 'strippable' backing or fibreglass carrier. The film is bonded using ceramic slurry or flame-sprayed ceramic. Leads are attached by spot-welding. A high degree of operator skill is necessary and so this task is best left to specialists.

Weldable strain gauges

For some applications, it is possible to spot-weld a gauge to a surface. The weldable strain gauge consists of a carrier (stainless steel or Inconel) with a gauge attached with epoxy–phenolic adhesive or ceramic cement. A capacitive–discharge spot-welder is then used to run a series of miniature welds around the edge of the shim.

This offers advantages in some situations. The necessary surface preparation is limited to basic grinding and degreasing and the installation time is greatly reduced. Many weldable gauges are pre-leaded and some are even pre-protected, so they require no further installation time after welding.

The disadvantages are obvious. Hitherto emphasis has been laid on mounting a gauge with a thin adhesive line, but we have now added a shim perhaps 0.1 mm thickness. This can reduce the sensitivity of the installation and reinforce thin sections. Further, it is only suitable for ferrous substrates. Gauges of this type are mainly used for stress analysis applications on large structures such as bridges, rails and oilrigs.

Special materials

Some materials need special treatment during surface preparation to ensure the bond will be successful and that the substrate will not be damaged.

Titanium must be heat treated to 175°C three times and degreased at each return to room temperature. Failure to follow this regimen has resulted in reports of the gauge falling off.

Copper and some copper alloys should not come into contact with ammonia, so the neutraliser stage should be replaced with thorough cleaning with distilled water.

Porous materials including cast iron should be locally warmed with a heat gun after surface preparation as some of the chemicals may still be in the surface layer and compromise the long-term stability of the installation.

Plastics and composites can be sensitive to solvents and so compatibility must be established. Unfortunately these materials may also be contaminated with silicone-release agents during manufacture and so the degreasing stage cannot be ignored. Isopropanol is a good choice for degreasing, followed by a thorough cleansing with conditioner solution that has been heated to 65°C. Some plastics, most notably polyethylene, are completely resistant to cyanoacrylate adhesives so epoxy adhesives must be used.

There are many other materials that need special attention, but the above examples are given to demonstrate common pitfalls. If there is any doubt, then the strain gauge supplier should be able to advise on surface preparation procedures and adhesive compatibility.

Tools and installation accessories

Some tools necessary for successful strain gauge installation can be found in a general-purpose toolkit, but many are specific to strain gauging and fine electrical work. Some general-purpose tools, especially some soldering irons, are simply not suitable for use in this environment.

Consumables

Chemicals

A form of degreasant is required, ideally in an aerosol container to avoid contamination of the contents. Isopropanol is an effective product that is safe on many plastics.

Surface conditioner is normally a weak phosphoric acid solution for effective contaminant removal.

Surface neutraliser for final surface preparation is usually a weak ammonia solution.

Rosin solvent is used for flux removal after soldering. Aqueous solutions are unsuitable for this, but a mixture of toluene and isopropanol is ideal.

There is a wide choice of adhesives (around 30 from one manufacturer alone), but a shortlist can generally be made with advice from the strain gauge supplier.

Abrasives

Surfaces can be ground, filed, polished, grit-blasted or hand abraded. Files are acceptable for removing large amounts of material but not for the final polishing stages. Here are some other recommendations.

Silicon carbide papers can be used wet or dry. Usually they are used wet to avoid clogging and to assist in the cleaning process. Conditioner is recommended as the wetting medium where compatibility has been ascertained. Various grades are available from coarse to fine, and generally three grades (220, 320 and 400) are useful for a variety of materials and surface finishes.

Grinders of many descriptions are available, ranging from angle grinders and drill attachments to hobbyists' tools. The larger tools are suitable for large rough structures but can be too aggressive. Hobbyists' drills, normally small 12 V powered hand-held devices, are available cheaply and have many attachments, ranging from small wire brushes and hard grinding wheels to abrasive flap wheels and flexible disks.

Cleaning materials

The chemicals used in installation will require, at various stages, to be dried from the surface or applied vigorously (described as 'scrubbing'). Clean tissues from a box can be sufficient, but gauze pads from medical suppliers are ideal. Cotton-tipped applicators are used extensively for scrubbing and should be of the wooden-handled variety.

Leadwires

A range of wire types is necessary for different applications. Solid copper enamel-insulated wire, 34AWG, is ideal for intra-bridge wires. Main leads should ideally be 4- or 6-core twisted and screened in either polyvinyl chloride or Teflon insulation for full bridge installations. When installing quarter-bridges, three-lead ribbon cable is convenient to use.

Solder

Solder is preferred to be of low melting point, normally lead based (60% lead, 40% tin, sometimes called '60/40') with a flux (rosin) core. Lead-free solders have a higher melting temperature (greater than 220°C compared with 183°C for 60/40) and require greater skill to use. Solders of smaller diameter are easier to work with: 0.5 mm is ideal for most if not all strain gauge work. Larger diameters are not generally required or recommended.

Terminals

All strain gauge installations require an interim terminal between the gauge and the main leadwires, though some gauges are already fitted with an integral terminal. Terminals are normally in a 'dog-bone' shape and can also be in a 'Y–I' configuration for three-lead quarter-bridge installations. They are available in many sizes, and can be cut to suit the application as they come in strips of eight or more. They must be bonded in a similar way to the strain gauges, usually during the same bonding process.

Gauge-handling materials

Suitable gauge-handling tape comprises cellophane for room temperature curing or Mylar for elevated temperature curing. The tapes must be of good quality and preferably from the strain gauge supplier to ensure they do not react with the adhesive or leave a mastic residue behind during gauge placement.

Paper drafting tape (low tack) is used for anchoring leads during soldering and other situations where temporary fixing is required. Masking tape is not recommended as this has a high tack and can leave residues on removal.

Clamping consumables include Teflon film, silicone gum (2–3 mm thick) and some form of backup plate such as 1 mm aluminium plate that can be formed easily by hand.

Gauge protection

A couple of protection materials can be useful to have available for general applications. Air-drying polyurethane or acrylic varnish is convenient and dries in approximately 20 min. Corrosion-free silicone rubber (RTV) is a single-component room temperature coating offering good mechanical protection. For more specialist applications, the strain gauge supplier will give advice.

Tools

Tweezers

Small tweezers are useful for handling at various stages of the installation process. The most obvious requirement is for gauge handling. Tweezers must be round-nosed, flat (i.e. unserrated) and undamaged. This is to ensure easy cleaning and that they will not damage the gauge during handling. Finer needle-point tweezers can be used for intra-bridge wiring, as can locking tweezers, but they must be used with care to avoid scratching or damaging the enamel insulation.

Layout marking tools

Once the position of the gauges has been decided, clear unambiguous lines must be marked on the surface. Ball-point pens that have run out of ink are good, as are 4H pencils. To locate the lines, a surface table with a height gauge, 'V' blocks and a dividing head is ideal, but vernier callipers and a steel rule are useful, though with a lower accuracy.

Cutters

Diagonal or end (flush) cutters are used for lead-wires. These can be used for wire stripping but purpose-made strippers are preferable to avoid damaging the conductors. Surgical scalpels are good for trimming and scraping operations if necessary.

Soldering

A good-quality temperature-controlled iron with a 1.5 mm iron-clad screwdriver tip is essential. The tip temperature must be adjustable by the operator for a variety of situations. General-purpose 'hobbyist' soldering irons are not recommended as the tip temperature can be extremely high. Fixed temperature controlled irons are usually unsuitable as the hysteresis (temperature variation) can be excessive. Portable gas irons should never be used.

Other hand tools

Miniature snipe-nosed pliers, small scissors (surgical shears) and dental picks can be useful additions to the toolbox as are spatulas, small camel-hair brushes and, for clamping during adhesive curing, a variety of spring clamps. A heat gun may also be useful when leads need to be spliced and protected with heat-shrink tubing.

Complete kits

For the dedicated strain gauge installer, complete kits are available that contain all the tools and accessories required in a variety of situations. These will generally also include adhesives, lead-wires, protective coatings and a high-quality soldering station.

Professional assistance

Strain gauge installation can be daunting but there are many resources available to assist. The costs to set up strain gauge installations can be significant and it may not be economic for only one or two installations. Therefore, there are many professional companies who can offer the full service, from recommending the products and gauge arrangement to providing a complete service from specification to a working product.

If installation is part of a continuous programme, there are training courses provided by both strain gauge suppliers and independent organisations. The British Society for Strain Measurement (BSSM) provides programmes as well as a certification scheme that gives evidence of competence in strain gauging. Additionally BSSM provide a s handbook showing many common types of installation (Pople, 1979) and a *Code of Practice* booklet detailing standard procedures for ensuring quality and consistency (British Society for Strain Measurement, 1992).

References

British Society for Strain Measurement (1992). *Code of Practice for the Installation of Electrical Resistance Strain Gauges CP1*. Newcastle upon Tyne, UK: British Society for Strain Measurement.

Pople J (1979). *BSSM Strain Measurement Reference Book*, Newcastle upon Tyne, UK: British Society for Strain Measurement.

Further reading

Adlington J, Mordan G, Chittey A (1996). *Resistance Strain Gauge Load Cells*. London: EAL Publications.

Vishay Measurements Group. *Application Note B-129–8: Surface Preparation for Strain Gage Bonding*. Basingstoke, UK: Vishay Measurements Group.

Vishay Measurements Group. *Application Note TT-603: The Proper Use Of Bondable Terminals In Strain Gage Applications*. Vishay Measurements Group.

Vishay Measurements Group. *Application Note TT-608: Techniques for Attaching Leadwires to Unbonded Strain Gages*. Basingstoke, UK: Vishay Measurements Group.

Vishay Measurements Group. *Application Note TT-609: Strain Gage Soldering Techniques*. Basingstoke, UK: Vishay Measurements Group.

Vishay Measurements Group. *Application Note TT-610: Strain Gage Clamping Techniques*. Basingstoke, UK: Vishay Measurements Group.

Vishay Measurements Group. *Application Note VMM-15: Quality Control of Strain Gage Installations*. Basingstoke, UK: Vishay Measurements Group.

Vishay Measurements Group. *Instruction Bulletin B-127–14: Strain Gage Installations with M-Bond 200 Adhesive*. Basingstoke, UK: Vishay Measurements Group.

Vishay Measurements Group. *Instruction Bulletin B-130: Strain Gage Installations with M-Bond 43–B, 600 and 610 Adhesive Systems*. Basingstoke, UK: Vishay Measurements Group.

Vishay Measurements Group. *Instruction Bulletin B-137: Strain Gage Applications with M-Bond AE-10, AE-15 and GA-2 Adhesive Systems*. Basingstoke, UK: Vishay Measurements Group.

Vishay Measurements Group. *Strain Gage Based Transducers: Their Construction and Design*. Basingstoke, UK: Vishay Measurements Group.

4

The measurement of displacement

Peter Ridgway Watt

Introduction

Displacement, in the context of tablet press instrumentation, generally involves such actions as the movement of a punch into a die or the rotation of a machine component. It may, therefore, be a linear movement of a few millimetres or perhaps an angular movement of some degrees. Such movements as these can be measured quite accurately in a variety of ways with commercially available transducers. There are, in fact, many types of displacement transducer on the commercial market, and it is possible to classify them in a variety of ways, but for the purposes of this chapter they have been grouped into those that generate an analogue signal and those that give a direct digital output.

In general, those in the first group provide a signal that is directly related to the position in space of some movable sensing element. The output may be described as 'absolute' in the sense that it uniquely defines that position, even if power to the device is interrupted and then reconnected. The conventional sliding potentiometer, used as a volume control in amplifier systems, has just such a characteristic.

In the second group are those transducers that generate a direct digital output rather than an analogue voltage. Some of these, like the analogue devices in the previous group, also provide absolute positional information. However, the simpler digital devices do not do this but only give incremental signals as they are moved. Thus each pulse of the output signal represents a given displacement of the sensing element from its previous position. Although the incremental type of digital transducer is generally much less expensive than the absolute type, it must be re-set whenever its power supply has been interrupted.

Both these forms of transducer measure the position of some sensing element in relation to a particular reference point. However, there is an alternative possibility of inferring the sensor position indirectly, and that is from continuous measurements of its velocity or acceleration during a known time. Some dynamic measurements of this kind have been reported in the general literature of tablet compression studies. For example, laser Doppler velocimetry was used by Depraetère *et al.* (1978) to study punch movement in a tablet press. Punch travel itself tends to be limited by the mechanics of the press to relatively short distances, seldom exceeding 50 mm. Transducers for that particular range were reviewed in 1979 by Garratt. In practice, the range of particular interest for compression studies is likely to be even more limited and will only involve those few millimetres in which the punches are in contact with the powder bed. Short-range, or micro-displacement transducers were discussed in 1972 by Sydenham in a paper that contains much useful information.

Displacement transducers with analogue output

The linear variable-differential transformer

The linear variable-differential transformer (LVDT) is probably the most widely used form of displacement transducer with tablet presses. As the name implies, the LVDT is a transformer. It is, however, one of special construction, with a straight tubular body that carries three sets of inductive windings (Figure 4.1). The windings comprise one central primary coil and two symmetrically disposed secondary coils, all in fixed positions. Along the axis of the system, within the tubular body, runs a guide rod that holds a short ferromagnetic core. It is the position of this movable core relative to the secondary windings that determines the variable differential from which the LVDT derives its name. When the core is precisely central with reference to the primary winding, the inductive coupling from that primary to the two secondaries is balanced. If the primary winding is energised by an AC waveform of a suitable voltage and frequency, the voltages generated in the two secondaries will be correspondingly equal and can be balanced against each other in a detector circuit.

The output signal itself may be considered as being of essentially infinite resolution and is likely to have a value of around 5 V rms (root mean squared) at the full-scale displacement of the core. The sensitivity and accuracy of the system as a whole must, nevertheless, depend on the external means used to detect that output signal. Using a modern digital AC carrier amplifier, it should be possible to achieve a sensitivity corresponding to better than 1 ppm of the total full-scale displacement. An analysis of the inductive sensor has been given by Hugill (1982).

The relation between signal and displacement is reasonably linear over the rated displacement range of the transducer, with an increasing non-linearity at each end of the scale. It is, therefore, important to check that the LVDT is installed correctly, with its mechanical and electrical zero points in reasonable coincidence. The maximum departure from true linearity within the specified range of movement depends on the form, the size and the cost of the device, but it is usually between 0.1% and 0.5%.

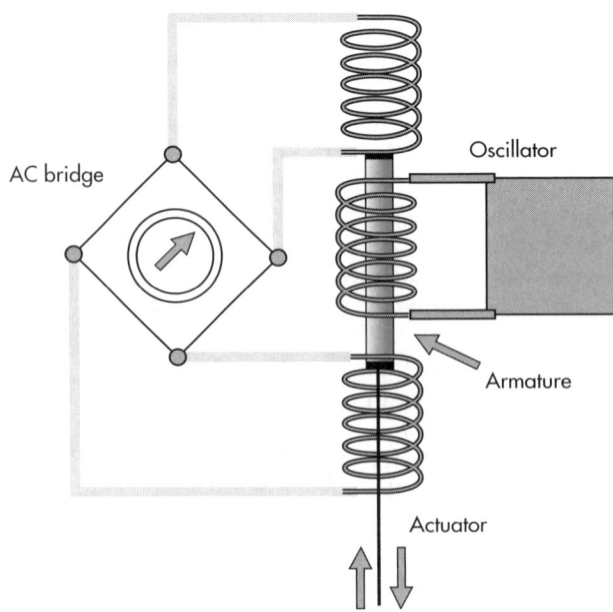

Figure 4.1 The linear variable-differential transformer (LVDT). The mechanism comprises three coaxial coils with a movable ferromagnetic core. The position of the core determines the distribution of the signal between the two outer coils.

Zero setting

In principle, there will be zero output when the armature is central, and the system precisely symmetrical. In practice, it is possible to adjust the zero point by offsetting the associated detector and amplifier, and this has the advantage of making it unnecessary to position the mechanical components with absolute accuracy. Sydenham (1972) has pointed out that a device capable of sensing to 0.1 nm need only be set to the nearest 20 μm.

Good-quality LVDTs with a range of a few millimetres are expected to have a measuring repeatability of the order 0.1 μm, and it is important to realise that repeatability may well be more critical than linearity. In modern systems, a departure from strict linearity can be corrected by the use of appropriate computing routines. Many articles on this can be found in the literature (e.g. Bolk, 1985). Lack of repeatability, however, cannot be treated in the same way. It requires a programme of successive measurements that can then be subjected to statistical analysis.

Thermal effects

Changes in temperature inevitably introduce errors into LVDT measurements, especially if the symmetry of the device is affected, for example by uneven expansion. The thermal drift of the zero point is likely to be of the order of 0.01% of total stroke for every degree centigrade. At the same time, there may be a change in sensitivity of about the same general magnitude. Although these error figures do not appear to be particularly large, it is worth remembering that the associated amplifiers, power supplies and measuring circuitry may themselves also have an element of temperature dependence.

Since the moving core that constitutes the sensing element does not need to be in physical contact with any part of the transducer body, the movement itself does not generate frictional heat. Nor does it absorb power from the mechanism being investigated. However, it is useful to remember that the coils themselves will inevitably dissipate some heat from the AC input, and that thermal expansion of the central armature rod could possibly introduce an error unless all parts of the system have reached constant temperature. When accurate measurements are to be made with an LVDT, it is, therefore, good practice to ensure that the whole measuring system is switched on well in advance and allowed to reach a steady equilibrium temperature.

Gauging transducers

There are some varieties of LVDT in which the core runs on linear bearings within the transducer body and is spring loaded to one end of its range of movement. These are called gauging transducers and are widely used in manufacturing practice to check the dimensions of mechanical components. As a rule, gauging transducers operate over a relatively short range of displacements, mostly from 0.5 to 5.0 mm. By comparison, the normal free-armature forms of LVDT may extend to displacements of 600 mm. Gauging transducers are designed for operation in hostile environments and are usually equipped with elastomeric gaiter seals that exclude dust or liquid contaminants.

If negligible operating forces, coupled with very high operating rates, are needed, then the free-armature transducers would be preferable. In case of doubt, it would be advisable to check with the transducer manufacturer.

Dimensions

When an LVDT is to be installed in some piece of experimental equipment, it is naturally important to have information about its length and diameter so that it can be allocated adequate space. As a very rough rule, transducers in the displacement range of a few centimetres tend to have a body length that is perhaps three or four times their measuring range, although the ratio of measuring length to body length is generally less for gauging transducers than for the free-armature types.

The diameters of these devices have traditionally been derived from Imperial sizes, and most of the free-armature models have been either of 0.75 inch (19 mm) or 1.0 inch (25.4 mm) gauge. The gauging transducers are often smaller, and here the most common size is 0.375 inches (9.5 mm).

There are, naturally, many exceptions to these simple rules; not surprisingly, the lower limits of diameter have been extended over the years. The miniature series transducers offered by Sigma, for example, are only 6.0 mm in diameter; however, at this size the measuring range is restricted to 1.0 mm. Some even smaller versions have been produced, although is it unlikely that they would be required for use with a relatively large machine such as a tablet press. Units in the 'XS-B' series from Schaevitz Engineering, for example, have a coil assembly with an outside diameter of only 4.76 mm: the overall length of these very small LVDTs is from 10.16 to 38.1 mm, depending on the measuring range, while the maximum measurable displacement is ±6.35 mm.

Excitation

Since the LVDT is a transformer, it requires an oscillator circuit that will provide an alternating voltage to excite the primary winding. In general, the excitation signal has a frequency within the range 1–50 kHz – although exceptions exist – and is only a few volts. Ideally, the optimum frequency for the signal is one at which there is negligible phase shift between the waveforms in the primary and the secondary windings. Under these conditions, the system is most stable to temperature changes. A commonly used value is around 5 kHz; the power level is small and is represented by some 3–10 V at approximately 10–35 mA.

Demodulation

Apart from the oscillator, the LVDT also needs to have a demodulator unit that will accurately turn the AC output into DC levels for display or analysis. Certain types of LVDT have both the oscillator and the demodulator built directly into their housing, giving a very compact construction. They provide a DC output from a suitable DC input and, therefore, have the description of 'DC–DC' transducers. These self-contained DC–DC transducers are undoubtedly convenient and may be of good stability and sensitivity. They are inevitably less adaptable to the miniature format, however, and so are not usually offered in diameters below the industry standard of 19 mm. If space is at a premium, it may be necessary to use a conventional AC–AC transducer with an external oscillator and demodulator.

Virtually all the LVDTs with diameters of less than 19 mm are of the AC–AC pattern. Apart from the sub-miniature devices mentioned above, both the spring-loaded, sealed, gauging transducers and the free-armature versions are available with body diameters of 9.5 mm or less. The SM series from Sangamo Schlumberger, for example, are of this diameter: the length of the SM1 body is 15 mm, and the moving core weighs only 200 mg. At this size, the total displacement range is ±1.0 mm, while the linearity may be either 0.3% or 0.5% of the total stroke, depending on the grade of the device.

The linear inductive displacement transducer

We have seen that the classical LVDT has three sets of windings; one primary and two secondary. It is however, possible to omit the central winding so that only the two symmetrical secondaries remain. The physical disposition of the components is then otherwise almost exactly as in the conventional LVDT, but the mode of operation is slightly different. In this instance, the two windings form adjacent arms of an AC measuring bridge, which is energised from an external oscillator and may be balanced by adjustment of the other two bridge arms in the normal way. After the bridge has been balanced, any change in the position of the iron core will de-tune the circuit, generating an output signal that is proportional to the displacement of the core. Some commercial transducers of this type have trimming controls with which their sensitivity can be adjusted. In effect, these form a second movable core that alters the inductance of windings in series with the main measuring coils. If an installation contains several such transducers, the adjustment facility can be used to set each one to the same level of sensitivity. In some instances, this two-coil construction has given an improved operating range for a given body length.

A useful discussion of the different approaches to inductive transducer design may be found in the 1975 textbook by Neubert.

The variable turns-ratio transformer

A low-cost device for displacement measurement has been described by Dann (1984). In this, single-layer coils of copper wire were wound on to rigid plastics tubes of different diameters, so that one coil could be moved axially within the other. The inner coil, used as a primary, was excited with a sinusoidal signal at a frequency of 1 kHz, while the resultant signal was measured from the outer coil with an AC voltmeter. It was found that the output was linearly related to the position of the inner coil to within better than 1%. Clearly, this does not approach the performance of a standard commercial transducer, but the interest here lay in the very high proportion of useful range to body length, where the device described had a length of 1080 mm and could apparently give an output at displacements up to 1000 mm.

The linear variable capacitor

In the LVDT and its half-bridge variant, movement of a ferromagnetic core alters the relative inductance of windings along a non-magnetic former. The same general principle can be extended, by appropriate constructional modifications, to provide changes in other electrical properties. For example, if cylindrical electrodes are used instead of wire-wound coils, the axial displacement of a third, smaller electrode can be used to vary the relative capacitances of the first two electrodes in a reproducible manner. Sensors using this principle were compared with inductive devices in a review by Hugill (1982).

Variable capacitance transducers are commercially available. One type, from Automatic Systems Laboratories Ltd, is sold under the name 'Super-linear variable capacitor' or SLVC. Like the LVDT, the SLVC is of infinite resolution. It can also be capable of an accuracy (defined in this case as the maximum departure from an ideal straight-line calibration) of better than ±0.00065 mm. The system non-linearity, taking both transducer and signal conditioner together, may be better than 0.04% of full scale for the best grade of SLVC.

The physical size of these transducers is somewhat less than that of their inductive counterparts, and the ratio of working range to body length may also be favourable. For example, the ASL model 3074 has a quoted range of 0–100 mm for a body length of only 220 mm.

The stability and general accuracy of SLVCs is very good indeed in most applications. A tendency to seize in use was reported by Clabburn (1983), but this was in a specialised field where the transducer had been required to follow the deflection of some explosively deformed plates. There are few devices in this field that will respond reliably at such speeds and, of course, that response would not usually be required in tablet compression studies.

A general account of the development of capacitive measuring devices for displacement up to some 10^{-2} mm was given by Jones and Richards (1973). The ranges involved were quite short in terms of tablet punch movement, but the article itself makes interesting reading, showing how sensitive capacitance transducers can be, since some of the devices described were stated to detect movements of less than 10^{-10} mm.

The linear magnetoresistive transducer

There are some materials that, in the presence of a magnetic field, exhibit appreciable changes of resistivity: this is the so-called 'magnetoresistive' effect. The effect can be maximised in certain semiconductors by appropriate doping. For example, indium antimonide, doped with tellurium, shows the effect quite well. The increase in resistance, produced by the presence of the magnetic field, is proportional to the square of the field intensity.

We have discussed above how the LVDT and the SLVC can be constructed as semifrictionless systems. The magnetoresistive effect can be used as the basis of an analogous form of displacement transducer, in which electrical resistance is the variable quantity. In this device, two similar resistors of semiconductor material form adjacent arms of a Wheatstone bridge. They are mounted in line and are symmetrically within the magnetic field of a small annular magnet attached to the moving plunger of the transducer. If the plunger is displaced, the magnetic field becomes correspondingly asymmetrical; the two magnetoresistive elements change in resistance, and the bridge is unbalanced.

Linear transducers operating on this principle are commercially available: the Blue Pot™ range, from Midori Inc., for example, covers displacement ranges from 0.60 to 8 mm. A typical midrange unit, the LP-5 UYW, has an output of approximately 1.0 V per 1.0 mm displacement, with a linearity within 1% of full scale. The maximum sticking friction is approximately 0.06 N.

In general, most of the devices that depend on magnetic field effects tend to show rather non-linear characteristics. However, they can be both sensitive and repeatable, so their output signals can be linearised in the data handling circuitry.

The Hall effect transducer

We have seen that certain transducers, such as the LVDT, require no physical contact between their fixed components and their moving element. A transducer of this type can be regarded as being essentially frictionless, provided that there is no ingress of particulate matter between the moving parts. However, since the internal clearances are often very small, maintenance of the original frictionless state requires much care and attention to detail.

Some transducers are much less demanding in their environmental requirements, since they operate remotely and have no moving parts to be affected by dust, grit or moisture. One such device is the Hall effect transducer, which has found various industrial applications in recent years. The Hall effect is produced when a current-carrying conductor passes through a magnetic field. Lorentz forces deflect the charge carriers in a direction at right angles to both the current and the magnetic field vectors, and a voltage develops across the width of the conductor (Figure 4.2). In commercial devices, the con-

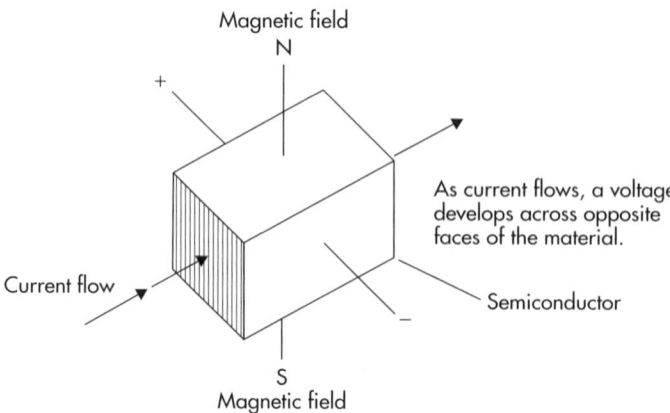

Figure 4.2 The Hall effect. In the presence of a suitable magnetic field, a current passing in one direction generating a voltage signal in a transverse direction.

ductor itself is usually made from a suitably doped semiconductor material, such as indium antimonide, in which the voltage output is as large as possible.

In order to utilise the Hall effect for displacement measurement, it is necessary to have a means whereby relative movement between two components gives a reproducible change in magnetic field at the site of the conducting element. This can be achieved by using a permanent magnet on one component, and a soft iron armature on another, provided that relatively short distances are involved. If a magnet is attached to a rotating component in a machine, a non-contact Hall effect switch can be used to generate a pulse at each revolution to establish the rotational speed.

The linear variable resistor

If a material of uniform resistivity is formed into a straight track, the addition of a sliding contact produces a potential divider, or potentiometer. Linear potentiometers have been known for very many years and can be of low cost; however, the early models were not suitable for use as accurate displacement-measuring devices on account of their poor electrical and mechanical properties. The resistive element was either in the form of a wire winding – which limited the available resolution of the system – or was a carbon composition, which was both electrically noisy and physically abrasive.

The development of improved resistive materials has made it possible for linear potentiometers to be produced specifically for displacement measurement. Usually, these devices have elements made from conductive plastics on an insulated backing such as Kapton polyimide film. The sliding contacts are of precious metal and are divided into several flexible sections in order to ensure a sound, continuous electrical connection. The toughness and grain-free character of the conductive plastics film are probably its main advantages over the earlier composition elements, since its electrical properties often still depend on the use of carbon.

The conductive plastics film provides infinite resolution; additionally, it has a long working life and does not damage the sliding contacts. Some devices of this construction are rated for up to 100 million operating cycles, which represents a comparatively long time in research and development, although a large production tablet press could run through this in 100 h of operation.

The temperature coefficient of resistivity of a conductive plastics film element is in the region of $\pm 400\,\text{ppm}\,°C^{-1}$. That of a modern carbon composition resistor is approximately $\pm 1200\,\text{ppm}\,°C^{-1}$, while good-quality resistance wire materials have a much lower value, which may be as little as $\pm 20\,\text{ppm}\,°C^{-1}$; a typical cermet resistor would have an average value of $\pm 100\,\text{ppm}\,°C^{-1}$. Thermal variations of this predictable form may not affect the function of a potentiometer that is energised at constant voltage, providing that the temperature changes causing them are uniform over the resistive element. Localised changes would naturally introduce considerable errors into the inferred position of the slider.

Linear potentiometers are available in an extensive range of sizes: the electrical travel may be as little as a few millimetres or as much as 2.0 m, and even with the larger potentiometers, the claimed repeatability can be better than 0.0125 mm. The acceptable stroke velocity varies from make to make, but it is usually of the order of $1.0\,\text{m}\,\text{s}^{-1}$; special models can be supplied for operation at approximately $3.0\,\text{m}\,\text{s}^{-1}$.

On these various grounds, the resistive devices would appear to be closely competitive with the LVDT and the SLVC. However, there is an important characteristic of the resistive potentiometers that must be taken into account during the choice of a displacement transducer: each of them requires a substantial axial force to move the shaft that carries the sliding contact. Apart from a few of the smaller units, most linear motion potentiometers are rated at 4.0 to 7.0 N. Although a force of this magnitude would scarcely affect the operation of a heavy tablet press, it will most certainly produce frictional heat. If the transducer is to be used for quasi-static measurements, the generation of frictional heat can probably be ignored. However, if it is intended that the device should be a permanent feature of a press used at production rates, then there is much to be said for the choice of an inherently frictionless transducer such as the free-armature LVDT.

Apart from the generation of unwanted frictional heat, there is a second problem for any sliding transducer that is permanently connected to a cyclically moving component such as a tablet punch: this is the repetitive nature of the movement. The production of a batch of tablets, even on an experimental basis, involves the punch in successive movements over what is, in effect, exactly the same range of travel for each compression. If the transducer is of such a type that there is physical contact between the actuator and the stationary components, then the track wear will be at its worst under these conditions.

The strain gauge conversion transducer

There is a second type of linear transducer in which the stationary and the moving components are in contact with one another, but in this instance frictional losses are kept to an acceptable minimum by the use of precision ball races. The housing of this transducer contains a pair of beams, cantilevered from one end of the system. The beams are provided with a four-arm bridge, comprising four foil gauges of $350\,\Omega$ nominal value. Down the axis of the device runs an actuator shaft that carries precision ball races in contact with the beam surfaces (Figure 4.3).

When the actuator moves, the ball races cause small deflections of the twin beams, so unbalancing the bridge and giving the familiar electrical output. Accuracy of these transducers is claimed to be good; they are commercially available with a non-linearity of $\pm 0.05\%$ of full scale, and a repeatability of ± 0.02 of full scale. Once again, this is a system with infinite resolution. The rated effect of temperature changes on zero and span is typically $\pm 0.005\%\,°C^{-1}$.

One useful consideration in relation to the use of the strain gauge conversion transducer is that it requires exactly the same external electronics – power supplies, bridge circuitry, and so on – as do the strain gauges used for force measurement.

The remote inductive transducer

Linear transducers such as the LVDT can be both sensitive and accurate, and they have been invaluable in fundamental research under laboratory conditions. However, if it becomes necessary to use displacement transducers in the presence of dust, grit or corrosive chemicals, there may well be problems in ensuring that the moving components continue to slide freely and that accuracy is maintained.

Some commercial transducers have no moving components and may be sealed in stainless steel

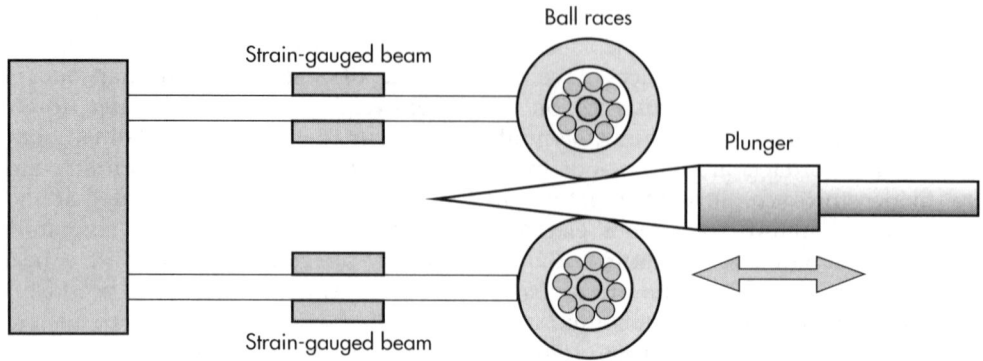

Figure 4.3 The strain gauge conversion transducer. Movement of the plunger causes controlled bending in one or more thin elastic beams. The amount of bending is monitored by strain gauges on the beams.

shells, and so these are well adapted to operation in hostile environments. One such device is the remote inductive proximity transducer. This usually comprises a resin-encapsulated coil assembly within a chemical-resistant outer case. It may be either an inductive half-bridge with two coils, or a quarter-bridge with a single coil, and it will be powered from an oscillator at a frequency in the range 5–150 kHz. The presence of an electrically conductive metal surface near to the coil assembly affects the impedance of one or both coils and gives rise to an output signal. In some versions of this device, the signal results from an imbalance between two inductances, while in the single-coil units, it is the generation of eddy currents in the metal target that provides the necessary variation. Other associated designs may operate on the principle of magnetic reluctance, in which displacement of a ferromagnetic plate changes the air gap in a high-permeability magnetic circuit. The principles involved in these different forms have been discussed by Hugill (1982).

Inductive transducers of this kind have inherently non-linear characteristics by comparison with laboratory devices such as the LVDT. However, after the signal has been demodulated, linearised and amplified, the final output signal can be entirely satisfactory. Taking, as an example the 'multi NCDT' system offered by Micro-Epsilon Messtechnik, linearity is better than 0.2%, temperature stability is better than $0.02\% \, °C^{-1}$ and, since there are no moving parts, there is a frequency response up to 100 kHz. Sensors from this manufacturer are available for measuring ranges that extend in 12 models, from 0.5 to 80 mm: the smallest units of these have a body diameter of only 2.0 mm. Units of this type are completely unaffected by dust, oil, water and general dirt.

The remote capacitive transducer

While the inductive proximity detector measures changes in the electromagnetic properties of one or more coils, the capacitive transducer monitors changes in the capacitance of an electrode system. The two devices perform more or less similar functions, and both are suitable for use in hostile environments. There is one notable difference, however; whereas the inductive devices depend on the presence of conductive metal targets, the capacitive device is affected by a wider range of materials including glass, ceramics, water and many plastics. It is, of course, necessary to shield the electrode system against stray capacitances, to check the physical condition of the electrode surfaces from time to time and, as before, to linearise the output signal. Apart from these points, the capacitance transducers can be very sensitive over fairly small measuring ranges. Industrially, both the inductive and the capacitive systems find their widest use in remote switching rather that in measurement, when their inherent non-linearities can then be ignored.

The magnetostrictive transducer

The displacement transducers that we have considered so far have generally operated over a restricted range of movement. Indeed, for force–displacement measurements within a tablet press, a maximum of 8–10 mm would probably be more than adequate. However, outside the die itself there could be movements of much greater amplitude that, at some time, might need a convenient means of measurement.

These comments are by way of introduction to an interesting form of displacement transducer with a particularly long measuring range: the magnetostrictive device.

Just as the presence of a magnetic field alters the electrical resistance of certain materials, so it may affect their physical properties. If a voltage is applied for a very short time across the ends of a ferromagnetic conductor, a pulse of current flows along the conductor. In the presence of a localised magnetic field around part of the conductor, the current pulse gives rise to mechanical stress as it passes the field boundary, and a corresponding shock wave passes along the conductor from the boundary region at sonic speed.

In a practical embodiment of these considerations (Figure 4.4), the conducting element is a length of nickel–iron wire, which exhibits the magnetostrictive effect well. The alloy wire is supported within a rigid non-magnetic tube, and it remains stationary in use. The moving head travels along the outside of the tube and carries an annular magnet. A signal generator is connected to the magnetostrictive wire and sends pulses along it at intervals. These pulses produce a local annular magnetic field around the surface of the wire; but at the position occupied by the movable magnet, there is a sudden discontinuity of the field. The rapid change in radial stress at that point induces a pulse of mechanical energy, and that pulse travels back along the wire at sonic speed to an ultrasonic detector at the transmitter end. From the measured transit time, and knowledge of the speed of sound along the wire (approximately $2800\,\mathrm{m\,s^{-1}}$), the transducer circuitry can calculate and display the position of the magnet. The claimed performance of this system indicates a good linearity, in the range 0.01–0.03% of full scale, repeatability of better than 0.01%; resolution with a 22.2 MHz clock is only to approximately 0.10 mm but is a function of clock frequency and can be varied. However, the particular distinction of the design is the great range over which it can apparently operate: standard units have a range of 3.0 m and larger distances have been suggested.

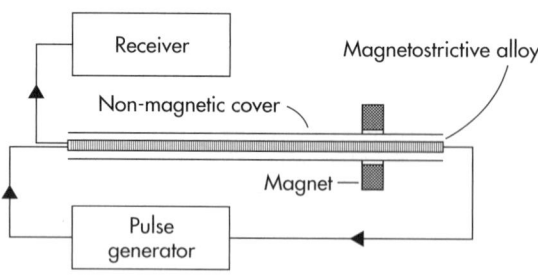

Figure 4.4 The magnetostrictive transducer. An electrical pulse is sent along the central conductor. When that pulse reaches the annular magnet on the moving component, there is a sudden localised deformation, and an acoustic wave travels back along the conductor at a known speed.

The magnetic sensor

The availability of linearising circuitry has made it possible to consider many systems that are inherently non-linear but reproducible. For example, Cooke and de Sa (1982) used the variation in magnetic field near a samarium–cobalt magnet as the basis for a device to measure displacements up to a maximum of 250 mm. A feedback fluxgate magnetometer was used as the sensing element, and its output, after conversion to digital form, was processed in a microcomputer. The authors claimed that the completed system was accurate to approximately 0.02% over the range 18–80 mm. Since DC magnetic fields can pass unattenuated through a range of materials such as glass, copper, aluminium and plastics, it was felt that the method might have applications in some experimental measurements.

The ultrasonic transducer

Although linearisation techniques are useful and practical, it is always convenient to have a form of transducer that is inherently linear in its operation. Measuring the transit time of a signal that is believed to travel at constant velocity is one way of approaching the linearity problem. That method has been used successfully in the magnetostrictive transducer described above, and has also been used industrially for liquid level measurement. Pulses of ultrasound can be generated conveniently by means of piezoelectric transducers; these are relatively inexpensive and are probably the most widely used transducer in commercial practice.

The major problem to be encountered is probably that of providing a suitable pathway for the ultrasound. Sound travels well in incompressible media such as water or steel, whereas it is attenuated strongly in gases – such as air – and is likely to be reflected at the interface between two dissimilar phases.

Modern ultrasonic devices for measuring liquid levels, therefore, often operate from within the liquid to be measured. They sense the position of the liquid interface above them, rather than looking downwards from above. It might seem possible to fit such a level sensor in a tube, partly filled with liquid, and to attach a narrow plunger to the moving component in order to produce an identifiable echo from its end. However, the currents induced in the liquid itself by the vertical movement of the rod would certainly affect the signal.

Ultrasonic 'sonar' methods can, in fact, be used in air, provided that the distances involved are small (a few tens of metres). The practical limit is the unknown air temperature, which changes the speed of sound and leads to potential errors. Otherwise, ultrasonics will travel over 30 m and back. The ultrasonic 'tape measure' is an example of the use of this method.

The principles and methods involved are described in some detail by Crecraft (1983).

The fibre optic transducer

Over the years, there has been a progressive increase in electrical noise levels, both within industrial premises and also within the electrical main supplies themselves. As a result of this, microprocessor-based systems tend to be at risk from spurious signals or high-voltage 'spikes' in the supply lines. Any system that can help to avoid these effects is likely to receive careful attention, and at present it seems that optical measurements hold out much promise.

There appear to be many areas in which optical transducers could replace the conventional devices that we have been considering in this chapter. One experimental device uses a composite fibre optic bundle, formed into a 'Y' shape. If the light is sent along one arm of the Y, it will emerge at the junction point; but if there is a reflecting surface close to the end face, a proportion of that light will return and travel along the other arm, where it can be detected with a suitable photosensitive element. Under appropriate conditions, the intensity of the reflected light can be used as a measure of the distance between the reflector and the fibre optic surface (Figure 4.5).

It is understood that this principle has been used in the construction of a fibre optic transducer with better resolution than that of an LVDT; but, as yet, such devices appear to have been made only as research tools. A paper by Giles *et al.* (1985) described a technique for balancing the losses inherent in the fibre optic leads in any intensity-based remote optical sensor, making use of an optical equivalent of the Wheatstone bridge.

In another design of optical transducer, the fixed and the moving component are represented by two toothed blocks that are clamped on either side of a fibre optic cable. If the blocks are brought towards each other, they deform the optical fibre from its original straight configuration into a more or less sinusoidal form. This has the effect of inducing transmission losses along the cable at points where the radius of curvature is small, and the measured loss can be used to infer the distance between the blocks.

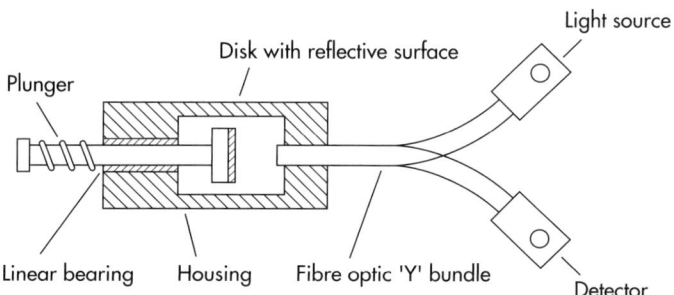

Figure 4.5 The fibre optic reflective transducer. Light is transmitted along one arm of a fibre optic 'Y' piece to a reflecting surface near the other end. A varying proportion of the light returns along the fibre bundle to a detector on the other arm, and the position of the reflector can be measured by measurement of the detector output.

An interesting discussion on the basic principles of fibre optic sensing is to be found in Culshaw (1983) and is to be recommended for general reading on this topic.

The optical wedge

Optical wedges are probably among the earliest forms of non-contact displacement transducer, having long been applied to the movement of isolated muscle strips in physiological laboratories. The moving component is usually a small glass plate whose optical density varies from one end to the other in a near-linear fashion; the stationary component is a fork sensor, with a light source and detector positioned on each side of the narrow gap through which the plate moves. Once again, this represents a frictionless system with good tolerance to misalignment. However, some care must be taken to ensure that the source and detector are not subjected to temperature variations, since these are likely to produce drift and instability in a single-beam arrangement of this kind.

If improved stability is essential, this can be achieved by the use of a differential absorption technique. In effect, the transmission of the wedge is measured at two wavelengths where its absorption is substantially different. The two readings provide the effect of a reference channel, thereby eliminating many of the fluctuations inherent in a single-beam system. A displacement transducer using this principle has been described by Theocharous (1985). Operated over the range 0–2.6 mm, the transducer had a measured resolution of $\pm 2.9\,\mu m$.

The optical potentiometer

Devices such as the optical wedge are useful additions to the range of frictionless displacement transducers and can be run at quite high speeds without the generation of frictional heat. However, the moving components still have appreciable inertia.

For some purposes, it may be essential to have a minimum inertial mass in the moving parts of the transducer, and one approach to this problem is through the light-sensitive potentiometer. This has two parallel elements, of which one is resistive and one is conducting, the space between the two being occupied by a photoconductive material. A voltage is applied across the ends of the resistive element. If the two elements are bridged by a narrow beam of light, the resistance of that material is locally reduced and the image of the light beam acts as a conducting but frictionless slider. The electrical potential of the conducting element can thus be varied without physical contact. Surface sweep speeds up to about $0.5\,m\,s^{-1}$ may be used with this particular device, and it is claimed it can resolve movements down to $1.0\,\mu m$.

Displacement transducers with a digital output

The optical grating transducer

Most of the transducers with analogue output have essentially infinite resolution; as a result their smallest resolvable increment is as much dependent on the signal-conditioning systems as on the transducer itself. Their output will, moreover, always be subject to some degree of amplifier drift.

Changing to digital operation provides a trade-off between sensitivity and stability. The smallest detectable increment now depends on the transducer and is almost independent of the associated electronics. Of course, the output from analogue transducers can always be displayed digitally through the use of an analogue-to-digital converter, but it is not actually produced in that way.

Once generated, the digital signal is in the form of pulses; they may represent a simple square wave, where the voltage is either 'on' or 'off', or alternatively, the signal may change between two voltage levels. In either case, the voltages employed can be chosen to be readily distinguishable from the ambient electrical noise. Digital transducers, as a group, have been reviewed by Woolvet (1982). They can take many different forms. By way of an example, we may consider a simple optical version in which a transparent plate, or grating, ruled with opaque lines at equal intervals is moved between a light source and a detector. As each line passes, the detector registers one pulse.

In practice, such a simple arrangement would be limited in its resolution by the line width necessary to interrupt the light beam. If two similar gratings, with their axes inclined to each other, are used, then the two between them generate moiré fringes of greater width than the lines themselves (Figure 4.6). One of the gratings now remains fixed between the light source and the detector, while the other one is attached to the moving probe. In this way an improved resolution is possible, with practical line spacings ranging from 25 to a maximum of 50 lines mm^{-1}. Above this, diffraction effects limit the available contrast, blurring the edges of the fringes.

For some purposes, an increment of 0.04 mm might be sufficient, but for most applications, it is usual to sub-divide the scale into smaller parts. This is done in two ways, of which the first step is to increase the number of detectors. Since the fringes are seen to move transversely across the grating when movement takes place, a row of equally spaced detectors can give several signals per fringe instead of only one. It is common practice to use four of them. At this point, the signal as received has a saw-toothed form, because the fringes do not exhibit clean square edges. It is, therefore, possible to use electronic level sensing in order to give the final sub-division into the maximum number of steps.

Clearly, using level sensing introduces an analogue measurement into what had previously been a purely digital system, but the likely error is normally only that of the uncertainty in the last digit. A grating ruled at 10 lines mm^{-1} can give a final resolution of 1 µm when these techniques are applied correctly.

The use of one grating gives an incremental system in which movements from a given zero point are monitored as long as the electronics remain in action. Appropriate circuitry keeps a running tally on the direction of the movements, so that they are added or subtracted as necessary.

It has, in the past, been customary to use a more complex construction when 'absolute' measurements were required. This called for a multiple-track grating, with several sections each of different spacings, so that any position could be coded uniquely. The same principles have been applied to the construction of rotary optical encoders, one track being necessary for each binary digit in the displayed value. Therefore, an encoder with a resolution of 1 part in 256 would require eight tracks. Using an absolute encoder system of this kind ensures that positional information is not lost if the power supplies are interrupted; consequently, it is good practice for many applications. However, the trend in modern electronics has been more and more directed to the use of components with very low power consumption that can be run from small backup batteries. Simple, low-cost incremental encoders can be run in this way, and although they do not have the capability of the true absolute encoder, they can certainly be run without losing positional information.

The use of gratings in metrology was reviewed by Sayce (1972).

The linear diode array

As its name implies, the linear diode array consists of a row of discrete photosensitive diodes mounted together in line at accurately known

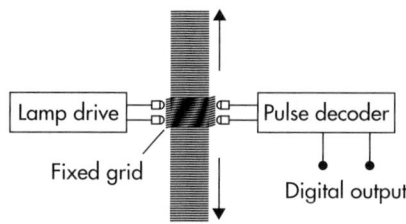

Figure 4.6 The optical grating transducer. If a transparent scale is passed between a source and a detector, the distance that it travels can be estimated by counting the number of times the beam is interrupted by the scale divisions. However when these divisions are very small, they may be too narrow to interrupt the beam. Using a second scale generates moiré fringes that greatly increase the sensitivity of measurement.

spacings in a standard integrated circuit package. The diodes usually have their own associated circuitry, which can interrogate them one by one and can transmit information from them to some external processing device. Linear diode arrays have many applications, and one that is relevant to displacement measurement is that of optical triangulation (Figure 4.7). In this technique, the image of an illuminated slit is projected at an oblique angle on to the surface of the object. The reflected image, having passed through a suitable lens system, is allowed to fall on a diode array. The height of the reflecting surface determines the position of the slit image, and hence which diode has the greatest illumination. This system will continue to follow subsequent changes as rapidly as the diodes and their circuitry can respond.

The resolution of this arrangement depends primarily on the number of diodes in the array – which could be anything from 2 to 4096 – while its sensitivity depends on the associated optics. The diodes in the array always show some variation in sensitivity, so the use of clean, sharp-edged images minimises positional uncertainty. It is also possible to use interpolation of the signal from adjacent diodes to gain further resolution. The light source in current models is usually a laser diode running at 675 nm, which gives a spot size of approximately 1.0 mm in diameter and a claimed resolution of 0.01 mm.

If only two diodes are used, the arrangement can be run as an analogue device, giving a continuously varying signal. A sensor using this principle was described by Tanwar and Kunzmann (1984).

Inductive element transducers

The optical transducers mentioned above are used extensively in industry for measurement in various gauging devices and some micrometers, though it is, of course, important to ensure that glass scales do not become contaminated with oils or other materials that might spoil their transparency. For some applications where this might be difficult, another approach is that of having a linear array of small ferromagnetic bar elements, embedded in a chemical-resistant matrix, in place of the optical components. The bars are detected by their magnetic properties, and interpolation is used as before to give enhanced resolution. This is a common construction for the sliding scales used in workshop callipers and machine tools, where resolutions in the range 0.01–0.001 mm are normal.

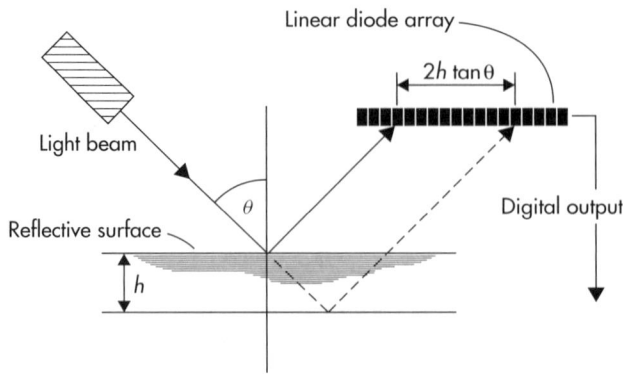

Figure 4.7 The linear diode array. The moving component carries a small reflector, which sends a sharp-edged beam of light towards the row of diodes. The vertical height of the reflector determines how many of the diodes are lit by the beam.

Optical interferometers

The principles of optical interferometry are well established, and models were constructed during the nineteenth century by Michelson. They are still capable of very accurate measurement and nowadays are able to take advantage of such components as the solid-state laser and fibre optics to provide a compact construction. Corke *et al.* (1984) commented that a modern interferometer can be used to detect displacements that may be as small as 10^{-14} m.

The basic concept of the interferometer is simple. Light from a suitable source passes into a beam-splitter and is divided into two separate beams. The two beams take different optical paths and are directed by mirrors or prisms to a second beam-splitter, where they are recombined and passed to a detector. If the two beams are in phase when they reach the detector, an output signal will be produced; if they are precisely out of phase, there will be no signal. For conditions in between, there will be a proportional signal from which the phase relationship can be calculated (Figure 4.8).

To use such a system for measurement, it is necessary to arrange that the path length of one beam will be modified by the parameter to be measured. One of the most sensitive ways of measuring displacement is one in which one of the mirrors moves in a direction normal to its reflecting surface; movement of the mirror then produces a double change in path length.

If a light source with a wavelength of 633 nm is used – this would be that of a helium–neon laser – then a half wavelength displacement of the mirror would produce one complete cycle at the detector for a movement of approximately 316 nm. In fact, very much less than one cycle can be detected with modern equipment, hence the sensitivity of better than 10^{-14} m quoted by Corke *et al.* (1984).

The limiting factor tends to be the frequency stability of the light source, since that is what provides the effective measuring standard. With a normal helium–neon laser, stabilities to approximately 1 part in 10^6 may be expected, although some specially arranged lasers are claimed to be stable to within 1 part in 10^{12}.

Dynamic measuring devices

The accelerometer

When a mass is allowed to hang freely from a spring, the spring extends and assumes an

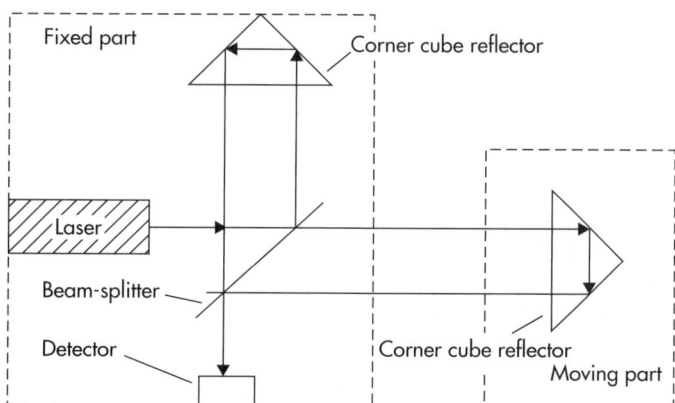

Figure 4.8 The interferometer. Light from a fixed source is split into two beams: one takes a path of constant length while the other takes a path of variable length before the two beams are recombined. The signal at the detector depends on the relative phase angles of the two combined beams. A displacement of one-quarter of a wavelength takes the signal from minimum to maximum.

equilibrium position at which the gravitational and spring forces are in balance: the system is then at rest. If the assembly is now caused to move in a vertical direction, the associated acceleration gives rise to a new force component since force equals mass multiplied by acceleration. The spring alters in length, in order to accommodate the new conditions. At any instant, the extension of the spring provides a measure of the force, and hence of the acceleration involved. If the acceleration and the elapsed time are known, the distance travelled is then calculable from Equation (4.1):

$$s = ut + \tfrac{1}{2}at^2 \qquad (4.1)$$

where s, distance travelled, a, acceleration, t, elapsed time and u, initial velocity.

This simple arrangement represents the principle of the accelerometer, and it may be extended to measurement along one, two or three axes. One possible advantage of the accelerometer as a means of measuring, or inferring, the position of a component is that it is unaffected by external contamination with grit, oil, moisture and so on. The LVDT is susceptible to these influences and must be shielded from them. We have seen that there are other possibilities for non-contact displacement measurement, but the accelerometer appears to be unique in measuring its own movement through space. It does not require a fixed reference surface, a plunger or any other form of physical zero point.

A practical accelerometer contains four essential elements: a seismic mass, an elastic suspension system, a force-measuring arrangement and a housing (Figure 4.9). The seismic mass is normally a block of dense material; using a dense material makes it possible to keep the physical size to a minimum. It is desirable to restrict both the size and the mass of the complete assembly, so that it has the minimum loading effect on the test structure. Its mass should be at least one order of magnitude less that of the structure being measured.

The suspension may be some form of flexible beam, constrained in such a way that the system is primarily sensitive along one axis of motion; at least one manufacturer uses an annular shear design, following the general layout of a modern load cell.

Force, as in our previous examples, can be inferred from a measurement of the movement in the suspension. If the expected movement of

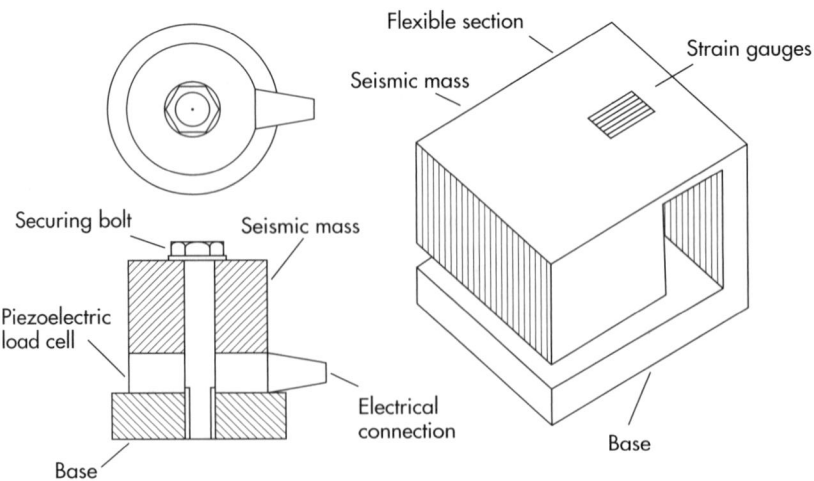

Figure 4.9 The accelerometer. If the small heavy mass at the heart of the accelerometer is caused to accelerate, a corresponding force must develop between the mass and its suspension system. Both units are sensitive to vertical acceleration. The force may be measured by means of strain gauges or piezoelectric elements.

the mass, relative to its housing, is more than a few micrometres, then it is possible to make this measurement with an LVDT. However, many commercial accelerometers are built around piezoelectric load cells, since this approach can provide a compact device with good frequency response. A third approach is that of using a servo-driven force balance; this is the method applied in some Schaevitz transducers.

The housing is designed to facilitate attachment of the accelerometer to the test structure; it may either be in the form of a block with a central bolt-hole or have a machined flange with several such holes.

Accelerometers of the piezoelectric variety, such as those from Kistler Instruments Ltd and Entran Ltd, can be very small and light: both these suppliers provide units with an all-up weight of less than 0.5 g. The upper limit of their frequency response is usually in the region 20–30 kHz, full-scale output is approximately 250 mV, while the charge sensitivity may be around $20\,pC\,g^{-1}$, where g is the acceleration from gravity. This can in the best circumstances give a maximum signal of $1\,V\,g^{-1}$: the measurement ranges are from approximately 0–5 g to some 100 000 g.

Servo-force balance accelerometers, although larger and heavier, are more sensitive. Minimum ranges are as little as $\pm 0.25\,g$, and full-scale output ± 5.0 V DC.

Piezoelectric accelerometers, like other piezoelectric devices, must normally be connected to a charge amplifier. Much care must be taken with the cable runs, and with the insulation resistance of all connections. If the mass of the device is not critical, it may be convenient to fit an integral amplifier within the transducer housing. This arrangement gives a low-impedance output and eliminates some of the problems associated with piezoelectric devices. Units in the 1010 series from Precision Varionics International contain a hybrid electronics package and provide a signal of some $25\,mV\,g^{-1}$; they weigh approximately 50 g.

Most of these transducers have a practically infinite resolution, limited only by the associated circuitry. The commercial display devices, sold for use with accelerometers, are scaled to read in units that are as small as 0.001 g, but the rated linearity of the transducers is not usually better that 1.0%. For the most accurate work, it would be desirable to calibrate the transducer on a suitable test rig.

Integrated micro electro mechanical systems

Integrated micro electro mechanical systems (abbreviated to iMEMS) is a form of accelerometer in which a silicon mass is micromachined on to the surface of a silicon chip. Analog Devices Inc make one-, two-, and three-axis sensors for the automotive market, and over 200 million have been produced to date. They make sensors for both low and high g values. Examples of sensors for low g are the ADXL203 (two-axis) and ADXL330 (three-axis). The latter device is in a package 4 mm × 4 mm × 1.45 mm, has three analogue outputs of $300\,mV\,g^{-1}$ with an accuracy of $\pm 10\%$, works over a range of $\pm 3\,g$ and has a bandwidth of 1.6 kHz. Power consumption is a miserly 180 µA at 1.8 V. A large range of devices is available, covering a range of both g values and accuracies up to $\pm 4\%$.

Velocity transducers

Optical velocity transducers

There may be times when it is useful to be able to measure such quantities as the angular velocity of the main shaft in a tablet press. It is relatively simple to make this measurement by optical means, using the intermittent interruption of a light beam. This was done by Ridgway Watt (1983) to measure the starting characteristics of a Manesty Betapress. A printed adhesive tape was applied as a complete band around the turret of the press. This tape has narrow black segments separated by narrow transparent bars. As the turret rotates, the surface of the patterned tape provides intermittent reflection of the light, and a stream of pulses can be detected.

Inductive velocity transducers

Inductively coupled devices are often used in industrial practice to estimate angular velocity or acceleration in machines where the absence of any electrical contacts may be desirable. If some toothed ferromagnetic component – such as a gearwheel – is conveniently accessible, a small inductive pick-up can be fixed in such a position that the teeth pass near to the pole-piece of the device while the machine runs. Each passing tooth varies the inductance of the pick-up coil and generates a roughly sinusoidal voltage across it. Some units nowadays incorporate integral electronics, in the form of a Schmitt trigger circuit, in order to provide a direct digital output that is largely immune to extraneous noise.

In this application, the inductive device merely acts as a switch: it has no linear measuring capability and depends upon the presence of the ferrous teeth on the gear wheel. There are, however, other forms of inductive velocity transducer that can provide a speed-related signal: for example, those in which a bar magnet moves axially inside two coils. An improved form of this was described by Cranshaw *et al.* (1976), who modified the central magnetic core by the addition of a narrow air gap at its centre. The modified device was said to have a constant sensitivity within 0.1% for displacements up to 4.5 mm.

Capacitive velocity transducers

Angular velocities can be measured with capacitive devices, just as they can by optical or inductive means. Rehman and Murti (1980) have used a construction in which a thin, slotted aluminium disk, attached to the machine under test, rotates within the air gap of a three-terminal capacitor. The periodical change in capacitance, introduced by the passage of the slots, is detected with a double-wound transformer ratio-arm bridge.

Laser Doppler-shift measurement

We have already seen that the laser has achieved wide application in the field of measurement, one reason being that it has good frequency stability. Using a laser as light source for an interferometer makes it possible to measure extremely small changes in the path length of one of the beams; however, the same equipment can equally well be used at constant path length to detect small changes in frequency. If one of the beams is used to illuminate a moving component, the scattered light from it will exhibit a Doppler frequency shift as a result of the movement, and the velocity of the object can be determined from the measured shift.

The techniques involved in laser Doppler shift measurement have been described in detail in a monograph by Drain (1980).

Smeets and George (1981) used the Doppler-shift principle in the construction of a velocity-measuring device that incorporated a phase-stabilised Michelson interferometer. The operation of the system is interesting. A point on the surface of the moving object is illuminated by the beam from a laser, so that the scattered light from that point exhibits a Doppler frequency shift. Some of the light, at its shifted frequency, is taken along an optical fibre to the interferometer, which is arranged to hold a constant phase at its two detectors. The constant phase is achieved by circuitry that applies a variable voltage across a Pockels cell. If the phase changes, the voltage changes within a microsecond and restores the balance. Monitoring the voltage across the Pockels cell then provides a direct measure of the velocity at the measuring point.

Miscellaneous methods of displacement measurement

Laser speckle photography

The transducers described above are, generally, capable of measuring displacements along a single axis. They could, for example, be used to measure the movements of a punch in the generation of accurate force–displacement curves. Sometimes, the requirements of research call for a different kind of displacement measurement. It may, for example, be of interest to know how the surface of a tablet, or a component, moves in three-dimensional space. Measurements of this kind must often be made in a non-invasive way, since the object of interest may not be accessible to conventional transducer elements; in such cases it may well be more practical to use photographic methods.

Using the phenomenon of laser speckle is one way of approaching the problem of measuring very small displacements, strains or distortions in more than one dimension. This may be done by simple double exposure to show movement after a given time delay or, more interestingly, in real time so that the movement can be observed as it takes place.

The speckle pattern itself is an effect caused by random interference between wavefronts, scattered from the surface of an object that is illuminated with coherent light. At any point in space where the fronts are in phase, there is reinforcement, or constructive interference. Where they are in antiphase, there is destructive interference. The net result of these random variations in phase and amplitude is that the illuminated object appears grainy or speckled: the intensity variations over the surface show no periodicity and have a generally similar appearance everywhere. If the speckle pattern is viewed through a lens system, the apparent size of the speckles is inversely related to the aperture used; consequently, a large f-number gives large speckles.

Although the speckle pattern is of a generally uniform quality, it is nevertheless derived from the structure of the surface on which it appears: if the surface moves, or is distorted, the speckle pattern moves with it. The mathematical interpretation of speckle patterns is beyond the scope of these notes, but even without a detailed analysis, they can be used in the assessment of small movements.

A practical method for displacement measurement was suggested by Leendertz (1970). For this technique, a half-silvered, inclined plane mirror was used to superimpose the images of two scattering surfaces; the composite image was then focused on to a high-contrast photographic plate. The plate was developed and then replaced carefully in its original position so that it acted as a shadow filter. If one of the two original surfaces then showed displacement or distortion, the local variations of phase shift resulted in the development of light and dark fringes over the image plane, which could be observed in real time through a suitable viewing lens.

Leendertz showed that it was possible to arrange the optics of the system in different ways in order to detect either normal or in-plane movements of the test surface. The method appeared to have some advantages over conventional holographic interferometry in terms of its relative ease of operation and its capability to eliminate the effects of normal movement when necessary. It is, of course, possible to carry out holographic interferometry in real time, provided that the hologram itself can be processed without movement from its original position: the so-called 'instant' holography systems provide this facility, but at a considerable expense. Speckle pattern photography is less demanding in its positional requirements and could, in principle, be carried out at moderate cost.

A review article with numerous references was published by May (1977).

Flash photography

In general, the laser speckle technique tends to be restricted to the measurement of very small displacements, which may be considered to be on a par with the movements detected by the use of strain gauges. When relatively large three-dimensional movements are to be analysed, it may be more practical to use conventional photographic methods. Quite frequently, the subject of interest may be large enough to see, but moving too fast for the eye to follow; here, the use of a standard electronic flash system may provide the necessary short exposure time to ensure the production of clear, sharp images. An electronic flash-gun provides exposure times that range downwards from approximately 1.0 ms – the exact exposure usually being dependent on the amount of light required for the prevailing conditions – and should be satisfactory for most work in this field.

Since the tablet press carries out repetitive cycles of movement, the action of some rotating part can often be used to trigger the flash at a selected moment in the cycle. For this purpose, the reed switch and magnet form a useful combination in which the gun may be fired by closure of the reed switch. The magnet would normally be sited at some point on the tablet machine, while the reed switch is supported on a stationary holder at a short distance away. The switch then closes once for each cycle of the machine. One advantage of this simple arrangement is that it is easy to make small adjustments to the triggering point by corresponding changes in the position of either the magnet or the switch. It is, therefore, possible to make successive 'stroboscopic' exposures that represent some operation of interest, taken over as many frames as may be needed for a detailed study of the movements involved.

For some purposes, this can be a practical and low-cost alternative to the use of a conventional high-speed cine camera. Furthermore, it is not limited to any particular format or film size. By comparison with the 8 mm or 16 mm cine format, the square 'still' frame in 35 mm or 60 mm format can hold a great deal of information. A typical black-and-white film such as Kodak Plus X or Ilford FP4 is capable of a resolution in the range 100–125 line-pairs mm^{-1} under good conditions; so, in principle, a standard negative measuring approximately 24 mm by 35 mm could hold well over 2000 line-pairs.

The possible significance of the quoted resolution is that it gives a rough guide to the size of single particles that might be identified in a photograph, with a given lens at a given distance. Successive single photographs on conventional film have been used by the writer to study the movement of particles as the punch entered the mouth of the die of a rotary tablet press. The expulsion of air from the die is seen to lift a rotating toroidal cloud of powder that then rolls up the outside surface of the punch. The effect is not visible to the naked eye.

In practice, digital cameras have almost replaced those using film for many photographic applications, but when equipment is to be chosen for a particular purpose, it is still worth considering what resolution will be required. If, for instance, the subject occupies a reasonable proportion of the visible frame, then in most instances it would seem quite satisfactory to use a run-of-the-mill digital camera of four megapixel resolution, but for the study of dust clouds, for instance, something better may be required.

The stroboscopic system outlined above can be most useful but it does have the disadvantage of showing different events on each individual exposure, rather than different phases of the same event. If this should be inadequate for some application, a reasonable compromise may be found in three-colour flash. For this procedure, Ridgway Watt and Harwood (1984) positioned three separate flash tubes in a common housing with a single exit aperture. In front of each tube was a colour filter, the colours being, respectively, red, green and blue. These colours can be fired in sequence by a programmable control unit. With this system, each frame of colour film is exposed to the three colours in rapid succession. Any stationary object appears normal, since the colours give white light by addition, but any object that moves is shown as three separate coloured images. As the images are all on the same piece of film, which remains stationary between the three exposures, the separation between them can be taken as a reliable indication of movement during the selected time period, both in magnitude and direction. The system was used to study the movement of small particles in powder clouds.

References

Bolk WT (1985). A general digital linearising method for transducers. *J Phys E Sci Instrum* 18: 61–64.

Clabburn EJ (1983). An inexpensive fast response displacement transducer. *J Phys E Sci Instrum* 16: 248–250.

Cooke M, de Sa A (1982). A new method of non-contact measurement of linear displacement. *J Phys E Sci Instrum* 15: 843–847.

Corke M, Kersey AD, Jackson DA (1984). Temperature sensing with single-mode optical filters. *J Phys E Sci Instrum* 17: 988–993.

Cranshaw TE, Lang G, Placido F (1976). A linear velocity transducer. *J Phys E Sci Instrum* 9: 9–10.

Crecraft DI (1983). Ultrasonic instrumentation: principles, methods, and applications. *J Phys E Sci Instrum* 16: 181–189.

Culshaw B (1983). Optical systems and sensors for measurement and control. *J Phys E Sci Instrum* 16: 978–996.

Dann MS (1984). A linear displacement transducer. *J Phys E Sci Instrum* 17: 184–185.

Depraetère P, Seiller M, Puisieux F (1978). Laser measurements of velocities of tablet punches. In

Proceedings of the International Conference on Powder Technology and Pharmacy, Basel.

Drain LE (1980). *The Laser Doppler technique.* New York: Wiley.

Garratt JD (1979). Survey of displacement transducers below 50 mm. *J Phys E Sci Instrum* 12: 563–573.

Giles IP, McNeill S, Culshaw B (1985). A stable remote intensity based optical fibre sensor. *J Phys E Sci Instrum* 18: 502–504.

Hugill AL (1982). Displacement transducers based in reactive sensors in transformer ratio bridge circuits. *J Phys E Sci Instrum* 15: 597–606.

Jones RV, Richards JCS (1973). The design and some applications of sensitive capacitance micrometers. *J Phys E Sci Instrum* 6: 589–600.

Leendertz JA (1970). Interferometric displacement measurement on scattering surfaces using speckle effect. *J Phys E Sci Instrum* 3: 214–218.

May M (1977). Information inferred from the observation of speckles. *J Phys E Sci Instrum* 10: 849–864.

Neubert HKP (1975). *Instrument Transducers,* 2nd edn. London: Oxford University Press.

Rehman M, Murti VGK (1980). A new capacitance speed transducer. *J Phys E Sci Instrum* 13: 655–657.

Ridgway Watt P (1983). Measurement of acceleration in a rotary tablet machine. *J Pharm Pharmacol* 35: 746.

Ridgway Watt P, Harwood N (1984). A novel photographic system for process machine analysis. *J Pharm Pharmacol* 36: 91P.

Sayce LA (1972). Gratings in metrology. *J Phys E Sci Instrum* 5: 193–198.

Smeets G, George A (1981). Michelson spectrometer for instantaneous Doppler velocity measurements. *J Phys E Sci Instrum* 14: 838–845.

Sydenham PH (1972). Microdisplacement transducers. *J Phys E Sci Instrum* 5: 721–733.

Tanwar LS, Kunzmann H (1984). An electro-optical sensor for microdisplacement measurement and control. *J Phys E Sci Instrum* 17: 864–866.

Theocharous E (1985). Differential absorption displacement transducer. *J Phys E Sci Instrum* 18: 253–255.

Woolvet GA (1982). Digital transducers. *J Phys E Sci Instrum* 15: 1271–1280.

5

Power supplies and data acquisition

Peter Ridgway Watt

Introduction

In earlier chapters of this book, we have considered the application of various transducers to measure force and displacement, and it is important to ensure that the information produced by the transducers can be used to the best advantage. This may involve manipulation of the signals received from one or more transducers so that some critical parameter may be available in real time for the operation of a control system; or it may be sufficient to have compaction data stored for record purposes. Regardless of how the transducer signals are to be applied, there will be two essential requirements. Firstly, that of providing appropriate power supplies to each transducer unit and, secondly, that of transferring the output signals from each one, without distortion, to the data processing system.

Most transducers require an electrical supply and correspondingly generate an electrical signal in the form of a continuously varying voltage: in other words, an analogue output, although there may be some exceptions to this generalisation. In any event, the first problem is that of bringing the electrical supply to the transducer.

Gauge excitation level

Strain gauge bridges – Wheatstone bridges – are usually powered with a low-voltage DC. We have seen in Chapter 2 that the output signal from a typical bridge circuit is very small, and it is a linear function of the applied or input voltage. Raising the applied voltage, therefore, gives a correspondingly increased signal, though at the same time it also increases the inherent self-heating that results from the current flowing through the circuit. It is clearly undesirable to raise the temperature of the gauge elements to any great extent; however, the heat generated is continuously conducted away into the substrate to which the gauge is attached; consequently, an equilibrium temperature is reached. The rate of heat loss depends on the area of the gauge, the thermal conductivity of the substrate and the thickness of the glue line. A good glue line should be as thin as possible: within the range 2.5 to 7.5 µm. Not only does this maximise the thermal transfer from the gauge, it also contributes to efficient transmission of strain. Nevertheless, the thermal conductivity of the substrate is the most critical factor, and there can be a 100-fold variation between that of conductive metals such as copper, on the one hand, and some insulating plastics on the other. An acceptable power loading for a gauge applied to a heavy steel substrate would be in the region of 1.6–3.1 kW m^{-2}, where the area concerned is that of the gauge grid. A mid-range value of 2.0 kW m^{-2} would be equivalent to 20 mW per 10 mm^2: half that would be acceptable on stainless steel, and twice that on copper or aluminium.

Modern strain gauge practice has made it possible to apply gauges of a few millimetres in length if this should be necessary, but where the available space imposes no limitation, there is clearly some advantage in using gauges with a somewhat larger area, perhaps in the region of 20–30 mm^2, since this will keep self-heating to lower values.

There used to be a simple rule of thumb to the effect that a gauge with an area of 10 mm^2 on a steel substrate could usually be run at a maximum of 2.5–3.0 V, though this would naturally have depended on the resistance of the gauge. Most of the earlier gauges had fairly low resistances of perhaps 120 Ω, but nowadays it is normal to use elements of 350 Ω or more, up to a maximum of 5000 Ω, often applied in battery-operated equipment; consequently, it is possible to use higher voltages or lower currents as the situation demands.

Rules of thumb aside, the most thorough way of determining the maximum voltage for continuous operation of a specific gauge on a given substrate is to connect a variable power supply unit (PSU), and to raise the applied voltage until thermal drift is detected. At this point the voltage can be reduced accordingly to a suitably safe level. It should then be possible to replace the variable PSU with one that provides a fixed output voltage at the desired value.

The power supply unit

Commercial bench or laboratory PSUs are readily available. They take AC power in at mains voltage, rectify and stabilise it, and provide a variable DC output; this can be accurately controlled and is usually displayed on one or two panel meters that indicate voltage and current.

A typical example is the Iso-Tech IPS 1810H from RS Components Ltd. The output of this PSU can be set anywhere in the range 0–18 V, and it can be operated at constant voltage or constant current, the selected values being displayed on analogue meters. At constant voltage, the line and the load regulations are stated to be 0.01% or less, while ripple and noise are less than 0.5 mV rms. Having a load regulation within 0.01% means that the output voltage should not vary by more than 0.01% of any changes in the load resistance. The resistance of a strain gauge bridge will usually be constant to within 1 part in 1000, so the stability of the applied voltage should be very good indeed.

Once the optimum voltage has been determined, the bench PSU can be replaced with an encapsulated PSU with an appropriate output. Many of these small units have some degree of adjustment so that they can be pre-set to the desired conditions. They are also provided with so-called 'crowbar' circuitry that protects them from accidental over-voltages and current limiting to protect the power supply from short circuits. Encapsulated units tend to come in two input groups: AC (mains) and DC. They also have three types of regulation: none, linear and switched mode.

However, before selecting a suitable module, it is worth checking whether the data processing equipment contains an integral power supply, as these are often included as part of the system and would be of good instrumentation quality. Equipment supplied by Vishay Measurements Group and Thames-Side Maywood are examples of such an arrangement.

Non-regulated power supply modules

Non-regulated PSUs are mainly of use when another regulator has been provided. It is sometimes convenient to make use of an unregulated 9–12 V DC power bus and locally regulate down to 5 or 3.3 V etc. as required by the electronics.

Linear regulators

Linear regulators usually have a series transistor that converts the excessive voltage into heat. The conduction of this transistor is controlled by feedback to maintain a constant output voltage. For current levels up to 1 A, it is often convenient to use linear integrated circuit regulators such as the LM7805 series (fixed voltage) and LM317 (variable output from 1.2 V). The two digits at the end of the number signify output voltage (e.g. the LM7805 is a 5 V regulator). It is necessary to ensure that the input voltage is at least 2.5 V higher than the output voltage but not too high, as the efficiency drops and heat dissipation can become a problem at higher currents. More modern low dropout regulators have a PNP ('Plug-N-Play') series pass transistor and only require a few hundred millivolts of headroom.

It is important to follow the manufacturer's recommendations for wiring. If two capacitors are not connected between its three terminals near the device, they may oscillate at up to 60 kHz. Note that the LM7905 negative voltage regulator series requires much larger value capacitors, and LDO regulators are particularly critical for not only the capacitors values but also their equivalent series resistance. A final point is that the power supply's output noise level is an amplified version of its internal reference voltage device; fortunately this is adequate for our application, as we are not interested in super-low noise. Zener diodes are fairly noisy, so instead these regulator devices use either a band-gap voltage reference at 1.2 V or a low noise process device called a buried zener.

Switch-mode regulators

Switch-mode regulators generate significant electrical noise at the switching frequency, which is often around 50 kHz. It is wise to add series inductors to the output as they are much more effective at removing these unwanted high frequencies than parallel capacitors across the output; electrolytic capacitors tend not to work so well at high frequencies since they may dry out and can reduce the stability of the output voltage. The advantages of this type of regulator are efficiency and cost. Some switch-mode regulators incorporate a transformer, which provides isolation between input and output terminals. The 'isolated' regulators are effective in removing earth loops.

Mains noise

Low-voltage signals are susceptible to electrical interference from a variety of external sources, which may include motors, refrigerators, thermostats, motorised valves and a whole host of electronic switches: in fact, any form of device that may give rise to airborne electromagnetic radiation. Apart from these, the electrical mains can be sources of a great deal of noise, which may itself be airborne or may be carried directly into mains-operated equipment. Mains noise may be manifest as a permanent distortion of the ideal sinusoidal waveform, although additionally it may occur in the form of random pulses or 'spikes' of short duration and high voltage. Peaks of several hundred volts are quite common, and some reach to more than a thousand volts. These can affect monitoring systems if they are not removed. Examples of noise from the mains are shown in Figure 5.1.

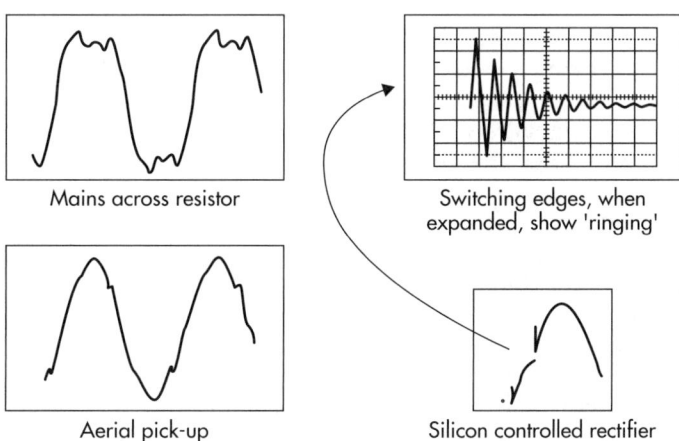

Figure 5.1 Examples of mains noise.

An interesting example of the effects that electrical noise can cause was encountered by one of the editors during an early attempt to interface an eccentric tablet press with a microcomputer (Armstrong and Abourida, 1980). As only limited computer memory was available at that time, it was decided to record the output of the upper punch force transducer only when the signal from it exceeded a small positive value. Otherwise the computer memory would mainly be filled with zeroes, since for much of the compression cycle, the upper punch was not in contact with the contents of the die, and hence there would be no detectable force. The computer was instructed to monitor the signal from the upper punch and to commence collecting data when this exceeded the threshold and to continue doing so until the signal fell back below the threshold. The tablets were to be made singly, with the die filled by hand and only one compaction event was to be recorded.

This arrangement worked perfectly when the press was rotated by hand. But when the press was driven by its motor, no data were collected. When the transducer output was viewed on an oscilloscope, it was noted with some surprise that when the 'start' button of the press was activated, a large electrical spike of very short duration was generated. The computer interpreted this spike both as the signal to start collecting data and also the signal to stop, so when the real compaction event occurred, no data were collected since the computer was programmed to collect data from only one event. It proved impossible to eradicate the noise from the 'start' button. The problem was circumvented by building a delay into the software so that data collection commenced only if the signal was still above the threshold some 20 ms after the threshold had been exceeded. In that time, the signal caused by the spike would have disappeared.

When a heavy inductive load such as an electric motor is switched on, the voltage across its terminals is temporarily reduced since the mains cables have a finite resistance. When the motor is switched off, the inductive energy stored becomes converted to a high voltage that will arc across the switch contacts, thus generating airborne radiofrequency noise and very high voltage peaks on the mains.

Computer switch-mode power supplies, unless of the power factor correction design, distort the mains by taking all their current at the peak of the mains sinusoid.

The effects of radiated noise can usually be minimised if the cables carrying the signal are screened with an external layer of conductive copper braiding, earthed at one end or possibly both ends. Electromagnetic interference suppression ferrites, manufactured by companies such as Steward, Chomerics, and Fair-Rite, are very useful in preventing high-frequency noise from entering a screened electronics enclosure. The cables are fed through ferrite rings or split rings are snapped over the cable to increase the cable's inductance. Clip-on flat ferrites are used with flat ribbon cables. PC video cables often have a bulge at each end, which is where the cable is fed through a ferrite tube to reduce radiation.

Ideally all signals should be sent and received differentially, so that common-mode voltages cancel out. Signal cables should always be kept as short as possible and laid well away from power cables supplying large motors. Earthing a cable at both ends may have undesired effects if the two ends are connected to local but separate mains earth points. If these two 'earths' should have a potential difference of, say, 1 V, and the cable's braid has a resistance of 1 Ω, then 1 A will flow and this can obviously cause a major problem.

As an example, this can happen if a motor does not have equal currents in its windings: any imbalance tends to flow through the bearings into the motor frame, which is itself connected to earth. Some variable-speed AC motor drives were designed to 'chop' the current at high frequency, and this technique led to premature failure of motor bearings, so more modern motor drives generate three-phase variable-frequency AC without the damaging high-frequency component.

Mains filters

Noise from the mains can be reduced considerably by the use of electrical filter networks. These are commercially available units that are connected between the mains supply and the mains-driven equipment. The simplest varieties contain a number of capacitors connected between the live, neutral and earth. The more elaborate ones also have inductive components in series with the supply. The characteristics of the simpler filters are detailed in British Standard BS613: they usually provide an attenuation of some 30 dB in the frequency range 600 kHz to 100 MHz, while having little effect on the 50 Hz base frequency.

In spite of the presence of input filter networks, high-voltage spikes may still pass through into mains-driven power supplies, emerging with the low-voltage DC output. Not only may they corrupt the signal output of the system powered by the PSU, but they may even destroy certain sensitive components such as CMOS devices, integrated circuits, and so on.

Transient suppressors

Solid-state transient suppressors may be used. These are PN silicon devices and are connected across the output terminals of the PSU. They are normally non-conducting at the rated voltage of the PSU, but at higher voltages they pass very rapidly into a conducting state. Their speed of response is measured in millionths of a microsecond, and they are able to shunt the excess power before serious damage results.

Battery power

So far, we have concentrated on permanent mains-operated supplies. In recent years, there have been alternative systems on the commercial market, running on batteries of various types. Using battery power enables short-term measurements of force and displacement on rotary presses to be made without the complication of introducing power from external sources.

In systems of this type, there have usually been on-board signal-conditioning electronics, so that data from the transducers could be transmitted by telemetry either continuously or as a short burst of stored data to a stationary receiver. The electronics also have the capability to stabilise the battery voltage, thereby avoiding some of the problems associated with mains-driven power supplies.

The disadvantage of the earlier battery-powered systems lay in their size, since it was necessary to occupy several die stations with the rather bulky components: the battery box, the electronics, the displacement transducer, and so on. Later versions improved on this to some extent by taking the electronics unit out of the die station and mounting it separately, reducing battery size by using lithium button cells, and so on. Even more recently it has been possible to miniaturise the complete system so that it can be contained inside an instrumented punch (see Chapter 6).

Power supply to and data acquisition from tablet presses

The supply of power to transducers fitted to an eccentric press usually presents few problems, and neither should data acquisition. Punches move in one vertical dimension and no difficulties should be encountered provided the electrical leads are given adequate support. The situation with rotary presses is different. Here the punches move in three dimensions, and the rotation of the turret precludes a fixed link between the power supply and the transducers, and also between the transducers and the data acquisition system. There are several approaches to solving this problem, but irrespective of the method that is chosen, attention must be paid to the electrical connections.

When punches are instrumented in the more traditional way, the gauged area passes into the punch guides during part of the machine cycle. The guides have sharp edges and it is important to ensure that the wires are not sheared off at the edge of the guide. The wires are often cemented into a shallow groove in the punch body so that they can be taken out from the end of the body in relative safety, though an axially drilled hole would probably be better.

Whatever means are used to lead the wires out from the punch, careful attention must be paid to the resident hazards around the turret. There are various fixed components on the press that can easily become entangled with the wiring from a moving punch, and it is worth spending some little time turning the machine by hand in order to check what is there. In our own instrumented Betapress, the wires were taken out at the end near to the punch tip, and were provided with 'Lemo' miniature locking plugs so that they could be connected to our rotary slip-ring unit. Using a Lemo plug reduced the risk of disconnection through centrifugal force as the turret rotates. Alternatively, the wires could have been taken out at the end near to the punch head: there are perhaps fewer obstructions at this end, but it is important to see that there is still sufficient clearance between the punch head and the punch guide at maximum compression.

Slip-rings

An apparently simple solution is to fit slip-ring gear to the press, using this both to supply power to the transducers and to take data out. This approach was reported by Ridgway Watt and Rue in 1979. The method itself was very satisfactory but was possible because it was used with a brand new Betapress that had been modified by the provision of an extended main shaft. The shaft was carried vertically upwards beyond the normal level and was machined with an internal cavity to accept electrical cables. Co-axially above the upper end of the shaft, a 12-element integrated slip-ring assembly of instrument quality (type PM-12-01-TC from IDM Electronics Ltd) was fitted that ran on its own bearings to ensure that it would not generate unacceptable levels of electrical noise. It had several brushes per ring and was loosely coupled to the Betapress shaft by means of an 'Essex' coupling so that the two rotated together but were mechanically separate (Figure 5.2). Holes were drilled obliquely through the turret and into the hollow shaft from points adjacent to the strain-gauged punches, both upper and lower, while corresponding holes were provided near the associated displacement transducers. Signal cables were then run up inside the central shaft and connected to the slip-rings above (Figure 5.3).

It had previously been suggested that slip-ring gear would not be suitable for such a problem because of the poor signal-to-noise ratios that might be expected. Sidor (1983) commented that 'slip-rings are far from reliable, and practically useless for permanent installation'. Indeed, this would have certainly been the case had an attempt been made to fit slip-rings to the main shaft itself, but using a separate self-contained unit of appropriate quality provided an entirely acceptable solution. The recorded noise level of this system after installation was $5\,\mu V$ per $1\,mA$, at a rotation rate of 3000 rpm, and this was felt to be satisfactory for the application.

Slip-rings represent one of the very few systems that can be operated on a completely continuous basis, with power going in and signals coming out without interruption.

Mercury seals

The widespread albeit inaccurate notion that slip-rings are useless for permanent installation has led several workers to look for alternative rotary contact systems, and arrangements using mercury as a liquid contact medium have been described by Marshall (1983) and by Walter and Augsburger (1986). This system was also used in instrumented capsule-filling equipment by Small and Augsburger (1977). In each case, one or more annular channels were machined into the upper surface of an insulating disk fitted concentrically above the turret head of the tablet press. The channels were partially filled with mercury and were electrically connected to the transducers below by means of metal screws let into the under side of each channel. Above the rotating channels was a stationary arm of insulating material that supported a set of conducting pins. As the pins dipped into the mercury, contact was established with the transducers below.

A major drawback to the use of mercury in this way is that its vapour is poisonous, and the use of stationary pins in a rotating mercury pool will, naturally, result in continuous renewal of the exposed surface. Even at room temperature, this would nowadays be considered an unacceptable hazard.

Radio-telemetry

The problem of retrieving information from rotating machinery is not restricted to the field of tablet manufacture, but has been well known in the context of general engineering for many years. As a result, there is a substantial body of knowledge on this topic, and there are commercially available telemetry systems that can be applied to existing presses. Some of the earliest examples of instrumented tablet presses were equipped with radio-telemetry links (Shotton et al., 1963; Wray, 1969).

The signal from conventional transducers such as strain gauges is usually in the form of a varying DC voltage, and the first step has generally been to convert this into an AC waveform that can conveniently be transmitted via a radio link or an inductive loop coupling. Signal conversion is accomplished by frequency modulation, in which the DC signal level is used to change the operating frequency of an AC oscillator. The oscillator will have a normal frequency of around 10 kHz, and can be swung through a small range of frequencies on either side of this central value by the application of positive or negative input signals. The output then appears as an AC waveform of fixed amplitude and varying frequency; it can be transmitted across a short distance, received on a suitable antenna and converted back into its original form by appropriate electronics.

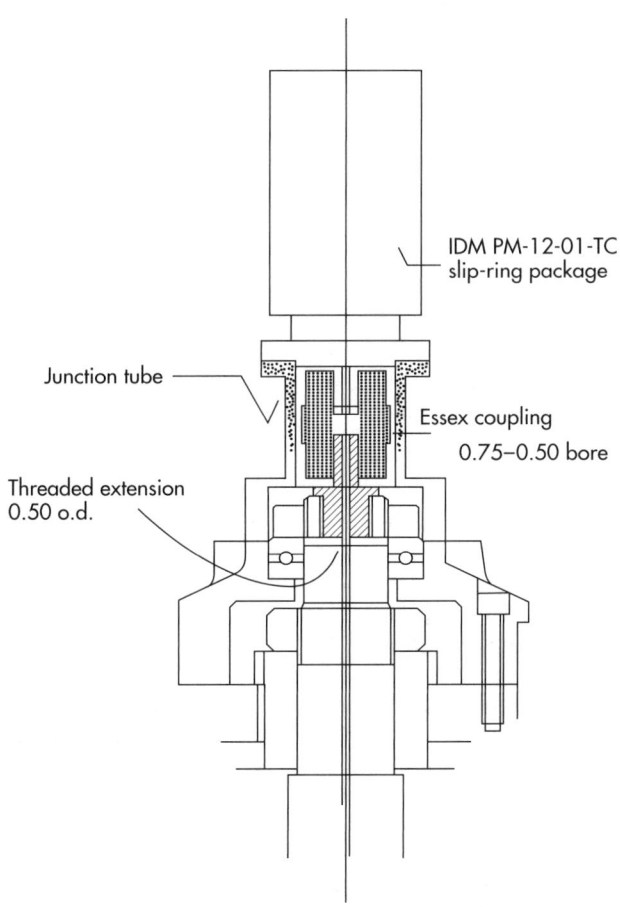

Figure 5.2 Schematic arrangement for a slip-ring system attached to a rotary tablet press.

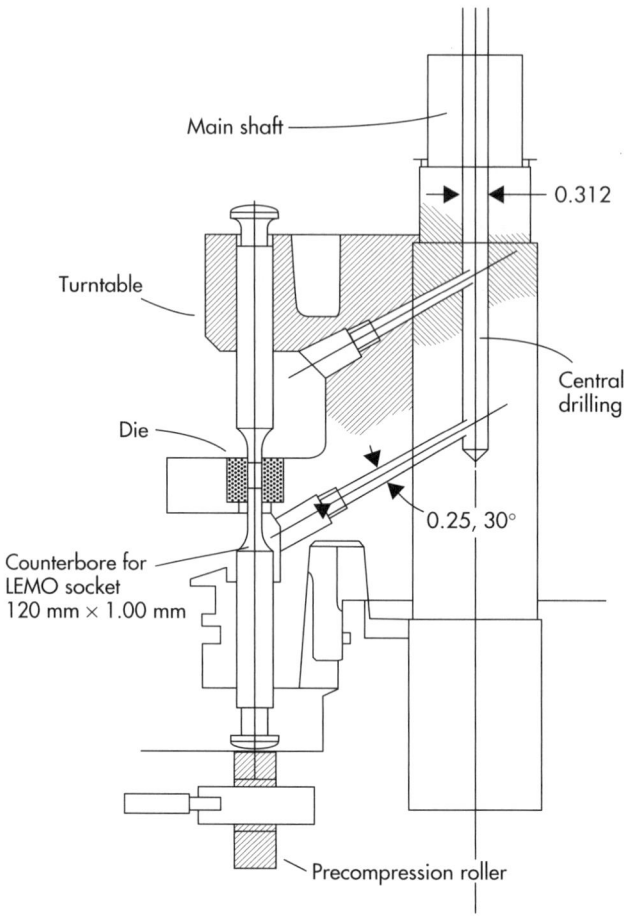

Figure 5.3 Schematic arrangement of drillings for signal cables involved in fitting a slip-ring to a rotary tablet press (not co-planar with the punches).

One disadvantage of the early telemetry systems was physical size. It was common practice to occupy one position in the turret for an instrumented punch, then to fit the signal-conditioning electronics into an adjacent position, and finally to use a third position for the battery pack that powered the whole system. Clearly, this meant that it was not possible to use the full complement of punches, and this might well have resulted in correspondingly abnormal behaviour of the press. For example, using fewer punches could lead to over-filling of the feed frame. This, in turn, could give rise to cyclic variations in tablet weight, because the fingers of the feed frame pass over the top of the die during the weight-setting operation and may produce a measure of extra compaction as they pass (Ridgway Watt and Rue, 1980).

Since then, the rapid development of micro-electronics has resulted in smaller and smaller devices with greatly increased ability and reduced power requirements. Voltage-to-frequency converters, for example, have been available as integrated circuits for many years: first as dual-in-line devices, later as surface-mounted packages. It is no longer necessary to occupy several stations with the telemetry equipment, which can be built into a block of minimum dimensions. Even the battery, once a fairly large component, can be replaced by one or more lithium button cells. It is also possible to use an inductive coupling to introduce power into a rotating device. A stationary coil on the outside, fed with AC, induces corresponding currents in a coil on the rotating component, the two forming a rotary transformer. The current so generated is then rectified and smoothed as necessary so that it can be used with the strain gauge amplifiers or displacement transducer circuits on the press.

Optical linkage

An alternative to radio-telemetry is to use a pulsed light beam to transmit information in digital form. Using frequencies within the infrared part of the spectrum reduces the interference that might be caused by daylight, but nevertheless provides a large capability for data transmission. One practical approach is that of using on-board electronics to operate a suitable light-emitting diode – an infrared LED – and to generate a stream of pulses that represent a digitised version of the transducer signals. This generally involves the use of an analogue-to-digital converter, a standard integrated circuit. Then the resulting data stream can be sent along a flexible fibre optic to a point in line with the main shaft of the rotating system, where the digital data can be read by a stationary photodiode.

Another method that has been used in commercial practice is that of storing the compaction data temporarily within the electronics block, which is mounted somewhere near the associated transducers. Then, as the turret rotates past an optical sensing head, the information that has been collected is sent out in a very short burst. When data from more than one transducer have been stored, their separate signals can be sent out in very rapid succession. Switching from one to another is known as 'multiplexing' and this is the method used, for example, to send many telephone messages simultaneously along a single cable.

A commercial system using this principle was available from Puuman Oy in Finland, under the name 'Portable Press Analyser'. As the name suggested, it was designed to be sufficiently small and light to be taken from one machine to another, so that the performance of each machine could be compared with others of a similar type. Indeed, it could be used with any machine employing the same type of punch, and facilities were provided for measurement both of punch force and displacement. The associated punch was made in two parts so that the strain gauges could be mounted internally, and this gave the gauges some measure of protection in use. The requirement for displacement measurement meant that one position in the turret was occupied by the instrumented punch, while the adjacent position held the LVDT transducer. A diagram of the Puuman Oy system is shown in Figure 6.19.

Signal display

With the increased availability of cheap computing power, the output of transducers fitted to tablet presses is now almost invariably fed into a computer for storage, display and arithmetical manipulation. This is discussed in detail in Chapter 8. Nevertheless, it is often advantageous to have some alternative means of displaying and recording the signal from a transducer, especially in the early stages of an investigation, and for calibration and checking purposes.

The method chosen for the display of a variable is likely to depend on the rate at which that variable may change. The time during which the punches of a tablet press are in contact with the die contents is of the order of 0.1 s. The inertia of a mechanical pen recorder is sufficiently high that the pen cannot follow the fluctuations of the signal. Neither can a digital voltmeter be used, since the eye cannot follow the rapidly changing digits. Nevertheless both pen recorders and digital voltmeters have been used in situations in which the signal changes more slowly, such as in the measurement of tablet-crushing strength.

Devices in common use for the display of variables at normal tabletting speeds include the oscilloscope and the high-speed ultraviolet recorder. Each of these produces a display in 'real time', that is to say as the event happens. It is also possible to produce a display in retrospect by the use of digital storage and subsequent retrieval. The data will of course pass into the store in real time; however, they can be retrieved at a much slower rate if necessary. The output from the data store can then be printed on a chart recorder, or an x–y plotter, to provide a hard copy of good quality.

The oscilloscope

If an oscilloscope is coupled to the output of the transducer amplifier, and is correctly set up, it will display the changing signal as a bright line on the screen. The duration of a single tablet event is short, and it is impossible to follow the moving spot by direct observation, so means have to be found to extend the time over which the trace can be viewed. One such means, often used in the early days of tablet press instrumentation, is that of photographing the trace as it appears with an oscilloscope camera.

In order to ensure that the oscilloscope trace is in a suitable position on the screen, it is necessary to have a triggering signal that will start the spot moving across the screen just before the compression event. This can be done with a reed switch, the contacts of which close when a magnet is brought near them. The reed switch is connected to the triggering contacts of the oscilloscope, while the actuating magnet is attached to some accessible part of the tablet press. The relative positions of the reed switch and the magnet are then adjusted until there is contact closure just as the punch begins to descend into the die. Reed switches and their magnets are often contained in a glass or plastic envelope and so are well protected against dust and dirt.

Some, but not all, oscilloscopes have an autotriggering facility, whereby the time base can be caused to start across the screen as soon as the signal to the y axis exceeds a set level. In this type, external triggering is not necessary, although some portion of the signal below the threshold may be lost.

The digital storage oscilloscope

Visualisation of the signal waveform is much simplified if the oscilloscope is of the storage type where the trace can be retained for some appreciable time after the event. Modern storage oscilloscopes are often equipped with digital storage systems, which represent a considerable advance over the earlier forms of analogue storage. The trace always remains clear and can be moved on the screen after the event. This makes it possible to study a waveform in great detail and to move it backwards or forwards across the screen.

Most of the oscilloscopes designed for serious laboratory work nowadays are of the digital storage type. Very-high-speed analogue-to-digital converters send data into memory, from which they are displayed on a cathode ray tube or LCD screen. The more expensive models use a colour display so that traces can be displayed on top of each other with no likelihood of confusion.

Data are fed continuously into the store until a trigger condition is detected. This causes a counter to count down from typically 50% of the store's sample size. Storing ends when the counter reaches zero. This means that it is now possible to see what happened before the trigger; this is known as 'pre-triggering'. Some digital storage oscilloscopes allow pre-triggering of 10, 50 or 90% of the store size, and have a variable time delay. For PAL/NTSC video signals, a digital storage oscilloscope can be set up to trigger on a given line or frame. Reed switch triggering, as described above, is no longer necessary.

The display is refreshed at a constant rate, so the brightness of the trace is constant and does not gradually fade away. Photographing a trace is simple, and many digital storage oscilloscopes have a floppy disk drive or memory card socket to copy data. Some have a much larger store memory and the stored waveform can be both shifted and expanded in the time axis.

Digital storage oscilloscopes have a comprehensive set of cursor functions that can measure voltage, time and frequency. Unlike older analogue instruments, the digital storage oscilloscope has an 'autoset' button that simplifies set-up.

A further development has led to the mixed signal oscilloscope, which combines two analogue channels with a logic analyser. A good example of this type of instrument is the HP 54645D. This has two 100 MHz analogue channels and 16 digital channels. One special feature is called Mega Zoom. Each trace has 1 Mbyte of memory though only 500 points are displayed. This gives the user the opportunity to zoom in and shift the stored waveform to a considerable extent.

The ultraviolet recorder

Whereas mechanical pen recorders are usually limited by the inertias of their components to a full-scale traverse time of approximately 1 s, ultraviolet recorders – or oscillographs – are not limited to the same extent. They use a moving ultraviolet light beam instead of a pen and produce a trace on photosensitive paper at writing speeds that are sufficiently high to record signals from the compression event. The movement of the light beam is controlled inside the instrument by means of a small mirror galvanometer. It is possible to select galvanometer movements with different characteristics to suit specific applications.

Because the duration of the compression event is so short, it is necessary to run the ultraviolet recorder at a correspondingly fast paper speed; otherwise the whole event will be compressed into an unreadably small space. Modern ultraviolet recorders can consume paper at a surprising rate, having a range of speeds from a few millimetres per second to several metres per second. In order to avoid a considerable waste of recorder paper, it is desirable to have a control system that will limit the paper run to the event being monitored. The ultraviolet recorder itself will often have an on-board timer that controls the duration of the run and is set to a value just exceeding the compression time.

In our own laboratories, we have once again employed the reed switch and magnet arrangement to generate a contact closure at the required moment. In a similar way, D'Stefan *et al.* (1985) used a magnet and a Hall effect switch to signal to their data acquisition system.

Once obtained, the traces from the ultraviolet recorder are usually reasonably clear to read; however, they may fade with age. For long-term record keeping, it may be useful to photograph the traces and store the photographs instead. Alternatively, spraying the trace with a suitable lacquer gives a record that does not fade for many years.

References

Armstrong NA, Abourida NMAH (1980). Compression data registration and manipulation by microcomputer. *J Pharm Phamacol* 32: 86P.

D'Stefan DA, Leesman GD, Patel MR, Morehead WT (1985). Computer interfaced instrumentation of a rotary tablet press. *Drug Dev Ind Pharm* 11: 83–100.

Marshall K (1983). Instrumentation of tablet and capsule-filling machines. *Pharm Tech* 6: 68–82.

Ridgway Watt P, Rue PJ (1979). The design and construction of a fully instrumented tablet machine. In *Proceedings of the International Conference of Pharmacy Technology and Production Manufacturers*, Copenhagen.

Ridgway Watt P, Rue PJ (1980). Tablet weight and density variation in a rotary press. *J Pharm Pharmacol* 32: 22P.

Shotton E, Deer JJ, Ganderton D (1963). The instrumentation of a rotary tablet machine. *J Pharm Pharmacol* 15: 106T–114T.

Sidor T (1983). Simple and accurate strain gauge signal-to-frequency converter. *J Phys E Sci Instrum* 16: 253–255.

Small LE, Augsburger LL (1977). Instrumentation of an automatic capsule-filling machine. *J Pharm Sci* 66: 504–509.

Walter JT, Augsburger LL (1986). A computerized force–displacement instrumentation system for a rotary tablet press. *Pharm Technol* 9: 26–34.

Wray P (1969). The instrumented rotary tablet machine. *Drugs Cosmetic Ind* 105: 58B–68B, 158–160.

6

Instrumented tablet presses

N Anthony Armstrong and Peter Ridgway Watt

Introduction

There are many ways in which force and displacement can be measured, and Chapters 2 and 4 covered a selection from the methods currently available. Some of the published work in tabletting has already been covered quite extensively in these chapters. The purpose of this chapter is to take some examples of instrumental methods applied to tablet presses of various types in order to look at some specific points in more detail, and to attempt to suggest possible future developments. Inevitably, this approach results in some repetition of detail here and there, but this has seemed justifiable on the grounds that it provides better continuity within each topic.

Measurement of punch forces

All tablet presses, of whatever design, embody at least one punch and die set, powered usually from a motor but occasionally by hand. In between the power source and the punch, there is a train of mechanical components through which the power is transmitted. Variations in the punch force will appear as corresponding variations in the transmitted force through each of the components, so any one of them could be considered, at least in principle, as a potential site for a measuring device of some sort.

To some extent, the choice of site for force measurements will depend on the purpose of the measurement to be made. This, in turn, will influence the degree of accuracy that is required, and the cost that can be justified by the measuring system.

Requirements of accuracy

In a research laboratory concerned with the physics of tablet compression, it is likely that punch force measurements will be needed at their most precise; say to 1 part in 10 000 or better. For much formulation work, a 12-bit accuracy – or 1 part in 4096 – is often quite acceptable, and this allows for convenient manipulation of the data by fairly simple computer systems. In a production area, concerned with the maintenance of tablet quality from a group of commercial tablet presses, a considerably lower standard, perhaps 1 part in 512, may well be acceptable. Finally, at the least accurate end of the measurement scale, would come routine monitoring of the tablet press as a check on its mechanical condition. Here, an appreciable departure from the normal power loading, for example, might be taken as a warning of a need to clean the punches and guides, or for general maintenance work on the equipment.

To put these considerations into perspective, it is probably useful to examine two stylised tablet presses, one eccentric and one rotary, and to look at the merits of various gauging areas in broad terms. On a real tablet press, these areas could be investigated in more detail by stress analysis and a specific gauging site could then be selected accurately.

The eccentric press

In the eccentric press, the upper punch is held in a sliding block that moves up and down above the stationary die. The block travels in a vertical slide-way and is driven from an eccentric at the end of the main shaft. The die and the lower punch remain in nominally fixed positions during the compression of a tablet. After the tablet has been compressed, the lower punch is then raised in order to eject the tablet through the upper end of the die. The whole system is usually powered from an electric motor, probably coupled through a flexible belt and pulley.

Measurement of the upper punch force

If the upper punch is considered first, we can see that its applied force is transmitted from the motor, through the drive belt, the main shaft, the upper eccentric, the upper arm, the punch holder and, finally, through the length of the punch itself. There will be corresponding reactions in the machine frame that will tend to separate the upper and lower punch faces as pressure is applied between them. Measurements related to punch force can be made at almost any of these sites, but the errors involved become progressively larger as the site becomes more remote from the punch.

The motor

Changes in the power absorbed by a machine are sometimes useful, if only very approximate, guides to the need for lubrication and mechanical maintenance work. If the power absorbed by an AC drive motor is to be checked for any reason, it is good practice to make the measurement with a sensitive wattmeter rather than a simple ammeter. Because of the inductive nature of the motor windings, the instantaneous values of current and voltage do not generally remain in phase with each other – the current usually lags behind the voltage – so that readings taken with ammeter and voltmeter may be misleading. Moreover, the amount of lag is not constant, since it is a function of the mechanical load on the motor. Variations from the normal, as distinct from absolute measurements, are more easily detected if the meter is provided with an electronic 'offset' facility so that it can be set to read zero under standard running conditions.

In some processes, such as wet granulation or mixing, the power consumption may be a good indication of correct running conditions and of the progress of the process. However, the requirements of tablet production are more finely critical, and measurement of absorbed power by itself is not a substitute for accurate determination of punch forces.

The drive belt

Variations in the mechanical resistance of a machine will induce corresponding variations in torque within the whole of the drive system. They will also be seen as changes in the tension of the drive belt, and these can be used to actuate simple warning devices if an overload should develop. For example, the belt tension can be sensed by means of a spring-loaded jockey pulley, which, in turn, can move against a microswitch if the tension rises above an acceptable level.

The main shaft and eccentric

The main shaft and eccentric both rotate when the press is in operation. Torque measurements taken from them would, therefore, have to be transmitted by slip-rings, or by some form of telemetry link. The possible advantage of such an area for gauging would be its facility for convenient total enclosure, away from dust and general exposure. Its disadvantages include the cost and complication of the telemetry link, together with the inaccuracies introduced by friction in the bearings and the slides. On balance, these disadvantages would usually outweigh the advantage of using this area.

The machine frame

Early instrumentation systems were designed to avoid the problems of telemetry by concentrating on those components that did not move. In 1954, Higuchi *et al.* applied Baldwin SR-4 strain gauges to the stationary frame of their Stokes model A-3 press in a system intended to provide force measurements within ±2% accuracy over the range 0–3000 lb (0–14 kN). The strain gauges were used in an AC bridge configuration running at 2.5 kHz and could be calibrated against a load cell directly under the lower punch assembly.

At the time, this arrangement was exciting and innovative, since it represented one of the very first instrumentation schemes that could make dynamic force measurements. It could be run at speeds up to 7 tablets min^{-1}. Almost all previous instrumentation work had been carried out with hydraulic presses at unrealistically low speeds without any consideration of dynamic variables, so even 7 tablets min^{-1} represented a great step forwards.

As instrumentation schemes began to develop, however, both the speed and accuracy of the Higuchi arrangement were seen as areas for improvement. Indeed, the non-linearity of the upper punch gauging system had been noted in Higuchi's original paper. The relation between frame distortion and applied force was commented upon by Shotton and Ganderton (1960). For their own studies on the effects of variations in compressive force, they abandoned the strain-gauged frame in favour of the strain-gauged punch. This was run in a Lehman press at 68 tablets min^{-1}.

Since that time, most of the published work relating to the instrumentation of eccentric presses has indicated a similar preference for measurement either at the punch itself or at the punch holder. It has usually been felt that the frame, however convenient as a site, has the disadvantage of remoteness from the action; moreover the signal that it provides rates rather poorly both for amplitude and for linearity. Once again, however, it is the application that determines how much effort and expense must go into an instrumentation system.

It has to be remembered that, in the past, press operators had to rely on the mechanical adjustment that simply controlled the depth of punch penetration. Asking the operators to change from this early but familiar system to one based on force measurement meant that the advantages of the proposed new method had to be demonstrated in a practical and convincing way. In the writer's laboratory, an old eccentric press was made available as a test-bed for experiments in instrumentation, and some of the following notes refer to work carried out to that end.

The upper arm

Provided that small errors in measurement are acceptable, it is both possible and convenient to use the upper arm as a site for strain gauges, bearing in mind that it will only display the maximum force during a compaction rather than the whole force profile. For some purposes such a measurement may be adequate, though for most research work it would be necessary to monitor the applied force continuously. Since the arm itself is readily removed from the body of the press, the gauging operation can be carried out within the workshop; both cleaning and handling are, therefore, facilitated. On a machine such as the Manesty F3, the upper arm is made as a casting. The bearings at each end are approximately parallel, but the surface of the metal between is largely in the rough-cast state, underneath layers of paint and filler. Before gauges can be attached, suitable areas must first be prepared. Using the machined ends of the arms as references, small flat recesses can be end-milled into the front and back surfaces of the arm so that they are equidistant from one end, parallel to each other and symmetrically disposed about the long axis of the arm. That is to say, they will lie along a straight line joining the centres of the upper and lower bearings. Using both the front and back surfaces in this way provides compensation for bending forces that may distort the arm during the tablet compression process and gives an acceptably steady, uniform baseline.

It is good practice to fit composite or rosette gauges in these positions, one element being oriented along the axis of the arm while the other is at right angles to it. The front and back gauges then form a compensated full bridge. They will be encapsulated after installation to ensure against mechanical damage or the ingress of moisture (Figure 6.1).

Since the arm moves in use, the connecting cable should be flexible, multi-strand wire, and it should be anchored securely to the arm at some point in order to avoid the transmission of stresses to the gauge pads. It should also be screened to reduce the effects of external electrical noise. If the signal from an unscreened bridge is displayed on an oscilloscope, the presence of generalised noise can often be observed as a

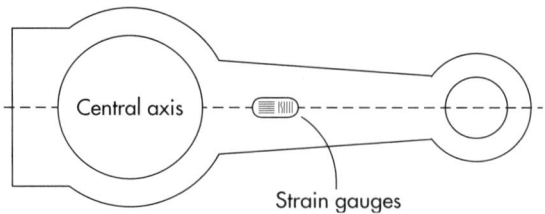

Figure 6.1 Upper arm gauging on an eccentric press, with the gauges positioned along the axis of the upper arm.

broadening of the trace, and a general 'fuzziness' of the waveform. These are indications of the need to provide suitable screening.

The upper arm of an F3 machine is not of constant cross-section, since it tapers slightly towards the lower end. Gauges fitted near that end will, therefore, show a fractionally larger signal for a given force, but the difference is not sufficient to make this an important factor. However, there may be some advantage in taking the electrical connections from a point nearer to the lower end, since there is rather less movement there and hence less risk of consequent damage to the leadwires.

The crucial test for this, and for any other of the force-gauging systems that may be used, is calibration against a transfer standard load cell positioned between the two punches. Since the load cell is likely to be much thicker than the average tablet, a suitably shortened upper punch should be prepared for this purpose. It is convenient to use such a length of punch that the upper arm is almost vertical during calibration, since this corresponds to the attitude of the machine at maximum compressive force.

For calibration, the machine is turned by hand while readings are taken from the two gauging systems over as much as possible of the likely range of forces to be encountered in practice. Using a restricted range of forces is not recommended, as it may fail to uncover errors in linearity or consistency. As the subsequent notes show, it is also important to check the output signal at several positions of the upper arm, preferably by cross-calibrating against a strain-gauged punch. This may reveal peculiarities not detectable at a single fixed position of the arm.

Errors involved in gauging the upper arm

Gauging an arm in order to measure punch forces involves various potential sources of error.

The cosine error must be considered. Since the arm is not always in a direct line with the punch, the force measured along the axis of the arm will show a cosine error by comparison with the punch force (Figure 6.2). However, this error disappears when the arm is vertical, at the instant of maximum compression. If the parameter of interest is just the maximum applied force during compression, then cosine errors can clearly be ignored. If, more elaborate relationships such as force–displacement curves are to be investigated, then the cosine error should be taken into account or another gauging area found. Even so, small angular variations do not introduce very large errors in the measurement. For example, at an angle of 2.5°, the cosine term is approximately 0.9990, and the error is approximately 1 part in 1000. At an angle of 1°, the error is reduced to approximately 1 part in 6500. Apart from being relatively small, the error is also systematic and calculable. Therefore, if the angular position of the arm is known, the signal may be corrected by a suitable algorithm in the data processing system.

The second possible source of error is the frictional resistance of the sliding components within the press. This cannot be dismissed but may nevertheless be held within acceptable limits for many applications by attention to the condition of the components involved and to their lubrication.

Thermal errors are always present to some extent in any system of this kind, but these again

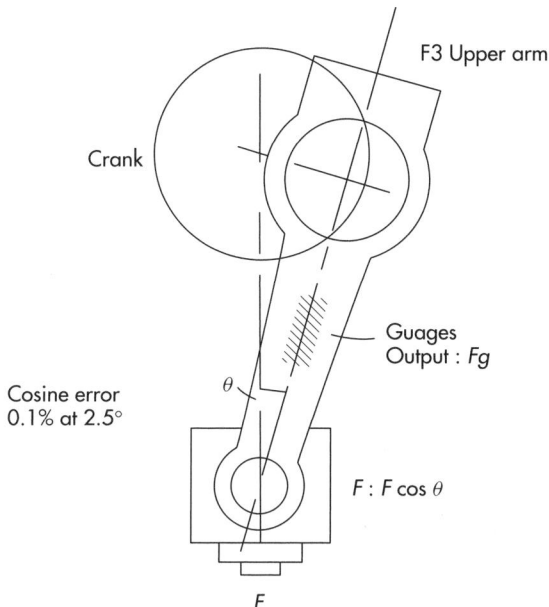

Figure 6.2 The cosine error resulting from the use of gauges mounted on the upper arm of an eccentric press if the arm is not vertical. The error is relatively small.

can be kept to low values, corresponding to better than $0.4\,\mu\varepsilon\,°C^{-1}$, by the use of composite STC (self temperature compensating) gauges in matched pairs.

We have previously noted that a well-used eccentric press, a Manesty F3, had been made available for experimental work in the writer's laboratory. In an initial experiment, we fitted a rosette strain gauge to the upper arm of the machine, gauging only its front surface and so producing a half-bridge. The gauged arm was then calibrated at the National Physical Laboratory in an accurate transfer standard compression-test machine. The calibration proved to be remarkably good and did not show any obvious departures from linearity when plotted out on graph paper. In other words, it would certainly have been within ±1%.

In order to check on the continued accuracy of the gauged arm, a transfer standard load cell (Type HS1, Graham and White Ltd) was obtained. An externally gauged punch with strain gauges in a balanced configuration was also prepared. Readings from both the arm and the punch could be compared against each other during the compression of a tablet, while corresponding readings could be taken from the transfer standard under static conditions. Compression curves produced from this arrangement were of normal appearance, and the maximum readings from both arm and punch were in agreement. The demonstration proved entirely convincing, and we were consequently asked to apply the same arrangement to a second but considerably newer press of the same type.

In this next experiment, the arm of the newer press was gauged exactly as before. Calibration against the load cell, in a more or less fixed position, again appeared to be satisfactory. However, when signals from the arm were compared with those from the instrumented punch through a complete cycle of punch travel, a different picture emerged. This time, the output signal from the arm showed what appeared to be a sine wave superimposed on the expected line (Figure 6.3). To investigate this unexpected result, the arm was disconnected completely from the punch holder and was allowed to hang freely from the upper drive eccentric. It was found that even with the arm completely unloaded in this way, slow rotation of the drive shaft nevertheless produced a sinusoidal signal from the strain gauges. It became apparent that there was a very slight lack of parallelism between the main shaft and the eccentric. This was rocking the upper end of the arm at each revolution and consequently it bent under its own weight. The effect was only apparent on a machine with reasonably well-fitting bearings. The looser bearings of the older press had been able to tolerate the misalignment.

The solution of this problem was quite simply to be found in the use of correct gauging methods. When both front and rear surfaces of the arm were gauged symmetrically, as they should have been at the outset, bending effects were cancelled out and the error disappeared completely.

In a third experiment, the upper arm of a practically new machine was fitted with strain gauges as before, the gauges being carefully balanced again so that there would be no errors resulting from out-of-plane bending. However, when the arm was tested against an instrumented punch, a new set of errors came to

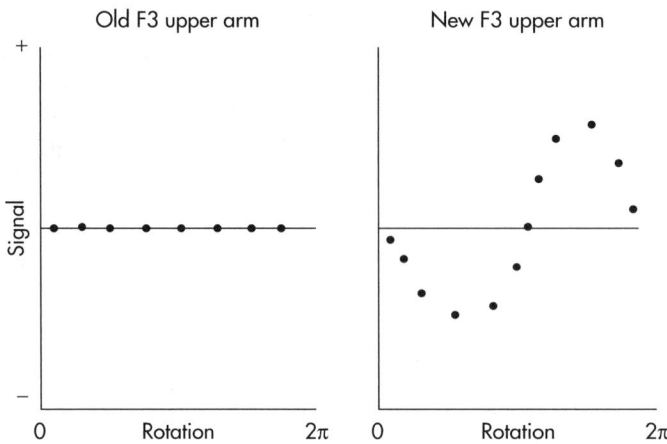

Figure 6.3 Bending errors in a gauged arm of an eccentric press caused by small errors in the machining of the eccentric cam. This is avoided by fitting the gauges symmetrically on the front and back surfaces.

light. This time there was marked hysteresis, such that the curves for increasing force did not follow those for decreasing force by several percentage points.

Eventually, this particular inaccuracy was found to result from tightness in the new bearing bushes at each end of the upper arm when under load. The tight bushes caused some in-plane bending of the arm that varied the strain sensitivity of the system as the arm rotated. Careful removal of high spots from the bearing bushes by scraping eventually eliminated this third source of error.

These examples emphasise that small physical differences, such as may be found between a new and a not-so-new machine, can sometimes have quite large effects on an instrumentation system.

The upper punch holder

The upper punch holder is certainly one site of choice for an accurate gauging system, as it is directly connected to one end of the punch and receives the same force as the punch itself without the interposition of other components. Moreover, a gauged holder can accept a whole range of different punches and can be used to compare one with another.

The main practical problem to be overcome is that of providing a force transducer that is sufficiently compact to fit in the rather limited space available. The most convenient solution is to use a load cell or a load washer in a machined recess at the upper end of the punch holder. If necessary, the recess may be extended into the body of the block that carries the punch holder in the vertical slide-way. However, this arrangement is not without possible sources of error.

The punches normally produced for tablet manufacture have a carefully formed tip, since that is necessary for the formation of the tablet, but in eccentric machines, the end of the punch shank may well be comparatively rough. If any rough-surfaced component is pressed against a load cell, it is likely to make contact only at a few points; consequently, there is a very high, localised loading. This may damage the surface of the load cell and cause inaccuracies in its output.

In order to improve the situation, it is customary to interpose a hard steel element between the load cell and the punch. This is often in the form of two separate washers with spherical surfaces that can accommodate angular misalignment between the punch and the load cell. It may also be useful to provide a flat disc of soft metal at the shank end of the punch to act as a cushion for any small, hard, high spots there. Brass or annealed copper is a suitable material for this purpose.

The punch securing screw may also cause difficulties. On simple presses such as the Manesty F3, the punch holder carries a securing screw that is tightened directly on to the shank of the punch. A small inclined flat area is machined at that point in order to provide a suitable bearing surface for the end of the securing screw, and the inclination of this flat area helps to lock the punch in its correct place. The screw system is entirely satisfactory for its normal purpose, but it can cause problems in an instrumented press, since it alters the local distribution of stresses. Therefore, if the punch is secured in this manner against a load cell in the punch holder, the force on the punch tip will not all be transferred to the load cell. Some will be directed through the screw and the body of the holder.

Apart from this, each change of punch involves the loosening and re-tightening of the securing screw. After every such operation, a different proportion of the punch force passes to the load cell. Marshall (1983) suggested that the punch-securing screws should be tightened while the punches are under load, with a tablet or a plastic plug between them. However, even this precaution may not overcome person-to-person variations in the tightening force that is applied. To some extent, the applied force on the screw can be standardised if it is tightened with a torque wrench, but the screw itself is a very imprecise component, and the torque on the screw head is not accurately related to the axial thrust that it produces.

Müller *et al*. (1982) reported on the same problem and showed that the output of their piezoelectric load cell was substantially dependent on the torque used to tighten the punch-securing screw, even though the screw itself had been made from brass. Calibration was restored when the depression in the punch was opened out to admit some movement, and the brass screw was replaced with one made from Makrolon (polycarbonate).

Lindberg (1972) fitted a flanged punch through the centre of an annular load cell (Kistler type 902 A), and held the shank end of the punch with an inclined securing screw. He commented that the screw had to allow some vertical movement of the punch within the holder.

In the author's opinion, it is desirable to have a means whereby the force transferred to the load cell remains completely independent of the way in which the punch is secured in its holder, unless there is to be a re-calibration after each change of punch.

One possibility is that of machining out the punch holder so that its central recess is appreciably larger than usual, and then fitting a separate inner housing into the enlarged recess. The punch is held within the inner housing by means of the usual securing screw, while the housing is retained against the load cell within the punch holder by an annular fastener that may itself be spring-loaded. Variations in the tightness of the punch-securing screw then have no effect on the way in which the inner housing bears against the load cell. It is, naturally, important to ensure that the inner housing cannot rotate with respect to the outer punch holder, since the punch will probably not be concentric with the axis of rotation. Apart from this, the assembly may well need to be used with punches that are not themselves round.

This approach was used by Jungersen and Jensen (1970) in the instrumentation of a Diaf TM 20 eccentric machine. In this instance, the upper punch was mounted at the lower end of a hollow steel holder that had been fitted with foil strain gauges (Figure 6.4). Somewhat similar arrangements were reported by Polderman *et al*. (1969) and by Duchêne *et al*. (1972). The latter group retained their punch by means of an annular cover, which fitted over it as far as the end flange and was itself screwed to the strain-gauged cylinder above.

Load cells, as already described, may either be of the strain-gauged variety or may be piezoelectric devices. At one time, it was difficult to find strain-gauged load cells that were small enough to fit into an upper punch holder; consequently, many laboratories elected to use piezoelectric cells. Apart from being compact, these cells have been claimed to be more sensitive, more rugged, more linear and less affected by temperature changes than their strain-gauged equivalents.

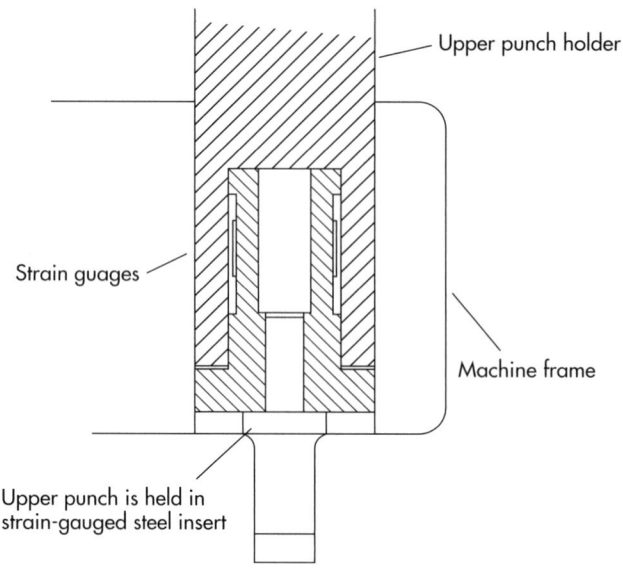

Figure 6.4 The upper punch assembly of the Diaf TM 20 eccentric press showing measurement of upper punch force using a system designed to overcome the asymmetrical effect of the punch-securing screw (Jungersen and Jensen, 1970).

But as noted above, some piezoelectric devices are less suitable for the measurement of static or slowly changing loads. Eccentric presses are often used for quasi-static compression studies, and in such cases, the possibilities of error through leakage of the charge must be borne in mind. This being said, the availability of piezoelectric load cells with integrated amplifiers has improved the capabilities of this transducer, although these cells have to be slightly larger to accommodate the internal electronics.

Marshall (1972) commented that strain gauges 'require a change in linear dimension of the member to which they have been bonded'. This is an interesting point, since it appears to imply that the piezoelectric load cells deform to a lesser extent under load. In fact, there often seems to be less difference between the two types of transducer than Marshall's comment would suggest. If we take a miniature piezoelectric cell in the Kistler range (type 9211) for example, its height is some 6 mm, while its quoted stiffness is approximately $400\,N\,\mu m^{-1}$ and its measuring range is 0–2500 N. On this basis, the device would contract by some $6.25\,\mu m$ at full load, which represents approximately $1000\,\mu\varepsilon$. Going next to a cell of much larger capacity, the M 195 Q/300 piezoelectric force washer from the Meclec Company has a quoted stiffness of more than $15\,kN\,\mu m^{-1}$, a height of 15 mm and a range of 0–300 kN. Here the contraction under full load would be $20\,\mu m$, or $1330\,\mu\varepsilon$. Conventional foil or semiconductor gauges can be successfully operated at this general strain level, and $1000\,\mu\varepsilon$ at full load would be a good design target for a transducer system.

From these figures it would appear that there can be an almost equal stiffness for some examples of the two types of transducer, and that this parameter by itself would not be sufficient to justify the use of a piezoelectric device.

The punch

From some points of view, the punch itself represents the best site for measurements of punch force. It is completely direct in its association with those forces, and it is independent of any shortcomings in the tablet press as a whole. The instrumented punch, like the load cell, can be removed from the press at any time, and measurements can be compared on similar presses at different locations. The obvious disadvantage of this approach is that it complicates the business of changing punches. Each one must have its own strain gauge bridge and its own individual calibration. Forces on different punches cannot be compared directly, but only through the transfer standard. Additionally, the wire connections to the gauges on the punch can be fragile and must be protected against accidental damage. In spite of these objections, there is a good deal to be said for having a few gauged punches in hand, if only for their interchangeability between machines.

The practical aspects of gauge installation have been discussed in earlier chapters. Before a gauge can be installed, however, the gauging site must be selected with some care. The literature of pharmaceutical technology, for example Marshall (1983), often contains a statement in relation to the gauging of punches: it is that, for maximum accuracy and a minimum of spurious signals, the gauges should be fitted 'as near as possible to the punch face'. So far, practical considerations have limited the accessibility of the punch face itself as a gauging site. For example, if the gauges are to be fitted to the outer surface of a punch, they must allow free entry of that punch into the die and must, therefore, be sited sufficiently far from the punch tip to leave any wire connections clear of the die.

Leaving such details aside, it may be reasonable to ask whether the punch face is really such an ideal site for gauging since it appears that, under some conditions, this could give rise to inaccuracies of measurement. We have already seen that it is desirable to use carefully matched strain gauges, fitted in a symmetrical configuration in order to balance out unwanted signals from bending or thermal drift. This is, of course, good practice and is generally accepted, but for complete accuracy, it is important that the distribution of strain on the punch tip should also be symmetrical. If the surface of the compact is uneven, then the stresses at the punch tip may themselves be uneven.

When any rigid body is pressed against another, contact between the two is necessarily incomplete. There may be several points of contact, but at the microscopic level, these contact points effectively separate the two surfaces (Figure 6.5).

Stress analysis of the two faces would show localised areas of high stress around the points of contact; in three dimensions, these areas would be seen as 'pressure bulbs' with concentric surfaces of equal stress. At some distance from the faces, however, the stresses become progressively more uniform and, from the engineering point of view, better distributed for measurement. It may be argued that compression of the compact

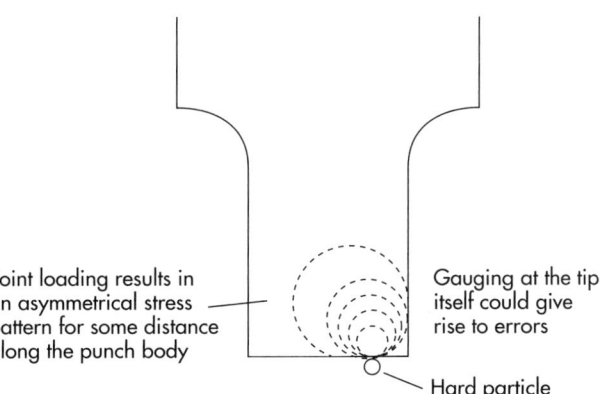

Figure 6.5 Localised punch tip loading if the tip is in contact with hard, irregular material. Further from the tip, stresses become more uniform across the width of the punch.

will spread the loading across the whole face of the punch, so that point contacts no longer exist. This is, of course, true to some extent; but the act of compression alone does not necessarily result in a completely even distribution of the forces across the punch face, or of the material within the resultant tablet. If the original material is unevenly packed in the die before compression, there will be at least some residual traces of a corresponding unevenness in the completed tablet. After compression, these traces may be reflected as variations in the amount of relaxation in the surface of the tablet.

The magnitude of any such effect must depend on the nature of the original material, its particle size, its surface hardness and its uniformity of filling. Nevertheless, such effects do exist, and in the opinion of the writer they should not be entirely ignored during the search for measurement techniques of greater and greater sensitivity. It has been suggested that the use of a gauging site near to the punch tip will minimise inaccuracies that result from longitudinal distortion of the punch during compression. In principle, that distortion should be constant over the parallel section of the punch. Moreover, if inaccuracies do exist, they should be determinable at the calibration stage.

From the engineering point of view, it would appear to be sound practice to mount the strain gauges well back from the punch face, at a position where the stresses in the metal have become largely uniform over the cross-section of the punch.

One of the best ways of reducing the effects of asymmetrical loading in a rod that is to be used as a transducer is to gauge along its neutral axis. This is the technique adopted in the commercial manufacture of strain-gauged bolts and clevis pins, and it involves the installation of gauges on the inside of a relatively narrow hole machined along the centre of the rod. Clevis pins are usually fitted with gauges that sense the shear forces produced by the two sides of a roller, whereas gauged bolts are arranged to measure tensile or compressive loads. The same technique can be used in the instrumentation of a punch. A punch of this kind is normally made in two parts, the tip section being gauged before the head end is fitted. A slightly more complex system was described by Schmidt et al. (1986), who constructed a two-part punch that carried a central axial strain-gauged element but also incorporated a miniature piezoelectric force washer. It was used in a Fette P2 press.

Mounting the gauges internally, along an axial cavity, has the double merit of allowing the gauged area to be as near to the tip as is considered optimal, while providing the greatest possible protection for the gauges themselves. Naturally, the connecting leads still represent a potentially weak point, as they must remain undamaged during the compression cycle. Neutral axis gauging has been utilised in the 'Portable Press Analyser' that was sold by Puuman Oy in Finland and by 'The Punch' sold by SMI Incorporated. Both of these systems are described later in this chapter.

Measurement of the lower punch force

The general considerations that we have noted in relation to the upper punch will, to a large extent, also apply to the lower punch. However, there are two points of difference, namely the range of measurement, and the physical configuration.

Firstly, the gauging system must not only be able to measure compression forces in the range, which may be tens of kilonewtons, but must also be sufficiently sensitive to measure the much smaller forces, perhaps only a few hundred newtons, involved in ejecting the finished tablet.

Secondly, though the upper and lower punches themselves may be quite similar, there is somewhat more room for instrumentation around the area of the lower punch. This facilitates the provision of a suitably gauged punch holder, so making it less necessary to have large numbers of gauged punches.

The lower punch holder

There have been three basic approaches to the problem of measuring forces at the lower punch holder.

One method has been that of applying strain gauges to the outer surface of the holder, at a level below that of the punch. This was done by Shotton and Ganderton (1960) in their studies on tablet strength, and illustrations of a similar nature have appeared in various subsequent papers, for example Schwartz (1981). The punch holder is machined with an annular groove at the gauging level so that the reduced cross-section provides improved sensitivity as well as some physical protection for the gauges. As usual, gauges are applied in a symmetrical configuration in order to minimise spurious signals from bending: Shotton and Ganderton, for example, fitted three gauges, equally spaced around the holder. The three gauges were connected in series and formed an active quarter-bridge.

A second approach has been that of using an external load cell underneath the lower end of the punch holder. This method was used in some of the earliest work, such as that of Higuchi *et al.* (1954). Using a commercially available load cell in this way enabled Higuchi's group to make force measurements quickly and easily. The cell could itself be removed for deadweight calibration at any time, and there was no immediate necessity for the group to acquire the skills of precise strain gauge installation. The lower punch holder had to be modified, but the modification was simply in the form of an extension rod that transmitted vertical thrust from the holder to the button of the Baldwin load cell.

Small load cells have been inserted within the punch holder itself, in direct contact with the punch shank. Williams and Stiel (1984) removed a portion of their holder and interposed a sealed piezoelectric element, together with an integral amplifier, between the two halves. Using an integral amplifier goes some way to overcoming the low-speed characteristics of the piezoelectric devices. Piezoelectric load cells were also used, for example, by Müller *et al.* (1982) and Graffner *et al.* (1985).

The engineering techniques used for the installation of a piezoelectric load cell are themselves equally applicable to load cells of the strain-gauged type. However, some care may be necessary in the selection of a sufficiently compact unit, since, in general, strain-gauged cells are slightly larger.

Of these three approaches, that of using the external load cell is probably the least satisfactory. Although it is convenient to employ a standard commercial unit in this way, the adaptation of the press is such that it no longer runs in a normal manner. The instrumentation of a press should always be 'transparent' in the sense that there is no change of operation or behaviour, otherwise any conclusions drawn from its use will be suspect.

Strain gauging the punch holder is more satisfactory, but may give a fairly low sensitivity because of the large cross-sectional area of the region that can be gauged. For that reason, it is worthwhile considering the use of semiconductor gauges. There may also be some doubt, just as there may be with the upper punch holder, as to the way in which stress distribution is affected by the tightness of the punch-securing screw.

The inserted load cell has once again the advantage of being a standard commercial unit, and it should give good results if care is taken with its installation. Clearly, if the whole upper part of the punch holder is mounted on top of the load cell, then effects from the securing screw will not be of any great significance. If the cell is fitted inside the punch recess, then means must be found to ensure that the punch can be clamped in a reproducible manner.

Jungersen and Jensen (1970), also recognising the possibility of uneven stress distribution, used a strain-gauged metal cylinder as a load cell and fitted it axially inside the body of their lower punch holder between two hard steel balls. The upper of the two balls remained in direct contact with the lower end of the punch and transmitted compressive forces from it to the load cell. This arrangement prevented any tendency to side loading (Figure 6.6).

A compromise is possible by constructing a replacement punch holder that is made in two parts: the inner portion carries strain gauges and acts as an integral load cell (Ridgway Watt, 1983a). As with most other eccentric presses, the Manesty F3 has a lower punch holder that is in the form of a substantial metal cylinder, around 210 mm in overall length. Its outer surface is threaded so it may be positioned with appropriately threaded collars in the lower arm of the machine; at its upper end there is a machined recess to hold the punch. Our construction

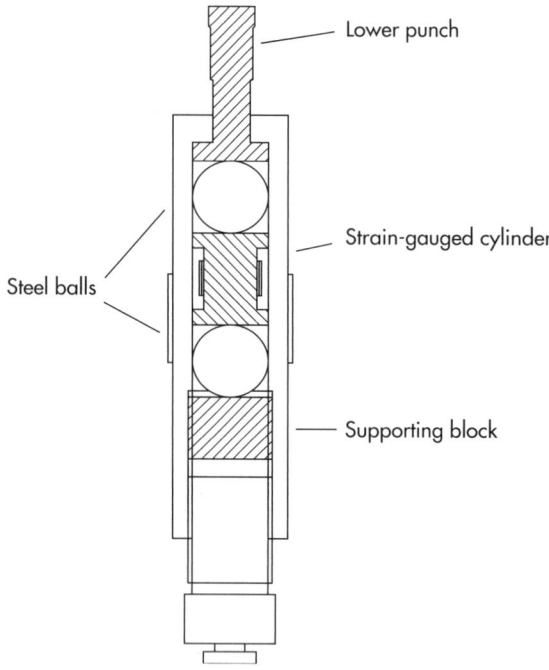

Figure 6.6 The lower punch assembly of the Diaf TM 20 eccentric press showing measurement of lower punch force. A load cell is loaded between two steel balls to prevent off-axis loading (Jungersen and Jensen, 1970).

very good protection against accidental damage since the gauges were within the steel body of the holder. We found that it was convenient in practice to retain the usual concentric drilling for the punch ejector rod, and to do so at its normal diameter of some 8.0 mm.

It is also worthwhile emphasising two points in relation to the materials of construction. Most steels have a very similar Young's modulus of elasticity, E. Whether the steels are mild, alloyed or stainless of various grades, they will usually have an E value somewhere in the region of $21 \times 10^{10}\,\mathrm{N\,m^{-2}}$. The lowest value for a standard steel would probably appear in type 316 stainless, while the highest might be in a chrome–vanadium alloy; but there is not more than approximately 10% difference between these. By comparison, both the elastic limit and the yield point can vary by as much as 10-fold from one steel to another. In designing a component for strain gauging, therefore, it is most important to check that the material chosen will

reproduced the standard form (Figure 6.7) but it now consisted of two components that fitted telescopically, one inside the other. The inner column was 22 mm in diameter and was a sliding fit inside the threaded body. It carried a punch holder of normal dimensions at its upper end and had a full, symmetrical, strain-gauged bridge, centred at a height of some 50 mm from its lower end. The inner column was bedded down on a soft annealed copper washer and secured by means of three M3 screws. The electrical leads from the strain gauge bridge were taken along a slot in the wall of the outer component, and then brought out at the lower end of the assembly.

The advantages of this arrangement were that it provided improved sensitivity for the gauges, since they were on a column of reduced diameter, that it positioned the gauges in a region where the stresses were uniform, and that it gave

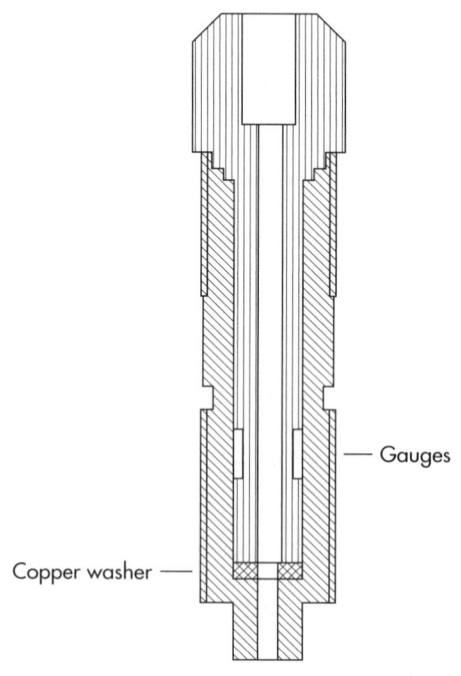

Figure 6.7 Measurement of lower punch force using a two-piece punch holder (Ridgway Watt, 1983a).

not only give the required strain rate per unit load but also that it will stand up to the largest predictable load without permanent deformation or catastrophic failure. It is usual to allow a safety factor of at least two: in other words, the maximum stress should not take the component more than half way to its elastic limit. In an unguarded moment, the writer once calculated the required cross-section necessary in a lower punch holder in order to provide a good signal from the ejection force, failing to remember that it would also have to stand the full force of compression. The result of this failure was a corresponding failure of the component itself, which slowly buckled under load and had to be re-constructed with a larger cross-section.

The lower punch

The procedure for gauging a lower punch will be virtually the same as that for an upper punch, apart from the need for measurement of ejection forces. In order to provide the necessary improved capability for handling small signals, one can conveniently use semiconductor gauges in this application. Their greatly increased sensitivity, by comparison with that of foil gauges, makes them very suitable for use with lower punches. Their increased cost is trifling when seen in relation to that of the installation as a whole, and their thermal characteristics nowadays can compare well with those of foil gauges. Their dynamic range can extend to well over 100 000 to 1, since the lower end of their practical scale is likely to be somewhere around $0.01\,\mu\varepsilon$, while the upper end will be well over $11\,000\,\mu\varepsilon$. For a rough estimation of what this means in practice, let us assume a punch area of $1\,cm^2$ and a Young's modulus of approximately $21 \times 10^{10}\,N\,m^{-2}$. The proportional change in length for a given force can be calculated from the expression strain = $1/E \times$ force (in newtons per metre squared). A force of 1 N should then result in a dimensional change of approximately $0.05\,\mu\varepsilon$: 2 tonnes would produce $1000\,\mu\varepsilon$. The maximum recommended load for a flat, shallow or normal concave punch of that cross-sectional area is approximately 8 tonnes, though it is unlikely that such a large force would normally be used in practice.

Rotary tablet presses

Almost all commercial tablet production is carried out on rotary presses, some of which can turn out as many as one million tablets per hour. The rotary press achieves these high production rates by the use of many punch and die sets, fitted round the periphery of a turret. The dies are filled at one or more points on the machine, and, as the turret rotates, the punch heads are brought, in turn, between pairs of stationary metal rollers. Passing between the rollers, the punch heads are forced together, and exert a compressive force on the granulation in the dies between them.

The action of the rotary press differs from that of the eccentric press in several ways. In the former, both punches move into the die together as the tooling passes through the rollers, whereas in the latter, it is only the upper punch that moves to any extent. Furthermore the two machines have different force–time and force–displacement characteristics, since one is driven by rollers while the other uses an eccentric.

Because both these characteristics can be critical in the formation of a satisfactory tablet, materials will not necessarily perform equally well in the two types of machine. Furthermore, though it is somewhat easier to instrument an eccentric press, the formulator must be able to work with a machine that represents production equipment. This means solving a number of instrumentation problems, most of which are associated either with the rotation itself or with the large numbers of punch-and-die sets.

We have previously noted the train of components that runs in the eccentric machine between the drive motor and the punch. The rotary machine has an analogous train, which includes the machine frame, the roller support shafts or 'roll-pins', the rollers themselves, and the punches. There are, of course, no separate punch holders in this type of press. Once again, all these components have been used as gauging areas, and are worth examination from that point of view.

The machine frame

When a rotary press is operated, the punch heads pass between the upper and lower compression rollers, exerting a force between them. This force is resisted by the frame, which is designed to maintain the rollers at a constant separation, and varying amounts of strain are thereby induced within the elements that constitute the frame. Measurement of these strains has quite frequently been used as a means of inferring the forces at the punch tips, although the relation between the two quantities is not always straightforward.

One of the earliest studies in rotary press instrumentation was that by Knoechel *et al.* (1967), who worked on a Stokes model 540-35 and a modified Stokes model BB2-27. The design requirements for their system included the facility to use any punches and dies, the measurement of individual punch forces and the ability to transfer measurements from one machine to another. On the first two grounds, they elected to work with the machine frame as they felt that it was desirable to have detachable transducers. One solution was to take out some of the existing screws and bolts in the machine and to replace them with strain-gauged versions. Both semiconductor and foil gauges were used, according to the available strain levels at the points of measurement. The upper punch force was measured on the Stokes 540-35 press and the lower punch force on the Stokes BB2-27 press (Figure 6.8).

As the measurement was indirect, the values obtained included errors represented by the friction in the punch guides and in the overload mechanism, but the authors considered these to be negligible. However, the accuracy of measurement was limited by the fact that the output signals were read from the screen of an oscilloscope, on a scale of approximately $0.6 \, mm \, kN^{-1}$, so errors from friction in the mechanism could easily have escaped notice.

Additionally, it is perhaps worth noting that the compressive force on the BB2-27 press was only measured at the lower punch, on the assumption that the upper and lower punch forces would be equal. More sensitive methods have subsequently established that the forces on the upper and lower punches are not necessarily the same.

In the following year, Goodhart *et al.* (1968) published an account of work on the instrumentation of a Manesty 45-station Rotapress. In this study, various sites were tested for signal strength, uniformity and freedom from extraneous noise (Figure 6.9). Out of six sites evaluated, the authors concluded that the most satisfactory was the upper end of the compression column, just below the pin supporting the upper roll carriage. The surface of the column at this point was brought to a mirror finish, and strain gauges were then bonded directly to each side, in a balanced configuration. Even so, the measured levels at this point were only in the region of $20–40 \, \mu\varepsilon$, which corresponded to approximately one-tenth of the force at the punch tip. Other sites were found to produce poor or noisy signals.

The adjusting wheel eye-bolts, which support the lower roll carriers on the Stokes BB2 press, were also used by Wray (1969) and later by Salpekar and Augsburger (1974). The tie rod, which joins the upper and lower roll carriages, has been a popular site for instrumentation. So too, in some machines, has been the end of the compression column. In this last position, load washers can be used under the securing nuts.

The problem of locating the best site for gauges is underlined by these studies. As far as the magnitude of the signal is concerned, stress analysis of the machine frame could have identified suitable areas at an early stage. A technique such as SPATE could have provided, had it been available in those days, a full picture of the stress distribution. It would not necessarily have shown up high-frequency noise, though this would have been detected by the use of exploratory strain gauging. The noise itself may result from 'ringing' in the structure when one part strikes another – for example, when a punch head strikes the pressure roll – and may, in some instances, produce resonances in the overload spring system.

When the strain level in a component under tension is too low to provide a really adequate signal, there is always a temptation to consider increasing the strain by thinning the component. Marshall (1983) commented that if a tie rod is to be strain gauged, it must be 'necked

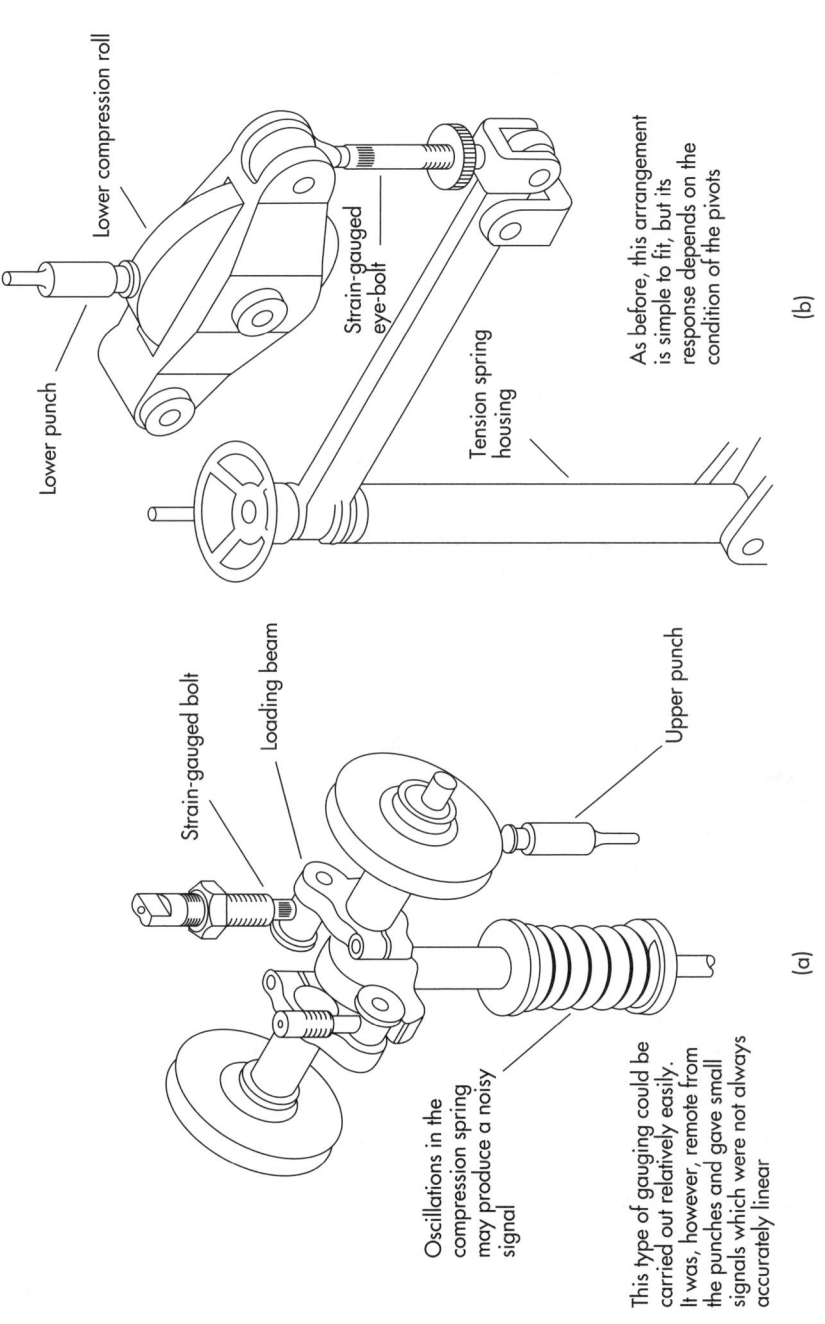

Figure 6.8 Measurement of force in two rotary presses using bolts fitted with external strain gauges. (a) Upper punch force measured on a Stokes 540-35 press; (b) lower punch force measured on a Stokes BB2-27 press (Knoechel et al., 1967).

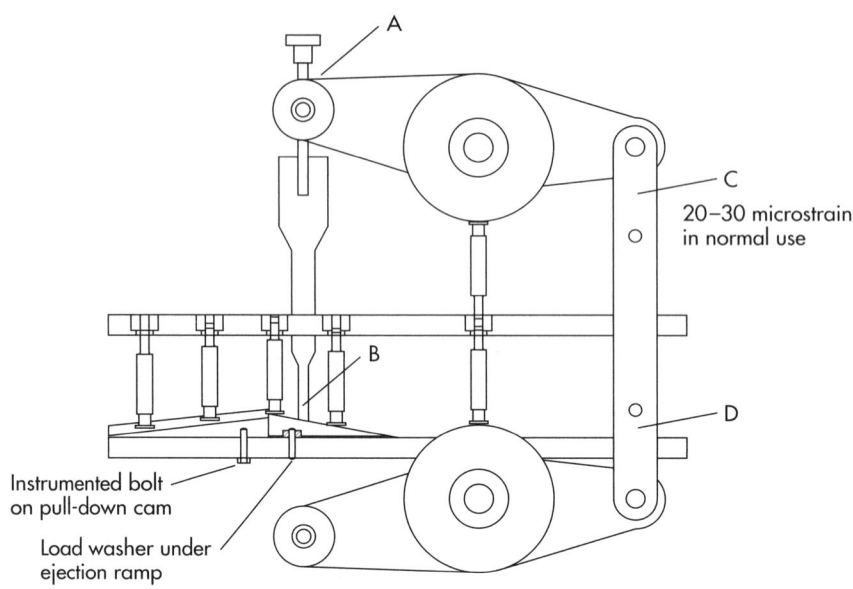

Figure 6.9 Measurement of force in a rotary press, using sites remote from the punches. A, top pressure link above eye; B, bottom pressure link at the narrowest point; C, upper section of compression column; D, lower section of column above table (Goodhart *et al.*, 1968).

down' to improve the sensitivity of measurement. Necking down, or thinning, in this way, results in a locally increased strain in proportion to the change of cross-sectional area, and hence to a larger signal. However, in some components it will alter the mechanical response of the machine and may, therefore, be thought inadvisable.

Where the component is relatively long, a possible alternative solution may be to introduce a strain shunt by attaching a bar of similar metal by the side of it, welding at two well-separated points, as shown in Figure 6.10. If that added bar is itself necked-down at its centre, there will be a local concentration of stress there when the main component stretches. A gauge at the thinned region will thus give an appropriately increased signal but will cause little change in the mechanical stiffness of the heavy component.

One problem with the early instrumentation systems was the relative crudity of their associated data-handing and read-out equipment. Measurement from a Polaroid photograph of an oscilloscope trace was an imprecise operation by modern standards, and it was often not too clear whether the basic procedure was giving adequate accuracy or linearity. Certainly, Ridgway (1982) has commented that 'easily-obtainable measurements can be made on the main frame or the pressure-roll carrier; these tend to lack accuracy.'

What these pioneering experiments do seem to show is that the careful selection of a gauging site can make an appreciable difference in signal quality. One would no longer expect to use the oscilloscope trace as the sole means of accurate measurement, but it is still a valuable indicator of noise on the signal and is certainly worth using from the diagnostic point of view.

The precise measurement of the forces themselves is, of course, likely to be achieved through microprocessor electronics that can handle signals of fairly low amplitude and linearity. In that context, measurements taken from the machine frame are still being used. Moller-Sonnergard and Kristensen (1978) used this approach in their instrumentation of a Kilian Prescoter, fitting semiconductor strain gauges to the surface

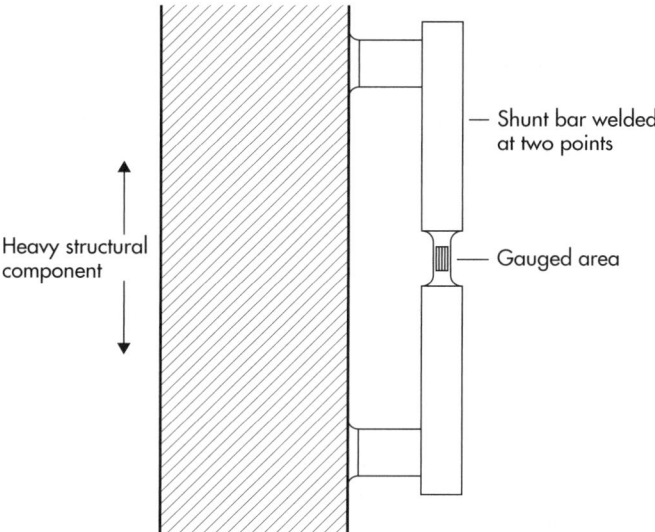

Figure 6.10 Use of a strain shunt to increase the signal from a heavy component.

of the central pillar that holds the upper pressure roll. D'Stefan *et al.* (1985) applied strain gauges to the tie-rod in a 25-station Manesty X-Press. This was used for studies on computer-interfaced instrumentation, in conjunction with a Hewlett-Packard 9825/T desktop computer, linked in turn to a DEC PDP 1170 mainframe computer. Using this arrangement enabled D'Stefan's group to store data within the 9825/T, and subsequently to plot compression curves of high graphical quality on a 7245B printer-plotter.

The roll-pin

The constant search for greater sensitivity and accuracy has led workers in the instrumentation field to examine parts of the rotary machine that are more directly associated with the punches. The roll-pins, on which the compression or pressure rolls are mounted, have been used for instrumentation schemes in two possible ways: either the pin itself has been gauged or the normal pin has been mounted on gauged supports.

To take an example of the first method, Ridgway *et al.* (1972) cut away a portion of the roll-pin in a Manesty Betapress and inserted a piezoelectric force sensor in the cavity. A suitably shaped saddle-piece was fitted in order to restore the circular cross-section of the pin, and the roll reassembled centrally over this. At the maximum working load of 6000 kg, the reported deflection of the load cell was only 0.5 µm. The pin itself was twice the normal diameter, so it was necessary for the roll carrier to be machined out accordingly.

It was reported that this system operated satisfactorily, and this would not be surprising, since the piezoelectric transducer was positioned well. However, there can, under some circumstances, be an inherent problem. If the machine carries a full complement of punches and is compressing a fairly porous granulation, there may be punch heads in contact with the pressure roll at almost all times. The load cell, therefore, would not return to a physically unloaded state between stations. Strain gauges would register, correctly, the fluctuating load as the punches passed through, but the piezoelectric cell, since it can be regarded as an essentially AC device, would see the minimum signal between stations as its effective zero level. It then appears, to the measurement system, that the peak values registered are too low. If the die fill is then adjusted to give a greater measured force, the punch heads travel through the machine at an even higher level, the error is compounded, and the system drives itself to destruction.

For some time, the piezoelectric sensing method was used by Manesty Machines as a basis for the automatic control of tablet weight. However, the problems of zero drift were such that the Manesty system, the 'Tablet Sentinel', was eventually redesigned using strain gauge sensors. Other tablet machine manufacturers, almost without exception, now also use strain gauges for choice in their own force measurement and control systems.

It has been said that hindsight is always 'twenty–twenty', and there is no doubt that looking back, one can often see that instrumentation schemes have not always gone according to plan. However, it can be instructive to examine past errors and to learn from them. For example, at one time we ourselves had the opportunity to acquire a completely new 16-station Manesty Betapress and to equip it with a comprehensive measuring system (Ridgway Watt and Rue, 1979). It seemed good practice to take force measurements from the compression and pre-compression rollers, and we were advised to use piezoelectric load washers for this purpose, since at that time the problem of zero drift was not widely appreciated. The upper and lower roll-pins were modified so that the bronze bearing bush at one end fitted into a steel block, mounted between two side pieces that constrained it to move only in the vertical plane. Each block was spring-loaded against a Kistler 902A piezoelectric load washer so as to maintain continuous contact between the two. In practice, the other end of the roll-pin was found to move, if only slightly, in its bronze bush; consequently, the effective length of the pin was not definable. It was, however, possible to overcome this particular problem by splitting the bush and then clamping it tightly around the roll-pin.

The pre-compression roller assemblies were mounted on vertical rods. Each was intended to be a sliding telescopic fit in an outer block that held a further load washer, and as before, the rod and the washer were spring-loaded together.

When the system was tested, one problem became apparent at once: the pre-compression rollers did not give the expected signals. As noted above, the physical movement required to provide full-scale output from a piezoelectric load washer is extremely small. That movement was not sufficient to overcome the sticking friction between the sliding parts, and a proportion of the force consequently failed to reach the washer. When the telescopic arrangement was replaced with one in which the rollers were mounted on a bending beam, using strain gauges to monitor the bending forces, much more satisfactory results were obtained.

If we were to carry out this exercise again, we would probably opt for the use of a commercially made load cell. This could be of the shear beam type, with the roller support block bolted directly on to it. Not only would this provide a virtually frictionless movement for the roller block, but it could be removed for easy calibration. Alternatively, an internally gauged clevis pin could be used in place of the plain roll-pin. As with the main compression rollers, this method has the particular advantage of requiring no engineering work on the press itself, since the original plain pins can be replaced directly with gauged pins of the same diameter.

In order to verify the performance of the main compression roller system, a strain-gauged punch was first prepared to establish that it had a reliable and linear output when measured against a transfer standard load cell. When the signal from the punch was compared with that from the load washer, however, a curious fact emerged. The calibration curve from one operator was reasonably linear, while that from a second operator was not. Investigation showed that one of the two took readings quickly, while the other took more time over them, and in that time the charge was leaking away. No such problems were ever encountered with the strain-gauged components and their associated amplifiers, all of which gave consistent results.

Under steady running conditions, corresponding to those of industrial production, the piezoelectric load cells also gave consistent results, and other workers appear to have come to the same conclusions. A review by Moldenhauer et al. (1980) made the same point. Müller et al. (1982) worked with both strain gauges and piezoelectric transducers. They favoured the piezoelectric transducers and published comparative curves that certainly demonstrated a noisier trace from the strain gauges. However, in our own laboratories we have seen widely differing noise levels from otherwise identical transducers. In these instances, the source of the noise was traced using an oscilloscope and was rectified by appropriate screening. Of course, the connections between a piezoelectric transducer and its associated charge amplifier must always be of the highest electrical quality, since the signal is of such low power. The cable must, therefore, be of a low-loss type with complete screening. The wiring between a strain gauge bridge and the bridge amplifier must take power in, as well as a signal out, and is perhaps more susceptible to extraneous noise. It may not always be realised that care in the screening of these leads can be just as important as with any other type of transducer.

No doubt the controversy over the relative merits of piezoelectric cells and strain gauges will continue. But for an experimental or semi-experimental press, to be used either dynamically or statically, in research or development, we would prefer to use strain gauges at all measurement points.

It will be apparent that the system outlined above had disadvantages for use in static measurements, though it also involved major modifications to the Betapress itself. A mechanically simpler approach would have been to substitute internally gauged roll-pins for the original solid ones, and very little engineering work would have been necessary. Strain gauges are fitted at two regions in order to sense the shear forces applied by the pressure roll, as shown in Figure 6.11.

This approach is, in fact, used by some tablet press manufacturers for original equipment; it is the method adopted by Frogerais, Korsch and others. Apart from this, there are other advantages to be gained by using strain-gauged transducers, in so far as it is then possible to standardise on one type of signal amplifier instead of two.

The pressure roll

The difficulties that were experienced in the development of the early frame-gauging arrangements led to investigations of the roll-pin and its supports as alternative transducer sites. Work has also been directed to instrumentation of the roller itself. Deer et al. (1968) designed a modified pressure roll that was partly cut away to allow for some amount of distortion under load. In essence, the roll was divided into a stationary inner hub section, with two horizontal spokes, and a rotating outer rim: between the two was a Skefco 7.5 inch roller-race, which allowed free rotation of the outer part. When a load was applied to the pressure roll, the flexible spokes deflected and allowed the whole outer portion to lift through a short distance: the actual deflection was quoted as being 0.0025 inches (0.06 mm) per ton, and this movement could be detected optically by means of a moiré grid, or by the use of strain gauges. The modified roller was used on a Manesty D3 machine.

With regard to this particular device, Thacker (1979) commented that 'while this offered an excellent research tool, it was not sufficiently robust for use under production conditions'. It

Figure 6.11 The internally gauged roll-pin, with strain gauges fitted to two regions to sense the shear forces applied by the pressure roll. The roll-pin has an axial cavity in which are fitted shear-sensitive strain gauges. These are mounted inside the grooved regions shown.

would also be true to say that the modified roller was inevitably more springy than its production counterparts, and it might, therefore, have produced slightly different results on the compression of some materials. At any rate, the instrumented roller was eventually overtaken by the instrumented roll-pin.

Another point to consider in relation to the construction is that ball and roller races do need to be provided with rotary seals if they are to be run in a dusty environment. Under the right conditions they are remarkably good, but grit can destroy them fairly quickly. If grit cannot be excluded, then 'plain' bearings, such as bronze bushes, tend to have better lasting properties than those of an unprotected roller race.

The punch

As with the eccentric press, strain gauge instrumentation of one or more punches of a rotary press provides a very direct means of assessing forces at the punch tip. Even if the basic measurement is usually to be taken from the roll-pin, it is extremely useful to have at least one gauged punch for the purposes of calibration under running conditions.

Working with the rotary press, however, carries some additional complications. Since the punch now travels in a guide that is not completely frictionless, the gauges must be fitted in such a position that they are clear of the guide during compression, that is between points A and B in Figure 6.12. Even more attention must be paid to accurate balancing of the applied gauges to compensate for bending forces, and it must be possible to run the press without damaging the strain gauges, or ripping the wires off the punch.

The first problem should present no real difficulty, since it is merely a matter of measuring the height of the punch guide above the die, and fitting the gauges at some point within that distance from the punch tip.

Gauging the punch at its tip diameter should give the largest signal, but from the point of view of protection, it has been usual practice to provide an area of reduced diameter in the main

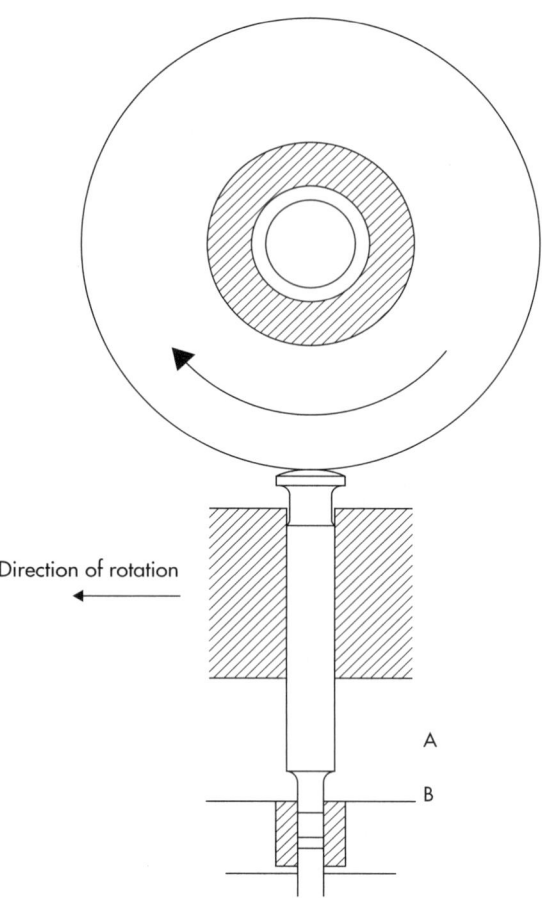

Figure 6.12 The strain-gauged punch. Since the punch is subject to sliding friction within its guide, gauges are applied to the region that will be exposed when compaction takes place (between points A and B in the figure).

body of the punch so that the installed gauges can be covered with a substantial layer of epoxy resin (Figure 6.13). After machining, the recessed section can be milled with four accurately positioned flats to receive the gauges, as this can facilitate good compensation for bending moments. We have noted, in relation to the eccentric press, the need to effect this compensation, and also to gauge at a point where the stresses are relatively uniform, since the surface of the die contents may not be completely flat. The situation in the rotary press is similar, but may be more critical since centrifugal forces can

The measurement of displacement in tablet presses

Eccentric presses

The measurement of punch displacement has been carried out with surprisingly little change since the earliest days of tablet machine instrumentation. Naturally, there have been improvements in the methods and equipment, but most of the published studies have followed the original lead of Higuchi et al. (1954) from the

affect the uniformity of packing. Naturally, this can only happen if there are voids within the die after filling, but this may happen when the press is set to produce thick tablets at high speeds. Alternatively, the gauges may be fitted internally, and the connecting leads may be taken out at either end.

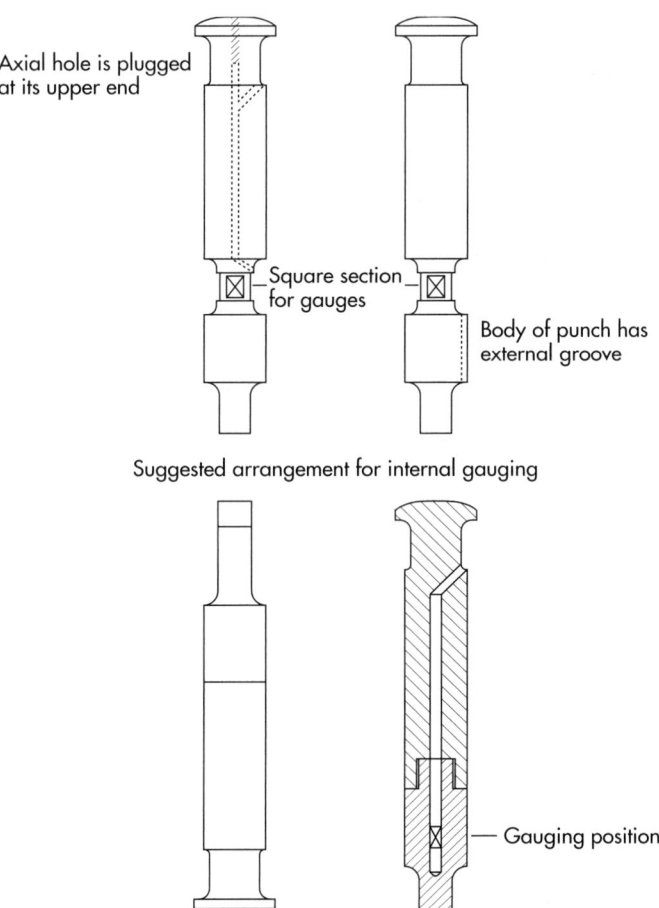

Figure 6.13 Punch gauging methods. Gauges are usually applied to a thinned region of the punch or may be fitted internally for maximum protection and resistance to bending errors. Leads may be taken out at upper or lower ends.

University of Wisconsin. Higuchi and colleagues adapted a Stokes Model A-3 eccentric press to measure upper punch displacement through the addition of a linear variable-differential transformer (LVDT; Schaevitz Model 500S-L). The body of the LVDT was attached to the frame of the press in a brass mounting assembly; the core was similarly linked to the upper punch holder through a cross-arm that projected over the LVDT body.

The design requirement for the measurement of punch travel in this set-up was that it should be accurate to within 0.002 inches (0.05 mm). Even in those days, it is probable that the LVDT itself would have been capable of much better accuracy and resolution than this, but the read-out systems available then were of limited capability. In this instance, the transducer signal was taken to a strain gauge amplifier, and thence to an oscillograph. The trace from the latter was scaled to show 10 mm for 0.44 mm of punch displacement, so fine distinction would have been difficult.

Clearly, the movement of the armature would be influenced by distortion of the machine frame, and by the elastic deformation of the punches themselves. In order to make allowance for these factors, the Wisconsin group brought the punches face to face, applied varying loads and observed the apparent displacement from the LVDT. With flat-faced punches of 0.375 inches (9.5 mm) diameter, the correction was found to be approximately $0.02 \, \text{mm} \, \text{kN}^{-1}$ of lower punch force; only the lower punch force was, in fact, measurable.

Of course, with no powder in the die, the upper and lower punch forces would have been virtually identical. Had a tablet been produced, the upper and lower forces would have been appreciably different as a result of die-wall friction and the correction would, therefore, be less accurate. Modern data handling systems, unlike the oscillograph, would have been able to display this difference in legible terms.

Since 1953, the LVDT has been widely used for the measurement of punch displacement, and there have been only minor changes in physical layout. As we have seen, the modern LVDT is capable of giving readings that, in some examples, are reproducible to better than $0.1 \, \mu\text{m}$. The device itself is of essentially infinite resolution, and therefore its ultimate accuracy is dependent on its power supply, amplifier and read-out facilities. It should be possible to resolve the output signal to better than 1 part in 20 000 or more. One implication of this is that there is a trade-off between resolution and scale length. Consequently, it is desirable to select a transducer that just covers the particular range of interest without any unnecessary extension beyond that.

Choice of scale length

In the example of the eccentric tablet press, the upper punch itself may have a vertical travel of several centimetres; indeed, it must have, on the one hand, at least sufficient movement to clear the feed shoe on each machine cycle. On the other hand, the part of that movement that requires precise measurement is likely to be the much shorter distance over which the punch is in contact with the powder bed. Using an LVDT with the reduced scale length of just those few millimetres will give a correspondingly larger signal for a given displacement. For example, an AC-energised LVDT with a nominal linear range of ± 2.5 mm can have an output of $180 \, \text{mV/V} \, \text{mm}^{-1}$; a similar unit with a linear range of ± 10 mm has an output of only $36 \, \text{mV/V} \, \text{mm}^{-1}$. The larger signal will provide better linear resolution.

If a short-scale transducer is used to provide the best resolution, reasonable care has to be taken with the alignment of its two components. The total travel of the punch may well be more than the length of the body of the LVDT, and if this is so, the armature may pass completely out of the body at one end of its stroke. When it returns, it must also re-enter without touching the sides, or it will cause damage. Fortunately, the modern LVDT is available in versions with a fairly wide radial clearance between armature and body. In the Schaevitz HR range, for example, the internal diameter of the body is 9.5 mm, while the outside diameter of the core is 6.35 mm. The radial core-to-bore clearance is consequently approximately 1.6 mm. Such a tolerance should be well within the machining capabilities of a laboratory workshop, although it is also necessary to ensure that the armature support rod will not vibrate excessively when the tablet machine is running.

For some very demanding conditions, it is now possible to buy types of LVDT with an increased radial clearance. Within the Schaevitz range, for example, units in the XS-A series have a bore diameter of 38.1 mm for a core diameter of 6.35 mm. Such a large radial clearance, 10 times that of the standard HR model, has its application in areas where the core and the body need to be isolated from each other by some non-magnetic carrier such as glass or Teflon tubing. Surprisingly enough, the output signal is within 20% of that from the normal LVDT.

If a transducer with a fairly close core-to-bore clearance is used, it is important to ensure that the gap does not become blocked with deposits of powder. In dirty industrial conditions, displacement measurements are often made with the sealed, spring-loaded type of LVDT; the so-called 'gauging transducer'. Such a device can be mounted on a machine in such a way that its probe makes intermittent contact with some part of the moving component, and it will remain hermetically sealed. It is necessary, however, to check carefully that the conditions of use do not give rise to 'contact bounce', which would superimpose oscillations on the output signal.

Requirements of accuracy

The Wisconsin group were able to define their own need for accuracy in displacement measurement as being within 0.05 mm. It would be much more difficult to produce a corresponding definition nowadays, since the purpose of measurement is often to generate force–displacement plots that show the work done in tabletting or to measure tablet thickness and hence calculate *in situ* porosity. The inherent problem is that, as compression proceeds, the force on the punch increases sharply; at the same time, the punch displacement decreases towards zero. Multiplying the one parameter by the other, in order to estimate the work done, is open to appreciable error if the value of either one is in doubt. The best that can be said in this respect is that the signal itself can be read to approximately 1 part in 20 000 (−9999 to +9999) in a standard four digit display. Its ultimate accuracy must then depend on the accuracy of the corrections applied to it.

Problems with machine distortion

When it became apparent that much useful information could be obtained from an instrumented press, many laboratories decided to build their own systems. Since there were difficulties associated with the instrumentation of a rotary press, such systems were usually based on an eccentric machine. In some cases, only punch forces were measured, but in others, the facility to measure the movement of the punches was included. This enabled the relationship between punch movement and applied force to be investigated, leading, among other developments, to the construction of force–displacement curves and the measurement of tablet porosity while the force was being applied.

There are many examples of such developments in the literature, such as Moldenhauer *et al.* (1980), Armstrong and Blundell (1985) and Ragnarsson and Sjögren (1985). In principle, these have usually followed more or less similar general schemes for the measurement of displacement. Punch displacement was measured with an LVDT, one component of which was joined to the punch holder and the other to the machine frame; relatively long arms were used so that the integrity of the die table could be maintained.

A typical example is illustrated in Moldenhauer *et al.* (1980), who commented that 'an inductive displacement transducer can be easily attached to the punch guide'. From the mechanical engineering point of view, no doubt it can; however, the apparently simple may well conceal some possible sources of error.

While Moldenhauer and his colleagues were preparing their own paper, Ho *et al.* (1979) published a most interesting communication. They mounted an LVDT on a Manesty E2 eccentric press in three different ways (Figure 6.14).

For the mounting system A, the casing of the LVDT was bolted to the machine frame, while the armature was operated by a bracket bolted to the sliding top punch carrier. For system B, the

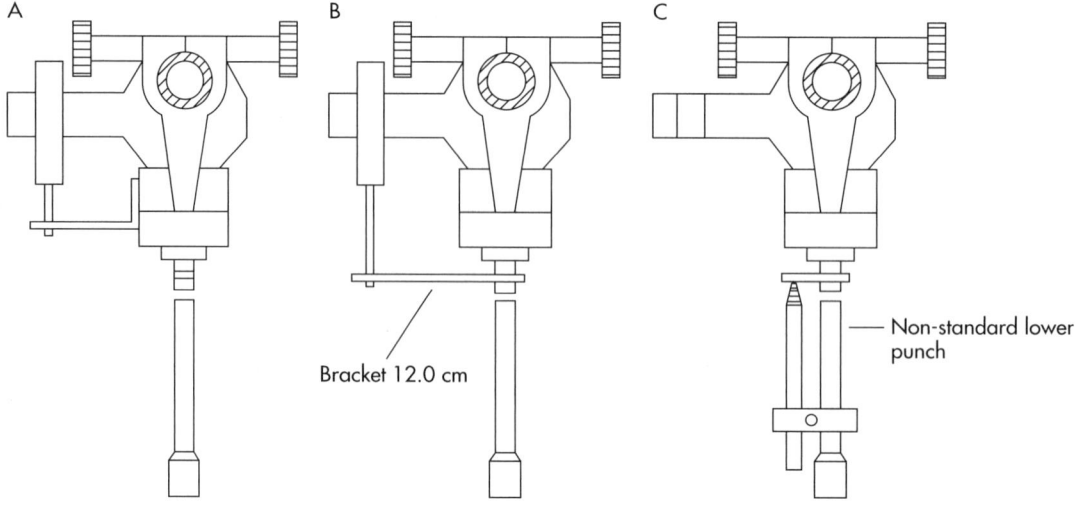

Figure 6.14 Three different positions (A–C) to place displacement transducers on a Manesty E2 press (Ho et al., 1979).

outer case of the LVDT was clamped as before, but the armature was now operated by a bracket, 120 mm long, attached to the top punch. In system C, the bottom punch and holder were replaced with a non-standard assembly that brought the case of the LVDT parallel to the modified lower punch at a distance of approximately 20 mm. An actuator block was attached to the upper punch so that the transducer, passing up through a hole in the die table, effectively bridged the punches.

The press was equipped with Kistler load washers, type 9021, and could be run under power with all three of the mounting systems. The punches were allowed to close on a steel disc within the die, the disc having a diameter of 10 mm and a thickness of 6.31 mm in the unloaded state. The application of forces up to 200 MPa had no permanent effect on the thickness of the disc.

The results given in Table 6.1 clearly show how the choice of mounting affected the accuracy of the results at a range of applied forces. If the transducer was attached to any moving part of the press remote from the punch tip, it would only read accurately so long as there was no force acting on the punches. Whenever there is material under pressure between punches, the situation can be changed through some or all of the following effects:

- the punch can shorten under the applied force (Ho et al. (1979) made no allowance for this)
- the upper and lower punch holders can be forced apart
- the frame of the press can twist through asymmetry in the eccentric and upper arm
- the upper punch holder can tilt to accommodate free play in the slides.

Whether any one of these effects is of importance must depend on the accuracy expected from the displacement measuring system.

The results show that the use of a bracket at some distance from the axis of the punches and the die can give rise to large non-linear errors in the estimation of tablet thickness. Unfortunately, the system that gave the greatest accuracy in these experiments did not represent a practical configuration for a working press, since the body of the LVDT had to be passed through the die table.

It is interesting to compare these results with those of the earlier workers in this field. Even from the start, it was apparent that the applied force would affect the length of the punches,

Table 6.1 Comparison of three displacement transducer positions (Ho et al. 1979)

Applied pressure (MPa)	Mean apparent thickness (mm)	Error (%)
System A		
0	6.31	
50	6.08	3.65
100	5.97	5.39
150	5.87	6.97
200	5.78	8.40
System B		
0	6.31	
50	6.31	0.00
100	6.29	0.32
150	6.16	2.38
200	6.08	3.65
System C		
0	6.31	
50	6.30	0.16
100	6.31	0.00
150	6.30	0.16
200	6.29	0.32

and the checks carried out by Higuchi et al. (1954) were very similar to those reported by Ho et al. (1979) and later by Krycer et al. (1982). Higuchi's correction factor was essentially linear, representing approximately 20 μm per kilonewton of force on punches of 9.5 mm diameter. As the punches were operated well within their elastic limits, a linear relationship would be expected if the only potential error was the shortening of the punches. The experiments of Ho et al. (1979) revealed a substantially non-linear error, since the discrepancy at 200 MPa pressure was not, in any instance, twice that recorded at 100 MPa. This indicated a non-linear distortion in the frame of the press and perhaps a twist in its structure.

Most tablet presses are designed to have a massive frame construction, while the working ends of the punches are relatively narrow. One might, therefore, expect that the major part of the observed strain would be associated with the punches, rather than with the frame. However, the distribution can be inferred from the published figures by looking at the dimensions of the punches used. The punches used by Ho et al. (1979) had an area of approximately 0.78 cm². These might have been expected to exhibit something like 1000 με at the maximum pressure level of 200 MPa used in these studies. If the effective length of the punches – between the LVDT body clamp and its probe – was approximately 70 mm, then a contraction of approximately 70 μm would be expected under maximum load. The errors determined by Ho et al. (1979) for mounting systems A and B were 530 μm and 230 μm, respectively. One must conclude that there appear to be much larger movements in the frame than in the punches, although some of the errors could result from tilting in the actuator brackets. It would have been interesting if these workers had used LVDT brackets and operating arms of different lengths, since this would have shown up any angular misalignment.

Clearly this work casts doubt on the reliability of any displacement measurements that are taken exclusively from the upper punch guide.

It would seem possible to achieve a compromise between accuracy and convenience by using more transducers. Almost all reported instrumentation schemes, including that of Ho et al. (1979), have employed an asymmetrical layout, with a single LVDT at one side. If the punch-carrying block should be a less than perfect fit in the vertical slide-way, it may well tilt to left or right as the main drive shaft rotates, thus moving the end of the actuator bracket up or down. The magnitude of the effect at the LVDT will then depend on the length of the bracket. It is also likely to be a non-linear function of the punch force. Fitting a second LVDT on the other side of the machine, as a mirror-image of the first one, would make it possible to average the two signals and at least reduce, if not eliminate, the effects of tilting. Such an arrangement was employed by Belda and Mielck (1999a), who used two displacement transducers mounted on short arms on either side of the punch. They used this to carry out a critical examination of the errors that might arise in the measurement of punch separation and the consequences on the estimation of parameters related to the tabletting process such as porosity and work expenditure (Belda and Mielck, 1999b).

By the same token, a similar system could be applied to the lower punch holder. In this way, both punch movements could be referred to a common reference, such as the surface of the die table, and it would still be possible to use standard punches. The extent of machine distortion will to some extent depend on the type of press being used. It might be expected to be greater in those presses such as the E2 that have an open front, and less in those with a more closed construction. In either case, one would expect the maximum distortion to be in the plane of the slide-way.

The accuracy of the system could be checked in the same way as before. If residual errors can be established to be repeatable and reasonably invariant, it may be acceptable to apply appropriate corrections by software for the sake of retaining an otherwise normal tablet press with interchangeable punches and dies. Those corrections, additionally, will vary in detail from punch to punch, according to the cross-section used and the force applied.

Cook *et al.* (1988) followed up the pioneering work of Ho and colleagues. They sought to fit a displacement transducer to an eccentric press while retaining the latter's normal mode of operation. In their system, the displacement transducer was mounted close to the upper punch. Its armature passed through a small hole drilled in the die table and made contact with a set screw clamped to the lower punch ejection height adjustment screw. This enabled the press to operate in its normal way, and tablet thickness could be measured from the relative positions of the two punches. The actual movement of each punch could not be measured. Cook *et al.* (1988) found that at forces in excess of around 5 kN, a linear error in punch displacement was obtained.

Other displacement transducers

The LVDT has been by far the most widely used type of displacement transducer in tablet press instrumentation, but there are certain disadvantages in its application to eccentric presses such as the Manesty F3. Ideally, the displacement transducer would be fitted as closely as possible to the axis of the punch and its holder to minimise bending errors, but the problem here is that the feed mechanism sweeps across the die table at each operating cycle of the press and would damage any transducer element standing in its way. For this reason, it has been customary to mount the displacement transducer at a safe distance from the moving parts, and some workers have chosen to fit the LVDT on extended arms of quite considerable length. It will be apparent that any errors introduced by tilting of the punch holder and distortion of the machine frame will be exaggerated by this construction. Nowadays there are several types of remote displacement transducer that could be used in principle to overcome the problem of the feed mechanism, since they do not require actuator rods or other projecting components.

Laser Doppler measurement is one possibility. If a laser beam, which is at a fixed frequency, is reflected from a moving surface, the reflected beam undergoes a change in frequency according to the well-known Doppler principle. Such an arrangement was reported by Depraetère *et al.* (1978).

There are two reasons for using this means of measuring movement in tablet punches. One is that the reflector, mounted on the moving component, can be of very small mass and small size. The mechanical effect on the machine is, therefore, minimal while the theoretically attainable precision is high. Of course, like all such systems, the laser Doppler method only provides information on the relative movement between two elements; in this case, the light source and the reflector. Once again, the choice of positions for the optical components is critical, and, as with any optical system, it will be important to ensure that the light path is not obscured by dust from the material being compacted.

The second is that, unlike the LVDT, the Doppler system provides a measure of speed, rather than of instantaneous position. Therefore, because the signal is proportional to punch speed, it can be used without differentiation to produce plots that display the integral of the deformation energy.

It is common practice nowadays to use low-power lasers as sources of narrow, focused beams that can be used in levelling devices, cutting equipment and various other workshop applications. If such a beam is directed at an angle on to a reflective surface, the position of the reflected beam will depend on the relative height of the surface in relation to the source, and this can be measured by allowing the beam to impinge on a linear array of photodiodes. These arrays are available with large numbers of elements, and the accuracy of measurement can be extended beyond that of individual diodes by well-established interpolation techniques.

Velocity transducers might also be used. If it appears useful to plot deformation energy integrals, one can also measure velocity by electromagnetic means. Velocity transducers generate a voltage that is proportional to the speed of a small permanent magnet moving axially along the centre of a uniform coil. This type of device is externally similar to the LVDT and can be mounted in the same way. Such a transducer was used by Müller *et al.* (1982) for compression studies with a Fette Exacta 1 eccentric press. It does not, however, overcome the need for some form of actuator rod as a means of moving the magnet in relation to the outer winding.

The two following devices do not need any external components. The first is the accelerometer. One way of determining the position of a body in space is that of measuring its acceleration over a known time. Compact accelerometers are commercially available and usually contain a small dense element that is mounted on a flexible support of some kind. If the device is caused to accelerate in the direction of its axis, the inertia of this dense element causes it to lag behind. The relative movement between the moving element and the outer body can then provide a signal that can be interpreted as movement of the assembly. Some miniature accelerometers achieve this by the use of powered coils surrounding the moving element, as in the LVDT. Others are built around lead zirconate titanate sensors that generate their own signal, providing that the acceleration is alternating and not static. Such a device, fitted in close proximity to the punch holder, could provide information about its position without compromising the feed system.

The second device is the remote inductive transducer. The nomenclature relating to displacement transducers is not always used in a specific way; thus the LVDT itself is sometimes described as an inductive transducer – which, in a sense, it is – but in manufacturers' literature, the term is more likely to be applied to one of those transducers that operates by eddy current losses in the presence of a ferromagnetic, or sometimes a non-magnetic, target. Inductive transducers operate without physical contact, and they will work perfectly well under a coating of dust, so they have certain qualifications for tablet compression studies. In terms of resolution, these non-contact devices compare well with the more conventional LVDT. For example, the units made by Micro-Epsilon Messtechnik GmbH have static resolutions that go down to 50 nm in their smaller models. Their frequency response is static up to 50 kHz (the 3 dB point), and their temperature shift is approximately $1.0\,\text{ppm}^\circ\text{C}^{-1}$. Some space must be provided between the inductive probe and other closely adjacent ferromagnetic materials, but usually a clearance of some two probe diameters should be sufficient for normal linear operation. The diameter of the probes in this series depends on their measuring range, but as an example, a probe with a range of 2 mm has a body diameter of 12 mm. It would seem quite practical to use these transducers on a tablet press, perhaps employing the die table as the ferromagnetic target; nevertheless, we have not so far seen them mentioned in the tabletting literature.

Measurement of displacement in rotary tablet presses

The majority of the published reports on rotary press instrumentation have been concerned with the measurement of force or pressure, rather than displacement. Those that have included displacement transducers have tended to describe rather similar arrangements, since the layout is partly forced by the machine design.

As a rule, one of the punches is fitted with a short arm that can actuate the armature of a transducer, which is mounted in or near to the adjacent punch guide and is connected to a suitable telemetry system (Figure 6.15). An example of this approach was described by Walter and

Augsburger (1986) as part of the equipment for a Colton 216 rotary press, and appeared in the Portable Press Analyser from Puuman Oy, which is described more fully later in this chapter.

Measurement errors in the rotary press

There are various sources of possible error, which need to be examined. To begin with, the punch itself must always have some amount of free movement within the punch guide so that it can slide without difficulty; as a result it is capable of rocking slightly from side to side. An actuator arm, projecting horizontally from the top of the punch, will transfer this free play to the probe tip of the transducer to an extent that depends upon its length. It is, therefore, desirable to keep this arm as short as possible.

In the Manesty Betapress instrumented by Ridgway Watt and Rue (1979), the LVDT was mounted in a machined hole between two punch guides, rather than in an existing guide. In this way, it was possible to keep the effective length of the arm to a minimum. It was mounted near the head of the punch, being secured by a single screw that passed through the thickness of the punch. Making the arm readily detachable simplified the insertion of the punch into the machine.

When the punch carries such an arm, it must naturally be prevented from rotation. This can be done readily enough with a normal key-way, such as would be used for shaped punches, but here too, there must be an element of free rotational play. It is consequently important to check that the surface of the arm is flat, and accurately normal to the axis of the punch. This can be achieved by machine grinding the assembled arm, using the body of the punch as a reference.

It is usual to fit a gauging transducer with a rubber bellows seal to exclude powder from the

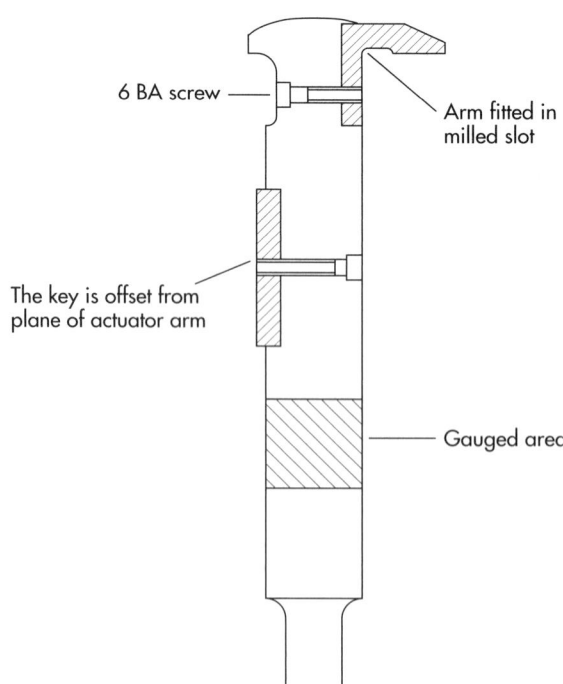

Figure 6.15 A punch from a rotary press fitted with a short arm to actuate a linear variable-differential transformer (LVDT). Keeping the arm as short as possible minimises errors resulting from the punch tilting in its guide.

mechanism, since the armature has to run in linear bearings. Here, as in the eccentric press, the problem is merely that of ensuring that the spring-loaded probe remains in contact with the actuator arm during measurement over the whole range of machine speeds and tablet thicknesses. If the signal from the device is free from 'steps', or obviously superimposed vibrations, it is likely to be in satisfactory contact. Our press carried a Schaevitz model D400, with a linear range of ±5 mm from its central position. This gave an output of 2000 mV for a mechanical displacement of 5.0 mm. The spring-loaded core had a mass of 1.0 g. Both upper and lower punch movements were monitored with exactly similar systems. Power to the two LVDTs was led in via a 12-way slip-ring assembly, and signals were taken out in the same way.

It might be necessary to retain balance in the turret to compensate for the mass of metal removed to provide a hole for the transducer. This can be achieved by drilling another hole diametrically opposite the first of such a size as to restore dynamic balance.

Measurement of ejection forces

In principle, the relatively small force needed to eject a finished tablet from the die can be measured in two ways: using a strain-gauged punch or using a gauged ejection ramp.

The strain-gauged punch

A gauged lower punch is normally used to measure compaction forces, but if it is provided with reasonably sensitive gauges, it should also be capable of measuring ejection forces to a sufficient degree of accuracy.

However, there are two important practical objections to this. One is that strain gauges deliberately fitted in such a way as to clear the punch guides will only register the force needed to push the tablet from the die. They will not indicate the additional force needed to impel the punch along the guide. It may well be desirable in a production machine to ensure that neither the guides nor the die have become sticky. The assessment of punch pull-up or pull-down forces is discussed later in this chapter.

The second objection is that ejection force measurements are likely to be required from all the sets of tooling in the press, since they will not be identical, whereas there is only likely to be one gauged punch. On these grounds, most workers have opted to measure ejection forces via the ramp or the ejection cam. The strain-gauged punch then becomes invaluable as a standard against which the ramp transducers may be calibrated.

The ejection ramp

Knoechel et al. (1967) approached the problem by using strain-gauged bolts to support the ejection ramp on their Stokes 540-35 rotary press. They fitted a substantial steel bracket directly underneath the ramp and held the lower ends of two gauged bolts in this. The upper ends passed upwards through clearance holes in the base of the machine and were screwed against the underside of the ramp. The existing holding-down bolts were slackened off just sufficiently to ensure that the ramp was supported on the upper ends of the strain-gauged bolts, some 0.05 mm above its normal position (Figure 6.16). A similar system, but with only one gauged bolt, was applied to the BB2-27 press.

These two systems certainly gave signals; nevertheless, they were quite difficult to interpret, because as the punch head moved along the ejection track, its loading on the gauged bolts varied continuously. A load of approximately 100 kg applied at the leading edge of the ramp on the BB2-27 produced a signal of 2 mV from the bridge circuit, whereas the same load, applied directly over the gauged bolt, produced only 1 mV. Apart from this, at any one time there were several punch heads in simultaneous contact with the ramp, so the electrical output represented a rather complex summation of forces.

In most later work where the ramp has been retained as a single component, it has been supported on three, widely spaced transducers

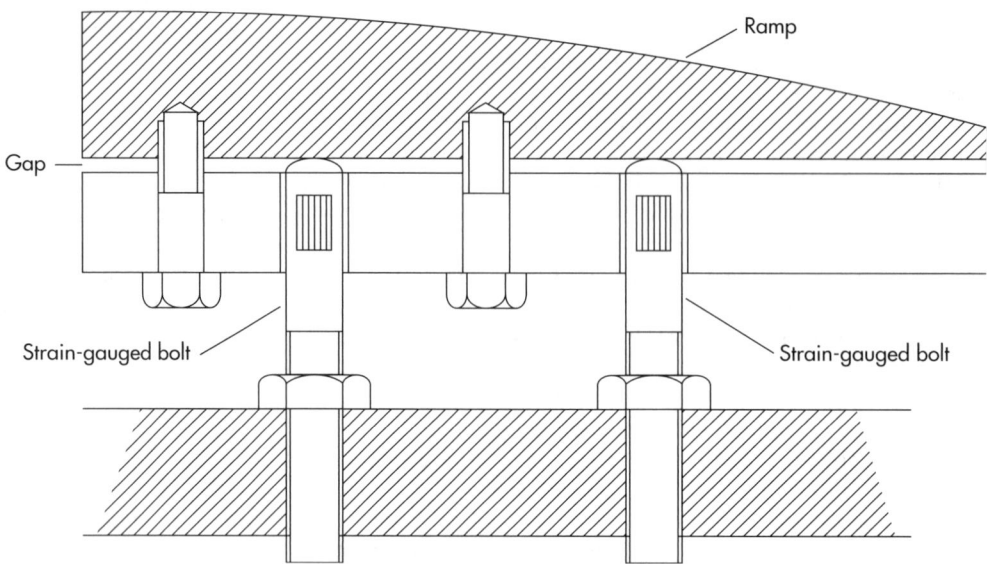

Figure 6.16 A gauged ejection ramp for a Stokes rotary press. The force on the ramp is taken by two strain-gauged bolts (Knoechel et al., 1967).

whose outputs are summed. This approach seems to overcome the problem of positional variability, since the sum of the three signals will be independent of the point at which the load is applied. An example of this arrangement is that of Williams and Stiel (1984), who used piezo-electric load cells to instrument the ejection ramp of a Manesty Betapress. The cells were inserted into three blind holes in the underside of the ramp, and their signal outputs were wired in series. The authors noted that the calibration was carried out 'statically, after installation', which perhaps raises a doubt or so in view of the known characteristics of the transducers. Some years ago, Ridgway Watt and Rue (1979) used a similar system and from the experiences with these load cells, we ourselves would nowadays prefer to calibrate dynamically rather than statically, and to use a strain-gauged punch as a reference.

The second drawback to Knoechel's original system was the complexity of the signal produced. If the machine was run with a complete set of punches, there were always several punch heads in simultaneous contact with the ramp. This problem could be side-stepped by using a reduced number of punches. This was the solution used by Williams and Stiel (1984), who ran their press with every other punch removed. However, using the rotary press with a limited number of stations raises further points that must be considered.

At the early stages of a formulation study, there may be some advantages in using a limited number of punches, since it is thereby possible to prepare tablets without using a great deal of material. The compression characteristics of the machine remain substantially the same however many stations are used, and useful information can be gained even from a single punch and die set. We ourselves used this method from time to time to compress individual tablets in an instrumented Betapress. For this procedure, the die was filled by hand at the feed-frame position, and the drive motor was then started so that a single tablet was made. Since the turret only makes approximately half a revolution between the feed frame and the main compression rollers, it was important to determine the speed of rotation at the time of compression and to check that this corresponded to the normal set speed of the press.

In fact, measurements showed that the Betapress, run with only one station of tooling, reached its full speed within considerably less than half a revolution (Ridgway Watt, 1983b). It seems safe to conclude that the punch velocities would correspond to those of steady running conditions.

For our own work on single tablets, the die was filled by hand. If the machine is run continuously, hand filling becomes impractical and so the die must then be filled via the feed frame in the normal way. However, when only a few stations are in use, less powder than usual is taken from the feed frame and as a result, the frame may overfill. When the lower punch travels up the weight-setting ramp at the filling stage, it expels the excess powder back into the frame underneath the rotating fingers of the feeder paddle. As the fingers pass one by one over the top of the moving dies, they partly obstruct the return flow of the powder and induce a small amount of pre-compression within the die. This can lead to cyclic variation in tablet weight, an effect that disappears when a full set of tooling is used (Ridgway Watt and Rue, 1980).

Modified ejection ramp designs

For many purposes, it is desirable to have an instrumented machine that can be operated under normal production conditions with all stations working. Ejection forces may themselves be measurable on single tablets, but it may take a comparatively long run to establish whether, for instance, the granulation is likely to build up on the working surfaces during production of tablets in commercial quantities.

The split ramp offers an alternative approach for obtaining readily interpretable ejection force readings. The design of the gauged ramp is altered so that only one punch head can possibly touch it at any one time. The generally accepted solution to the problem has been simply to reduce the width of the gauged section so that it becomes narrower than the distance between two adjacent punches. Wray (1969) described an early example of an arrangement in which part of the normal ramp was replaced with a tool steel cantilever beam, supported on an outrigger bracket and instrumented with foil strain gauges.

Systems of this type are now quite commonly used, although there is always a slight problem of signal linearity. Since the punch travels in a circular path around the machine, the applied load runs at varying distances from the support of the cantilever. The induced stress is correspondingly variable, just as in Knoechel's original device. Marshall, in his review article of 1983, suggested that the ramp may be sectioned into three parts, each independently mounted on load washers (Figure 6.17). This makes it possible to measure the whole ejection force profile, from initiation to final ejection, at any setting of the lower punch heights.

A possibility that does not seems to have tried so far is that of mounting the individual ramp sections on shear force load cells instead of cantilever beams. These would provide a constant signal, independent of the point of loading. Yet another method is that of linearising the cantilever output signal by reference to the known position of the punch head: this has been applied by D'Stefan et al. (1985) in the instrumentation of their Manesty X-Press. D'Stefan and his colleagues removed a segment from the standard ejection cam, or ramp, and constructed a cantilever beam from flat ground die steel to fit in the gap. The cantilever was held securely at its outer end and was gauged with Constantan strain gauges of the stacked rosette pattern (Micro Measurements CEA-06-061WT-350) on the top and bottom surfaces. It was arranged to deflect by 0.38 mm under a maximum force of 2 kN, and it was calibrated by dead-weight loading. Loading at different points around the locus of the punch gave variable signals, from which a correction curve could be constructed and then applied to the raw data from the strain gauges.

Measurement of punch pull-up and pull-down forces

After a tablet has been compressed, the upper punch must be withdrawn completely from the die to allow for ejection of the tablet and for the next cycle of die filling. So that this withdrawal may be conveniently achieved, the punch head is arranged to slide within a stationary ramp that

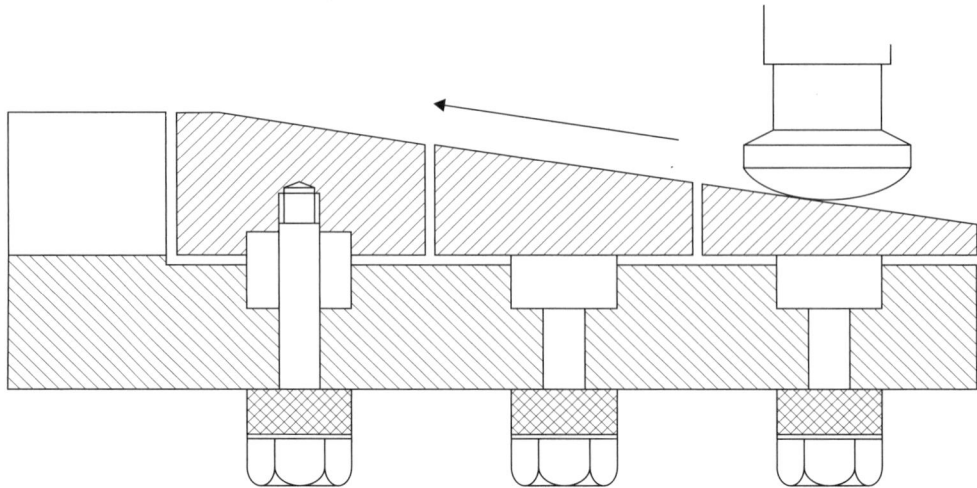

Figure 6.17 A segmented ejection ramp. More than one punch can be in contact with the ejection ramp at the same time, so the signals from the different punches need to be separated. Here the ramp has been divided into several parts with individual force washers; if there are less than three washers per segment, signal correction is required. The load washers may be fitted externally or internally.

lifts it as the turret rotates. The 'pull-up' force required to lift the punch is usually quite small, but if material accumulates in the sides of the die, and on the punch tip, then the movement can become relatively stiff, even to the point when the machine may suffer damage.

The development of a large pull-up force can indicate that the machine requires cleaning or some other attention, and it is, therefore, useful to warn the operator well before the force approaches a dangerous level. In general, pull-up forces on the upper punch and pull-down forces on the lower punch have been measured by instrumentation on the cam track. Goodhart et al. (1968), for example, achieved this with a Manesty Rotapress. They removed one of the three bolts that held the pull-down cam in position and substituted an internally strain-gauged 'Strainser' bolt, with a range of 0.45 kg, for the one nearest to the ejection ramp. Under good running conditions, the instrumented bolt gave a signal corresponding to approximately $6.0\,\mu\varepsilon$.

A much more sensitive arrangement, designed to measure both upper and lower punch tightness, has been described by Ho et al. (1983). In this system the cam tracks were reconstructed from Nylatron, which is a type of nylon with a low-friction molybdenum disulphide filling, and they were both equipped with semiconductor strain gauges (Techni-Measure type GB3-350). As the turret of the press rotated, the positions of individual punches were constantly monitored by proximity detectors. Should either of the strain gauges register an excessive tightness, the associated electronics could not only shut down the press but also indicate the number of the offending punch.

In order to provide a convenient means of calibrating the shut-down level, Ho et al. (1983). constructed a special punch that could be adjusted to varying degrees of tightness within the punch guide. This was achieved by the use of a small Nylatron plug in the side of the punch. The plug was set with an adjusting screw so that it required a force of approximately 150 N to move it along a standard guide. The machine was then run, and the electronics were adjusted to trip out at this level.

The facility to stop a production machine when punch tightness exceeds a predetermined level is nowadays offered as an option on some commercial presses. However, the tightness may sometimes be measured indirectly by monitoring a secondary variable such as the power absorbed by the drive motor. Bathe, in a short review from 1979, mentioned that this method has been used in the tablet presses made by Friedrich Horn in Germany.

Measurement of punch face adhesive forces

When a compressed tablet has been pushed out of the die by the action of the lower punch, it makes contact with the stationary sweep-off blade in a rotary press or the leading edge of the feed shoe in an eccentric press. It is then detached from the punch surface. At this point, sufficient force has to be provided to overcome the adhesion between the tablet and the punch. If a tablet sticks to the face of the lower punch, then the force needed to remove the tablet is increased. Sticking to the face of the upper punch can even cause rupture of the tablet if the adhesive force between the tablet and the punch face is greater than the cohesive forces holding the tablet together. Usually only a film forms on the face of the punch, but even this can be difficult to eradicate.

The first workers to study sweep-off forces were Mitrevej and Augsburger (1980). They fitted a strain-gauged cantilever beam in front of the normal sweep-off blade on a Stokes rotary press and used it to measure the adhesion of tablets made from various direct compression diluents. The instrumented beam was in the form of a narrow stainless steel blade, gauged with a complete bridge of foil gauges. The gauges were of the STC compensated type (Micro Measurements CEA-06-125UT-120) and were carefully fitted on each side of the beam, one of each pair being parallel to the major strain axis and one at right angles to it (Figure 6.18). The bridge was connected to a carrier pre-amplifier by means of a suitably screened cable. The instrumentation system indicated that the beam had a linear response for a given load and a given point of application.

Mitrevej and Augsburger (1980) reasoned that the adhesion force was the difference between the total force measured at the beam and that from the momentum of the tablet. The latter could be calculated from the mass and the velocity of the tablet as it travelled around the die table. It was necessary to apply two experimental corrections to the bridge signal. One correction was necessary to deal with the fact that the force

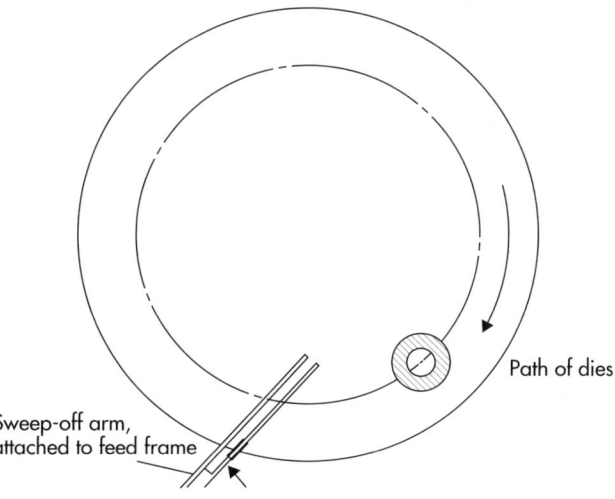

Figure 6.18 The measurement of adhesion between the lower punch and the tablet, using a gauged beam at the sweep-off point (Stokes RB-2 press; Mitrevej and Augsburger, 1980).

applied by the tablet is not exactly normal to the surface of the sweep-off arm, since the arm is set at an angle of approximately 52° to the radius of the die table. The second correction dealt with the inertia of the tablet being moved and was determined by experiment with a set of known suspended masses. The measuring range of the beam extended to a maximum force of approximately 1.2 N. Greater forces than this caused the blade to bend back into contact with the original sweep-off blade behind it.

Mitrevej and Augsburger (1980) noted that the adhesion force increased with an increase in compression pressure or a decrease in lubricant concentration. They also measured the force needed to eject the tablet from the die and found that differences in lubricant efficiency did not necessarily reflect differences in adhesion. A similar instrumented beam was used by Wang *et al.* (2004), who were able to relate adhesion forces to the intermolecular attraction between ingredients of the tablet formulation and the metal surface of the punch face.

Delacourte-Thibaut *et al.* (1984) measured the adhesion between tablet and punch face on an eccentric press. They found that because of adhesion, strain gauges on the lower punch showed negative force signals at various parts of the machine cycle.

Waimer *et al.* (1999a) adopted a different approach in that they measured the force required to separate the punch face from the tablet rather than measuring the sweep-off force. As the punch begins to rise in the die, adhesive forces cause the punch to be in tension before contact is lost between punch face and tablet. However, these forces are extremely low. For a compression force of 40 kN, the adhesion force is only 30–40 N, so a sensitive and wide-ranging system of measurement is needed. In a subsequent paper (Waimer *et al.* 1999b), the same workers studied the influence of punch engravings on adhesion. They did this by screwing small cones into the punch face and found that the angle of the cone had a major influence on the amount of adhesion.

Instrumentation packages

Introduction

As the use of instrumentation in tablet presses has increased, it is unsurprising that instrumentation packages have become available. These have varied in complexity, but essentially consist of a number of transducers, a power supply and equipment for amplifying and recording the output of the transducers.

To fit instrumentation to a tablet press requires that the user possesses or has access to mechanical engineering, electrical and electronic skills, as well as those relating to tablet formulation and manufacture. Such a broad range of skills may not be available. Therefore it is an attractive alternative to obtain an instrumentation package 'off-the-shelf' from a manufacturer who has already solved at least some of the problems associated with the selection and installation of transducers.

Hydraulic presses and testing machines

Much of the earliest research on the relationship between tablet properties and compressive force or pressure utilised a hydraulic press, often hand operated and almost invariably incorporating some form of pressure gauge. Typically a weighed quantity of the material to be compressed was loaded into a die closed at one end, a punch inserted into the upper end of the die and then force applied by downward movement of the upper platen of the press. Such a press is currently obtainable from Copley Scientific Ltd.

While this is a convenient method of applying and measuring force, the movement of the compressing punch is slow, non-uniform and cannot be said to imitate punch movement in a tablet press.

A later development was to use a press in which the punch movement was controlled and the applied load measured electronically, usually by means of load cells. Such devices are often referred to as 'universal material testing machines', and those made by the Instron Corporation have probably been used most frequently. Such machines have the advantage of moving the compressing punch at a predetermined uniform rate, but even so, the rate of movement is usually much slower than that of a tablet press. For example, in a study of consolidation mechanisms of pharmaceutical solids, Khossravi and Morehead (1997) used an Instron Model 5567. The crosshead of the press was programmed to move down at a rate of 5 mm min^{-1} until a pressure of 225 MPa was reached, after which the crosshead moved up again at the same rate. This compares with an upper punch speed in a Manesty F3 eccentric press ranging from approximately 140 mm s^{-1} to zero over the last 5 mm of its penetration into the die (see Chapter 9). Since it is known that the consolidation of some solids is dependent on the speed at which the compressing force is applied (Armstrong, 1989), results obtained by the use of such a low crosshead speed must be viewed with caution. However, as the tablets are made singly, the technique is useful if only limited amounts of solid are available.

Instrumentation packages

Instrumentation packages mainly comprise a power source, transducers and signal-acquisition devices, and several have been marketed. One of these, sold by Copley Scientific Ltd, was designed to be fitted to an eccentric press. Upper and lower punch forces were measured by piezoelectric load cells, and punch position by LVDTs. Transducer outputs were acquired with a computer, and compression parameters such as peak force, ejection force, tablet thickness and work of compression could be calculated from its associated software.

A similar system, the PC40, was devised by SMI Incorporated. Transducers measured upper and lower pre-compression forces, compression forces and punch movement, and it was designed for installation in any type of press. SMI also developed an instrumentation package called 'The Director' that could accept the output of up to eight transducers. It had modules for eccentric presses and rotary presses. The former permitted a detailed analysis of a single compression event, the calculation of porosity as a function of pressure, work functions and residual die wall stress. For rotary presses, in addition to the above, a statistical analysis of a production run could be obtained.

SMI also introduced an ingenious device that they called 'The Punch', which was described as portable press instrumentation on a single punch. Contained within the punch was a force sensor, an accelerometer, signal-conditioning and signal-amplification devices, an analogue-to-digital converter and limited computer memory. The punch was first connected to the input port of a computer and set up regarding sampling rates and the intended duration of the test. This information was downloaded to the microprocessor within the punch. The punch was then installed in the press turret and the press was run in the normal way. Compression data were collected, the punch removed and the data transferred to the computer, where it was analysed. Thus the punch was completely self-contained and did not need a telemetry system or other device to transmit the transducer outputs.

A perceived problem with this device is its use of an accelerometer to determine punch movement. Substitution of the relevant data into Equation 4.1 (p. 80) gives the distance travelled by the punch in a given time. However, unless there is some clearly defined starting position to act as a reference point, there is no way of knowing where the punch tip actually is.

A somewhat different approach was adopted by Puuman Oy with their Portable Press Analyser. In this system, upper and lower punches were equipped with force and displacement transducers, and a battery-powered data transmission and amplification units, a computer interface and associated software were also provided. The instrumented punches could be substituted for normal punches in a rotary press. The punches had short arms that activated the LVDTs. Transducer outputs were transmitted by an infrared data link to a data acquisition controller and from there transferred to a computer. The system was described by Paronen *et al.* (1994) and a diagram of the Portable Press Analyser is shown in Figure 6.19.

Matz *et al.* (1999) have evaluated the Portable Press Analyser. They found that force measurement was precise, but displacement measurement was affected by tilting of the punches when force was applied. If punch tilting was not considered, a deviation of displacement occurred of up to 110 μm from its true value. They also found that the deformation of the machine and the punches differed in dynamic or static conditions, and they suggested that correction for elastic deformation should be made from dynamic conditions. Errors from tilting could be reduced to approximately 18 μm by putting additional displacement transducers in adjacent turret positions.

Compaction simulators

Compaction simulators are designed to mimic the exact compression cycle of any tabletting press in real time and to record all the important parameters during that cycle. In 1972, Rees *et al.* described a device that used a mechanical tablet press to simulate the double-ended compression cycle of a rotary press. However, this could not simulate the compression cycle in real time.

The first true simulator was reported by Hunter *et al.* (1976) of ICI Pharmaceuticals Division. This device consisted of a frame supporting two hydraulic activators that were connected to a high-pressure hydraulic power pack. The activators could independently move punches within

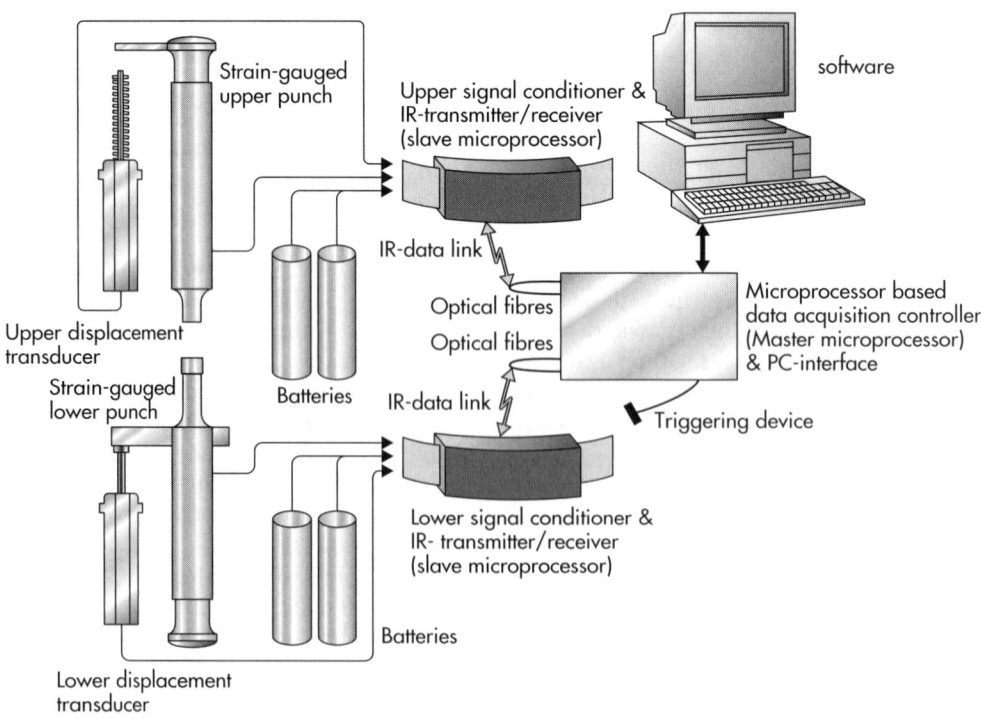

Figure 6.19 The Portable Press Analyser (Courtesy of Puuman Oy).

a die. Forces up to 40 kN could be achieved, the potential accuracy of the punches was better than 5 μm and a complete cycle could be performed in any time interval between 40 ms and 100 s. To simulate a particular tablet press, the displacement–time profile of the press was established by high-speed photography. The profile was then digitised. Load cells and LVDTs were attached to each activator, and their outputs were fed back to the control system to ensure that the input profile was being followed.

Until relatively recently, all compaction simulators were developments of the equipment of Hunter *et al.* (1976). For example, the simulator described by Celik and Lordi (1991) of the Pharmaceutical Compaction Research Laboratory at Rutgers University USA can apply loads up to 63 kN and move either punch at speeds up to 3.3 m s^{-1} at zero load and 1.6 m s^{-1} when under a load of 50 kN.

The use of simulators has been reviewed by Bateman (1988) and Celik and Marshall (1989). The latter have compared the equipment that can be used for tabletting studies (Table 6.2). Celik and Okutgen (1993) have suggested using a simulator to generate a data bank of information relative to the behaviour of materials under compression. The simulator would be used to develop standardised experimental conditions such as punch velocity, compaction pressure range, standardisation of tooling, etc.

One of the claimed advantages of the compaction simulator is its ability to follow any specified punch movement and, therefore, to imitate any given press running at any given rate. It is surprising how little use has been made of this facility. As Muller and Augsburger (1994) have commented 'Although compaction simulators have been designed to mimic the displacement–time behaviour of any tablet press, they have rarely been used in that fashion'. Instead, constant punch speeds are used, giving rise to a so-called 'saw-tooth' profile. Such profiles may well be selected because it may be difficult to decide on what other profile to use. The displacement–time waveform can be derived in several ways. The first is to use equations that describe the movement of the punches with time. These are discussed in Chapter 9 and depend on the dimensions of certain components of the press and its rate of rotation. However, as Armstrong and Palfrey (1987) have shown, such equations apply only if the force applied by the punches is zero (i.e. the press is operating with empty dies). If the die is not empty, the actual movement deviates from that predicted, the magnitude of the deviation being dependent on the applied force, the machine speed and the material being compacted.

Alternatively profiles can be obtained by examining the actual punch movement using, for example, high-speed photography (as did Hunter *et al.*, 1976) or high-speed video (Rubinstein *et al.*, 1993), though the application of such techniques can be difficult. It is, therefore, not surprising that so many workers have opted for saw-tooth profiles.

Table 6.2 Comparison of equipment used for tabletting studies (Celik and Marshall, 1989)

Feature	Single station press	Multi station press	Punch and die set	Simulator
Mimic production conditions	No	Yes	Maybe	Yes
Mimic cycles of many presses	No	No	Maybe	Yes
Require small amounts of material	Yes	No	Yes	Yes
Easy to instrument	Yes	No	Yes	Yes
Equipment inexpensive	Yes	No	Maybe	No
Easy to set up	Yes	No	Maybe	Maybe
Data base in literature	Yes	Yes	Some	No
Used in stress/strain studies	No	No	Yes	Yes

In recent years, a different type of simulator has become available. This is known as the Presster™ and is marketed by Metropolitan Computing Corporation. This acts as a high-speed single-station press, but it uses punches, compression rolls and pre-compression rolls that are identical to those of the press being simulated. A pair of punches and a die move linearly along a track, passing between the sets of rolls. The speed of the press is simulated by the speed at which the punches move along the track. Since the actual punches and rolls are used, the appropriate waveform of the press is followed and there is no need to make assumptions as with other simulators. The apparatus has been described by Levin (2004).

References

Armstrong NA (1989). Time-dependent factors involved in powder compression and tablet manufacture. *Int J Pharm* 49:1–13.

Armstrong NA, Blundell LP (1985). The effect of machine speed on the compaction of some directly compressible tablet diluents. *J Pharm Pharmacol* 37: 28P.

Armstrong NA, Palfrey LP (1987). Punch velocities during the compaction process. *J Pharm Pharmacol* 39: 497–501.

Bateman S (1988). High speed compaction simulators in tabletting research. *Pharm J* 240: 632–633.

Bathe P (1979). Recent developments in tabletting technology. *Mfg Chem Aerosol News*, 50: 33–41.

Belda PM, Mielck JB (1999a). The tabletting machine as an analytical instrument: qualification of the tabletting machine and the instrumentation with respect to the determination of punch separation and validation of the calibration procedures. *Eur J Pharm Biopharm* 47: 231–245.

Belda PM, Mielck JB (1999b). The tabletting machine as an analytical instrument: consequences of uncertainties in punch force and punch separation data on some parameters describing the course of the tabletting process. *Eur J Pharm Biopharm* 48: 157–170.

Celik M, Lordi NG (1991). The Pharmaceutical Compaction Research Laboratory and Information Center. *Pharm Technol* 15: 112–116.

Celik M, Marshall K (1989). Use of a compaction simulator in tabletting research. *Drug Dev Ind Pharm* 15: 759–800.

Celik M, Okutgen E (1993). A feasibility study for the development of a prospective compaction functionality test and the establishment of a compaction data bank. *Drug Dev Ind Pharm* 19: 2309–2334.

Cook GD, Duffield PJ, Oliver A (1988). Mounting an inductive displacement transducer on a single punch tablet machine. *J Pharm Pharmacol* 40: 119P.

Deer JJ, Ridgway K, Rosser PH, Shotton E (1968). A modified pressure wheel for the instrumentation of rotary tablet machines. *J Pharm Pharmacol* 20: 162S.

Delacourte-Thibaut A, Bleuze P, Leterme P, Guyot JC (1984). Étude du grippage et du collage, a l'aide d'une machine a comprimer instrumentée. *Labo-Pharma Probl Tech* 32: 673–681.

Depraetère P, Seiller M, Puisieux F (1978). Laser measurements of velocities of tablet punches. In *Proceedings of the International Conference of Powder Technology and Pharmacy,* Basel.

D'Stefan DA, Leesman GD, Patel MR, Morehead WT (1985). Computer interfaced instrumentation of a rotary tablet press. *Drug Dev Ind Pharm* 11: 83–100.

Duchêne D, Djiane A, Altglas F *et al.* (1972). Étude des comprimés. 5: Equipement d'une machine a comprimer alternative pour la determination des forces intervenant lors de la compression. *Ann Pharm Francaises* 30: 127–138.

Goodhart FW, Mayorga G, Mills M, Ninger FC (1968). Instrumentation of a rotary tablet machine. *J Pharm Sci* 57: 1770–1775.

Graffner C, Johnstone ME, Nicklasson M, Nyquist H (1985). Preformulation studies in a drug development program for tablet formulation. *J Pharm Sci* 74: 16–20.

Higuchi T, Nelson E, Busse LW (1954). The physics of tablet compression. 3: Design and construction of an instrumented tablet machine. *J Am Pharm Assn Sci Ed* 43: 344–348.

Ho AYK, Barker JF, Spence J, Jones TM (1979). A comparison of three methods of mounting a linear variable displacement transducer on an instrumented tablet machine. *J Pharm Pharmacol* 31: 471–472.

Ho AYK, Milham A, Lockwood L, Jones TM (1983). A system to prevent damage to rotary tablet machines caused by the tightening up of punches during tablet production. *J Pharm Pharmacol* 35: 112P.

Hunter BM, Fisher DG, Pratt RM, Rowe RC (1976). A high speed compression simulator. *J Pharm Pharmacol* 28: 65P.

Jungersen O, Jensen VG (1970). Studies on tablet compression. 1: Construction of an instrumented tabletting machine. *Dansk Tidsskr Farm* 44: 39–47.

Khossravi D, Morehead WT (1997). Consolidation mechanisms of pharmaceutical solids. A multi-

compression cycle approach. *Pharm Res* 14: 1039–1048.
Knoechel EL, Sperry CC, Ross HE, Lintner CJ (1967). Instrumented rotary tablet machines. 1: Design, construction, and performance as pharmaceutical research and development tools. *J Pharm Sci* 56: 109–115.
Krycer I, Pope DG, Hersey JA (1982). An evaluation of the techniques employed to investigate powder compaction behaviour. *Int J Pharm* 12: 113–134.
Levin M (2004). Tablet press instrumentation. In Swarbrick J, Boylan JC (eds), *Encyclopaedia of Pharmaceutical Technology,* 2nd edition suppl. New York: Taylor & Francis, pp 455–480.
Lindberg NO (1972). Instrumentation of a single punch tablet machine. *Acta Pharm Suec* 9: 135–140.
Marshall K (1972). Some observations on the elucidation and control of the compaction processes. In *Proceedings of the International Conference on Compaction and Consolidation of Particulate Matter,* Brighton.
Marshall K (1983). Instrumentation of tablet and capsule filling machines. *Pharm Technol* 7: 68–82.
Matz C, Bauer-Brandl A, Rigassi T, *et al.* (1999). On the accuracy of a new displacement instrumentation for rotary tablet presses. *Drug Dev Ind Pharm* 25: 117–130.
Mitrevej A, Augsburger LL (1980). Adhesion of tablets in a rotary tablet press. I: Instrumentation and preliminary study of variables affecting adhesion. *Drug Dev Ind Pharm* 6: 331–337.
Moldenhauer H, Kala H, Zessin G, Dittgen M (1980). Zur pharmazeutischen Technologie der Tablettierung. *Pharmazie,* 35: 714–726.
Moller-Sonnegaard J, Kristensen HG (1978). Instrumentation of a rotary tablet machine. *Pharm Ind* 40: 269–273.
Muller FX, Augsburger LL (1994). The role of the displacement–time waveform in the determination of Heckel behaviour under dynamic conditions in a compaction simulator and a fully instrumented rotary tablet press. *J Pharm Pharmacol* 46: 468–475.
Müller BW, Steffens KJ, List PN (1982). *Drugs Made in Germany* 25: 53.
Paronen P, Ilkka J, Wihervaara M *et al.* (1994). Application of portable rotary press instrumentation and compaction simulator to tablet development. In *Proceedings of the 13th Pharmaceutical Technology Conference,* Strasbourg, vol 1, pp 641–648.
Polderman J, de Blaey CJ, Braakman DR, Burger H (1969). Het gebruik van rekstrookjes in de farmaceutische technologie. *Pharm Weekblad* 104: 575–582.
Ragnarsson G, Sjögren J (1985). Force–displacement measurements in tabletting. *J Pharm Pharmacol* 37: 145–150.
Rees JE, Hersey JA, Cole ET (1972). Simulation device for preliminary tablet compression studies. *J Pharm Sci* 61: 1313–1315.
Ridgway K (1982). Pharmaceutical tablet making. *Chemical Engineer,* May: 169–173.
Ridgway K, Deer JJ, Finlay PL, Lazarou C (1972). Automatic weight-control in a rotary tabletting machine. *J Pharm Pharmacol* 24: 203–210.
Ridgway Watt P (1983a). Tablet press instrumentation 2. *Manuf Chem* 54: 42–45.
Ridgway Watt P (1983b). Measurement of acceleration in a rotary tablet machine. *J Pharm Pharmacol* 35: 746.
Ridgway Watt P, Rue PJ (1979). The design and construction of a fully instrumented rotary tablet machine. In *Proceedings of the International Conference of Pharmacy Technology and Product Manufacture,* Copenhagen.
Ridgway Watt P, Rue PJ (1980). Tablet weight and density variation in a rotary press. *J Pharm Pharmacol* 32: 22P.
Rubinstein MH, Petersen MA, Bateman SD, Stetsko G (1993). Measuring punch-time displacement profiles on a rotary tablet machine using a high-speed video system. *Pharm Tech Int* 5: 24–33.
Salpekar, AM, Augsburger LL (1974). Magnesium lauryl sulphate in tableting: effect on ejection force and compressibility. *J Pharm Sci* 63: 289–292.
Schmidt PC, Tenter U, Hocke J (1986). Force and displacement characteristics of rotary tabletting machine. 1: Instrumentation of a single punch for force measurements. *Pharm Ind* 48: 1546–1553.
Schwartz JB (1981). The instrumented tablet press: uses in research and production. *Pharm Technol* 5: 102–132.
Shotton E, Ganderton D (1960). The strength of compressed tablets. 1: The measurement of tablet strength and its relation to compression forces. *J Pharm Pharmacol* 12: 87T–92T.
Thacker HS (1979). Instrumentation of tablet machines. In *Proceedings of a Conference on Compaction of Particulate Solids,* University of Bradford.
Waimer F, Krumme, Danz P *et al.* (1999a). A novel method for the detection of sticking of tablets. *Pharm Dev Technol* 4: 359–367.
Waimer F, Krumme M, Danz P *et al.* (1999b). The influence of engravings on the sticking of tablets. Investigations with an instrumented upper punch. *Pharm Dev Technol* 4: 369–375.
Walter JT, Augsburger LL (1986). A computerised force/displacement instrumentation system for a rotary tablet press. *Pharm Technol* 1986: 26–36.

Wang JJ, Guillot MA, Bateman SD, Morris KM (2004). Modelling of adhesion in tablet compression. 2: Compaction studies using a compaction simulator and an instrumented tablet machine. *J Pharm Sci* 93: 407–417.

Williams JJ, Stiel DM (1984). An intelligent tablet press monitor for formulation development. *Pharm Technol* 8: 26–38.

Wray P (1969). The instrumented rotary tablet machine. *Drugs Cosmetic Ind* 105: 58B–68B, 158–160.

7

Calibration of transducer systems

Peter Ridgway Watt

Introduction

If an instrumentation system is to be of value, it must provide accurate measurements and must continue to do so in a reliable manner. Appropriate calibration procedures are, therefore, essential, both at the assembly stage and at intervals during the subsequent operation of the system.

An important consideration in this respect is that practically all present-day transducers depend on a chain of electronic sub-assemblies to manipulate and display their signals, so that there is no obvious or 'commonsense' relationship between the measured variable and its indicated value. In earlier times, the weights on a balance, or the graduations on a foot-rule, could be associated directly with the quantity being measured, but nowadays we have lost that directness. One consequence of this is that the digital read-out has acquired an aura of infallibility and will be believed implicitly even when the equipment behind it is manifestly on fire. If indirect measuring systems are to be used with confidence, they must be seen to give repeatable readings under known conditions, and recalibration should be such a simple matter that it becomes a normal routine. It should also be possible to relate the transducer output to some form of absolute standard from time to time as necessary.

Fundamental standards

In the SI system, now more or less universally adopted, there are seven basic units from which all others can be derived. These base quantities, with their units and symbols, have been described and defined in Chapter 1. Such variables as displacement, time and temperature can be referred in principle to the base units of the SI system. Variables, for example force, that are not among those seven fundamentals must be derived from base units taken in combination.

In practice, the base units are not equally accessible for everyday use; some of them can only be verified by the use of specialised equipment. It is, therefore, normal to approach these through the use of agreed sub-standards or derived units.

Practical standards

In the laboratory, economic considerations dictate that the equipment used for measurements should not be more complicated or expensive than necessary. The end result must justify the initial expenditure, and it is obviously important to assess the accuracy that will be acceptable for the work. In general, this implies measurements that are several stages removed from the fundamental standards noted above, and an extensive use of the methods that have become common in engineering practice.

The basic standards themselves are usually maintained in national institutions such as the National Physical Laboratory in the UK, the National Bureau of Standards in the USA and the corresponding bodies in other countries. As we have noted, these standards are, with the exception of that of mass, no longer physical objects but are quantities that, in principle, could be established by direct measurement anywhere in the world.

The same institutions also hold secondary standards, which are more likely to have a physical existence and can be used for comparison. Laboratory measuring systems can be checked against these secondary standards; those, in turn, may be validated by reference to the fundamental units. The following notes show how this process may be extended to the calibration of a transducer in a practical environment.

Force

The first problem in calibrating a force transducer is that of generating a precisely known force for comparison. Such a force can be produced, for most practical purposes, by a known mass in the Earth's gravitational field: that is to say, by a known weight.

Direct calibration

It is often convenient to have some removable components of the tablet press equipped with strain gauges so that they become force transducers in their own right. Gauged punches, for example, are useful in this respect for at least two reasons. Firstly, they can be transferred from one machine to another; secondly, they can be used during compaction, either dynamically or statically.

Generally, such components are detachable and portable; in this case, they can be taken to be calibrated by weight loading in a suitable metrology laboratory. There is a large deadweight testing machine at the National Physical Laboratory at Teddington, UK, for example, and this is operated by their calibration service for work of this kind.

The component is set up within the jaws of the test machine and is progressively loaded with accurately known masses up to the maximum expected force. Output readings are taken, first with increasing load and then with decreasing load in order to check for hysteresis. Ideally, the whole set of readings should be repeated after an interval to establish if there are thermal effects or measurable drift either in the component or in its power supplies and amplifiers.

Of course, the measured signal is always dependent on its associated electronics, and it is also desirable that they should be used in the deadweight test whenever possible. However, it may be difficult to move the electronics from the instrumented machine. In this event, it may be more satisfactory to calibrate a separate load cell for use as an indirect, or transfer, standard. The load cell, complete with its own electronics, can then be taken from place to place, or machine to machine.

Indirect calibration

Almost all the forces measured in an instrumented tablet press are related in some way to the forces at the punches. Consequently, it is a useful first step to obtain an independent instrument that can be used to measure the force between the punch tips, and which can itself be re-calibrated easily as a transfer standard.

A comparison load cell, with a range equivalent to that of the tablet press, is ideal for this purpose. Such a cell will be considerably thicker than the average tablet and cannot conveniently be interposed between the tips of standard punches. However, it is worthwhile shortening one or more punches – old or worn ones can be used economically here – to make room for the cell between them.

Most ordinary cylindrical load cells are designed for axial loading and may give inaccurate readings if the applied force is allowed to run off-axis. It is worth taking some trouble in the production of a suitable cell holder that will centralise the punches and the cell along the same straight line. The transfer standard load cell is of such a diameter that it will slide within a normal die cavity. It has a central hole into which various hemispherical heads can be fitted, and it is supported on a special lower punch with a spigotted upper end. A modified upper punch, with a concave end, ensures that the load cell is not subjected to off-axis loading (Figure 7.1).

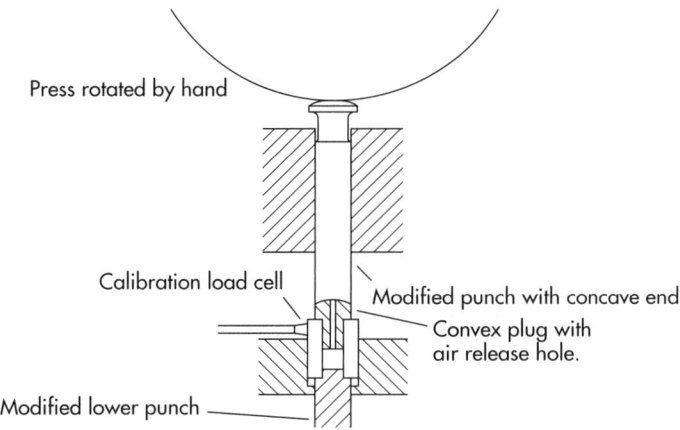

Figure 7.1 Force calibration in a rotary tablet press. The transfer standard load cell is of such a diameter that it will slide within the die cavity. It has a central hole into which hemispherical heads can be fitted, and the upper punch has a matching concave end to ensure the load cell is not subject to off-axis loading. As the press is rotated by hand, readings are taken simultaneously from the load cell and the instrumented pressure roll.

When the force transducers on the machine itself are to be checked, a die, the shortened upper punch and the cell in its holder are fitted carefully in place. The electronics are switched on and allowed to reach steady conditions. The press, whether eccentric or rotary, is then turned cautiously by hand in small increments, and simultaneous readings are taken from the two separate transducer systems.

As in the example of the deadweight testing procedure, all sets of readings should be taken with increasing and then with decreasing load, to check for hysteresis. From time to time, the load cell itself can be taken for separate deadweight calibration.

Potential sources of error

Ignoring, for the moment, inaccuracies in the load cell, we can see that the procedure indicated above may show errors under some circumstances. Some eccentric presses, for example, may have a strain-gauged upper arm as a means of sensing compression force. It was noted in Chapter 6 that this gives rise to a cosine error when the arm itself is not upright.

However, the reading of maximum force remains correct: whether the error is of importance, therefore, depends entirely on the purpose of the instrumentation.

In a press that has been fitted with piezoelectric load cells, there is a certain possibility of error through signal leakage during this particular calibration procedure. Since the machine must be turned by hand and is stopped for each reading, its rotation is slow by comparison with normal running, so caution is advised in the interpretation of results under these conditions.

In the author's opinion, it is prudent to acquire a pair of strain-gauged punches for use with presses that have piezoelectric load cells. The punches are first calibrated outside the tablet press, either by deadweight test or by use of a transfer standard load cell. The calibrated punches are then fitted into the press in the normal way, acting as the transfer standard load cell did before. This time, unlike the load cell, they are capable of being driven at machine speed and of compressing a tablet. Admittedly, the introduction of another link in the calibration chain puts the tablet press instrumentation at one further stage away from the fundamental standard, but the advantage of being able to make static, quasi-static and dynamic calibrations well outweighs the slight reduction in accuracy.

Telemetry

Using instrumented punches at machine speeds involves making some provision for the signals that they generate. In the eccentric press, where the punches only move up and down, a cable connection can be used. The rotary press presents more of a problem unless it has a slip-ring system or a radio-telemetry link, although more and more such machines now exist. Many workers have found it satisfactory to have the instrumented punch connected to a small transmitter that will fit into an adjacent punch guide, while a fixed antenna outside the turret picks up the radiated signal.

Other force calibrations

An instrumented tablet machine may have transducers arranged to measure not only compression but also ejection forces, die-wall stress and punch pull-up force. The calibration of all these variables is facilitated, as before, by the use of the instrumented punch. Of course, ejection forces are likely to be small compared with normal compression forces, and it may be desirable to have the lower punch equipped with semiconductor strain gauges for extra sensitivity. Many schemes for ejection force measurement on rotary presses give outputs that vary as the punch moves along the ramp, since its leverage is not constant in all positions. Calibration under these conditions then involves the construction of a suitable curve, from which all readings must be corrected by the data handling system.

Die-wall stress calibration is a much more controversial area. As a rule, gauged dies have been filled with some material that behaves as a liquid under pressure, and known forces have then been applied to the punches. The problem with the measurement itself usually seems to be a marked dependence of the signal on punch position, and this point is discussed in Chapter 9.

Displacement

Here we are concerned with the measurement of length, and of changes in length. There is a broad gulf between what can be attained in the standards laboratory – 1 part in 10^{10} – and what is normal in engineering workshop practice – say 1 part in 10^4. Turned components, for example, might typically be machined to tolerances of around $25\,\mu m$. The displacement transducer, often in the form of an LVDT, can in fact achieve realistic resolutions down to approximately $0.1\,\mu m$.

At this level of accuracy, probably the most widely used measurement device is the interferometer.

The interferometer

The principle of the interferometer has been outlined in Chapter 4; suffice it to say that it can be used to determine changes in optical path length to a maximum sensitivity of approximately one thousandth of a wavelength of the radiation employed. The light source is usually a frequency-stabilised laser, operating in the region around 600 nm, but even so, its stability is not perfect, and the uncertainty associated with the measurement is of the order 1 in 10^8. In general, it is a physical laboratory instrument, and its most practical application in tablet studies is probably that of checking the dimensions of gauge blocks, which are discussed below.

Gauge blocks

Gauge blocks move away from complex instrumentation and into the area more directly applicable to transducer calibration. These are simply sets of hard, dimensionally stable blocks with flat faces of defined surface finish and parallelism. A complete set may have between ten and a hundred or so pieces, and they can be used to make up various lengths as required by the use of one or more blocks together. The blocks may also be described as 'slip gauges', and they are used in engineering practice for a variety of purposes including the calibration of micrometers and – as we have noted – for the generation of accurately known lengths for comparison.

Table 7.1 shows, by way of example, the nominal sizes of block that are contained in metric set 2M (to the US Federal Specification GGG-G-15B). It will be apparent that very many specific lengths can be made from this set by taking one, two or more blocks in combination. Ideally the number of gauge blocks combined to produce a given length should be kept to a minimum by careful selection of appropriate sized units.

Gauge blocks are the subject of standard specifications such as BS 4311/68, DIN 861, GGG-G-15B and others, and they are available in various qualities that are given alphabetical or numerical codes. For example, the BS grade described as 'AA Special' has a length tolerance of $\pm 0.10\,\mu m$, while BS grade 'C' has an asymmetrical tolerance of $+0.5$ to $-0.25\,\mu m$. These differing accuracies are, naturally, reflected in the prices of the sets.

When the highest accuracy is needed, the gauges in a set can be individually calibrated and certified, either by the gauge manufacturers or by some independent body such as the National Physical Laboratory or the National Bureau of Standards. The measurements will probably be made with an interferometer. Gauge blocks may be made from a variety of materials, such as tool steel, tungsten carbide or chromium carbide, though alloy steel is the most popular choice. Since micrometers and many other measuring instruments are also usually made of steel, it is useful to have gauges with similar expansion characteristics.

The surface finish of these blocks, as manufactured, is normally better than $0.025\,\mu m$ centre line average (CLA), and in the higher quality sets may well be better than $0.01\,\mu m$ CLA. When two such blocks are put together under clean conditions, their opposed surfaces appear to adhere. This effect is known as 'wringing' and it is a good indication that the two blocks are both flat and clean.

The micrometer

Traditionally, length measurements in the instrument maker's workshop have been made with the micrometer gauge. In principle, this consists of an accurately made screw and nut. Rotation of the screw drives it forward through the nut in a precise manner, and the distance travelled can be read from a scale that indicates the amount of rotation.

The micrometer head is usually built into a rigid 'C'-shaped frame that holds the end of the moving screw in line with a fixed stop or 'anvil'. An object to be measured is held

Table 7.1 The nominal sizes of block contained in metric set 2M, made to US Federal Specification GGG-G-15B Set 2M (88 blocks)

First series: 0.001 mm increments (9 blocks)									
1.001	1.002	1.003	1.004	1.005	1.006	1.007	1.008	1.009	
Second series: 0.01 mm increments (49 blocks)									
1.01	1.02	1.03	1.04	1.05	1.06	1.07	1.08	1.09	1.10
1.11	1.12	1.13	1.14	1.15	1.16	1.17	1.18	1.19	1.20
1.21	1.22	1.23	1.24	1.25	1.26	1.27	1.28	1.29	1.30
1.31	1.32	1.33	1.34	1.35	1.36	1.37	1.38	1.39	1.40
1.41	1.42	1.43	1.44	1.45	1.46	1.47	1.48	1.49	-
Third series: 0.50 mm increments (19 blocks)									
0.5	1.0	1.5	2.0	2.5	3.0	3.5	4.0	4.5	5.0
5.5	6.0	6.5	7.0	7.5	8.0	8.5	9.0	9.5	-
Fourth series: 10 mm increments (10 blocks)									
10	20	30	40	50	60	70	80	90	100
One block: 1.0005 mm									

against the anvil, and the screw is turned by means of a knob or 'thimble' at one end until its end face touches lightly against the other face of the object. At this point, the thickness of the object may be read from a scale on the rotating thimble. In some constructions, the head may be mounted through a metal block for the measurement of depth or height.

Simple mechanical micrometers of this type are capable, at best, of reading the dimensions of rigid objects to the nearest 0.001 mm. Mechanical considerations limit the fineness of the screw thread that can be used to something in the region of two threads per millimetre (40 threads per inch in Imperial units), and this means that the practical sensitivity of the gauge will be limited by the scale on the thimble. It is not easy to read such a scale if it is divided into more than 50 parts, and at that level each graduation would correspond to a movement of 0.01 mm.

Greater sensitivity can be achieved by the addition of a vernier scale to the thimble, or by increasing its diameter. However, it is not easy to ensure that each measurement is being carried out at exactly the same pressure. Slight changes in pressure will cause variable amounts of distortion, both in the object being measured and in the frame of the micrometer. It is, therefore, common practice for the thimble to be fitted with a light friction clutch that, to some extent, limits and equalises the force that can be applied to the screw.

Nowadays, micrometers are available in versions with digital display. This may read from the rotation of the screw as in the mechanical micrometer, or it may run from a completely independent optical scale. In either case, estimation of the scale reading is greatly facilitated. Moreover, if the encoding system is digital throughout, it will not be affected by drift in the associated amplifiers. For digital micrometers in which the display operates from rotation of the screw, the limit of practical resolution still remains at approximately 0.001 mm, because the inherent errors in the feed screw are themselves of this order. The independent linear encoder could be made to a better accuracy but is, again, limited by the problems of variable pressure and temperature.

In work on tablet presses, micrometers are convenient for checking dimensions of punches, where accuracy to 0.001 mm may be thought more than adequate for the application. At this level of sensitivity, cleanliness is very important. The measuring faces must be kept free from dirt, moisture and particulate matter, and it is good practice to close them on to a clean lint-free tissue just before use. Care should also be taken to avoid holding the frame more than is necessary, since the warmth of the hand will induce local expansion.

Micrometer errors

From time to time, a micrometer should be checked against a known standard: it is convenient to use a gauge block for this purpose. Naturally, the screw pitch may be expected to remain constant over the working life of the instrument, but two sources of error may develop. Firstly, wear between the screw and the nut may give rise to some free play in the mechanism. In this event it is usually possible to remove the thimble, adjust a clamping ring that surrounds the nut section, and hence take up any free play. Secondly, the micrometer may indicate a departure from true zero when the anvil and the face are brought together. This situation can be remedied by careful adjustment of the thimble position, using a special tool provided by the manufacturers.

Calibration of displacement transducers

In order to calibrate a displacement transducer system, it is necessary to have some means of effecting a known amount of movement at the sensing point. Commercial transducers are generally supplied with a test certificate from the manufacturers, but it is naturally important to check that the original performance has not varied. Checks can be made on the bench, and on the tablet machine itself.

Displacement measurements are best carried out on some form of rigid, dimensionally stable surface that can hold the components in a fixed relation. In the toolroom, it is normal practice to use a heavy, flat slab of material known as a surface plate. Surface plates may be made of various materials, including granite, glass and metal, though there are advantages in the traditional cast-iron plate since it will conveniently hold magnetic clamping devices and V-blocks.

If the body of the transducer is held vertically and securely over the surface plate, its armature can be brought down so that it makes contact with the surface plate at some point near the centre of its travel, where its scale is at maximum linearity. Then slip gauges can be interposed between the armature and the plate, either singly or in combination, to check as many points on the transducer output characteristic as may be thought necessary.

It may be difficult to remove a displacement transducer for calibration once it has been installed in the tablet press. In this event, a technique must be found that facilitates some simple test that can be carried out *in situ*. There are two fairly straightforward procedures. In the first of these, the moving armature of the transducer can be displaced, as in the bench test, by the insertion of slip gauges. This method would be applicable where the transducer is of the 'gauging' variety and is lightly spring-loaded against an actuator bracket. The slip gauge may be positioned squarely between the actuator and the gauge head, or between the punch head and the compression roller. If the transducer is built into the sliding mechanism – as it might be in an eccentric press – and is itself inaccessible, it will be necessary to displace the whole punch holder assembly by inserting spacers between the punch faces. At this stage it may be important to distinguish between two problems: firstly, that of calibrating the transducer itself, and secondly, that of calibrating the whole system.

Calibration problems

As a general rule, calibration procedures of the kind outlined above are expected to provide reassurance that all is well with the system; nevertheless, there may be occasions when they reveal trouble. Both mechanical and electronic areas may have problems, and it is useful to develop a systematic approach to their solution. Some possible fault conditions are noted here.

No signal

Assuming that the system has in fact been switched on, the most likely cause will be a broken connection, probably at some component that is subject to vibration. Next may be a blown fuse in the transducer power supply: check for output voltage. Less likely will be a failure in the data collection circuitry, but it is as well to have a spare transducer amplifier and a spare data input module that can be interchanged as a check.

Low signal

There may be a high resistance in either the input or the output lines of the transducer bridge, reducing the working voltage. This could result from corrosion in a soldered joint through the presence of residual flux. The voltages should be checked with an electronic voltmeter such as one of the Fluke portable instruments, first at the transducer and again directly from the power supply module. As before, the amplifier and input modules can be replaced in order to establish if conditions are thereby changed.

Strain gauges, if attached to a substrate by adhesive, may become wholly or partly detached. This may not be easy to see, but it is worth examining the area with a $\times 10$ hand lens.

Signal clipping

What appears to be a normal calibration curve may flatten out at one end and the output signal would reach a maximum value even though the measured variable continues to increase. There may be several reasons for this effect, some mechanical and some electrical or electronic. Taking, as an example, the displacement transducer; it is possible for the body of this device to work loose in its housing through vibration from the operation of the press. If the armature itself remains connected, then the transducer signal will fall to low levels over part of its travel. This can be checked readily by inspection.

Signal clipping can take place in strain-gauged assemblies as a result of some discontinuity in the stress–strain relations of the parts. For example, a gauged component deforms linearly over a certain range, but then makes contact with another component. Now the two must move together, and the gauge output is likely to be reduced. In the experience of one of the editors, a worn bearing on a tablet press caused a problem of this sort, resulting in a force transducer calibration curve with two distinct linear portions. This type of fault can usually be checked by careful observation, although the movement concerned will, of course, be very small.

Electrical clipping is less easy to identify but can happen in an analogous way if the controls on the signal amplifier have been incorrectly adjusted. Most amplifiers have the facility to offset a signal so the amplifier output can be brought to an electrical zero. Ideally, the mechanical and electrical zero points would be the same, but it is easier to carry out the fine tuning electrically. The system may, therefore, have been set near the very limits of its working range. In this event it can only provide a small change in output before it reaches the end of its electrical travel.

Non-linearity

Some transducer systems inevitably generate non-linear signals, which must be corrected by the data handling equipment; examples here include some types of instrumented ejection ramp, and certain non-contact displacement transducers. In other systems, non-linearity may represent a fault condition. The charge amplifiers used to handle the very small signals from a piezoelectric load cell, for instance, are quite susceptible to variations in their input capacitance. A poor quality connecting cable, or an accidental splash of water, may give rise to distorted calibration curves.

8

Data handling

Alister P Ridgway Watt

Introduction

This chapter will discuss the various methods that can be employed to take meaningful information from a sensor and pass it into a computer. Each sub-system will be designed under a set of engineering compromises; therefore, in order to obtain accurate information about the process under investigation, the investigator will need to be able to check that each sub-system will meet its specific requirements. If one section does not have a sufficiently high bandwidth, for example, some of the information will be lost and the final data analysis could lead to incorrect conclusions, popularly described as 'garbage in, garbage out'.

It would be fair to say that there have not been many new electronics devices introduced since the early 1980s, but the modern devices have been improved very considerably. They are now generally faster, more accurate, cooler, cheaper and smaller, with much reduced power consumption. These factors have allowed fundamental changes in the method of data acquisition; for example, small surface-mount components to analyse, store and forward data may be placed much nearer the sensor than was previously possible.

Nowadays the typical computer has a Pentium 4 or AMD 64 processor, running under a Windows XP operating system. This is ideal for data analysis and the colour display may be used to advantage. Windows is not a real-time operating system, and problems may arise if data are continually streamed into the computer because the central processor unit does not give that task 100% of its attention. However, on the positive side, it comes with the Excel spreadsheet program, which is useful for both analysis and display of data.

Sampling system theory

Aliasing

Sensors typically generate an analogue signal that varies continuously as a function of the sensor's input. However, computer processing normally calls for the signal to be represented in numerical form as a stream of digital data. Consequently, it is necessary to have a means of converting one to the other. In order to achieve this, the varying signal must be sampled at regular discrete intervals, and those intervals must be carefully chosen in order to avoid a specific problem, generally known as 'aliasing'. A familiar example of aliasing is the observation that wagon wheels in old Western cinema films appear to slow down, stop and then rotate backwards as the wagon accelerates. When the spokes move by exactly one position at the frame rate of the movie camera, or multiples thereof, rotation of the wheel apparently ceases. Another example of aliasing is the use of a stroboscope to apparently freeze the motion of a rotating machine; the frequency of the flashing light exactly matches the speed of rotation of the machine. In these extreme examples, the effective sampling rate is that of the varying signal, and the 'frozen' image tells us nothing about the movement as a whole.

The mathematician Harry Nyquist proved that it is possible to sample a signal accurately provided that the sampling rate is at least twice the highest frequency present in the input. Radio engineers prefer a paraphrased version that states that a signal must be sampled at a rate equal to or greater than twice its bandwidth in order to preserve all the signal information. So for a telephone line with a typical bandwidth of 3300 Hz, digital samples must be taken at least every 1/6600 s: that sampling rate may be referred to as the 'Nyquist frequency' for that particular system, and it represents the minimum sampling rate at which no information is lost.

If the sampling rate is below this critical value, the measurements taken will lead to a distorted picture of the original signal waveform, and that waveform cannot then be recovered by subsequent filtering. The distortion of the signal is a further example of aliasing.

When high-frequency noise is present in the signal, it may itself be subject to aliasing, thereby adding to any noise in our first Nyquist zone, reducing the signal-to-noise ratio and introducing measurement errors. However, its effect will be dependent on the chosen sampling rate for the system.

Choice of sample rate

It is now possible to consider what is a suitable sampling rate for transducer signals from a tablet press. Consider an eccentric press producing one tablet per second. The cycle time of the press will therefore be 1 s. However, the compression event is much less than this, and in fact amounts to about 20% of the cycle time (this topic is discussed more fully in Chapter 9), and in this example will last for about 200 ms. It is good practice to detect up to the tenth harmonic, so for an event lasting 200 ms, the minimum Nyquist frequency should be one sample every 10 ms. In practice, a more realistic starting point would be 100 samples during the event, or one sample every 2 ms.

Similar considerations apply to other aspects of the tablet press cycle. The ejection event, for example, is of much shorter duration, perhaps of the order of 20 ms. Therefore, if ejection is to be studied, the equivalent frequency would be one sample every 0.2 ms.

If tablet production on a rotary tablet press is to be investigated, then the duration of the compression event must be considered anew. The movement of the punches of any rotary press can be calculated, as described in Chapter 9, and examples of the duration of the compression event are given in Table 9.6. That of a relatively slow rotary press such as the Manesty B3B is comparable to that of an eccentric press, and so similar sampling rates are appropriate. However with a high-speed rotary, the compression event may only last approximately 20 ms, so higher sampling rates must be chosen to take this into account.

The effect of choosing an unsuitable sampling rate is illustrated in Figure 8.1. Figure 8.1a represents the output of a displacement transducer attached to an object that describes a sinusoidal pathway with a frequency of 1 Hz; each complete cycle, therefore, occurs in 1 s. If the signal is sampled at the rate of 1 sample s^{-1}, then a constant output is detected (Figure 8.1b), the value of which depends on the point in the cycle when the sample is taken. In this series of diagrams, the first sample is arbitrarily taken when t = 0.1 s. The sampling rate is progressively increased to 2, 4, 10 and 20 samples s^{-1} (Figure 8.1c–f). It is only at a rate of 10 samples s^{-1} that a recognisable sine curve is obtained, but even then, samples are not taken at the points where the maximum and minimum signals are +1 and −1, respectively.

A representation of the force–time waveform from an eccentric tablet press is shown in Figure 8.2. The waveform is of approximately saw-tooth form at 2.5 Hz, and the tenth harmonic is 25 Hz. A low-pass anti-alias filter must be used to guarantee that signals above half the sampling frequency are attenuated. A rule of thumb is to select a Nyquist frequency of over 10 times the highest frequency that is required to be digitised, to simplify the filter's design. That leads to a sampling frequency of 1000 Hz, 200 samples being taken in a 200 ms duration of the compression event. Of course greater accuracy may be required, so the filters will need to be designed to settle to better than one 'least significant bit' (LSB), and the easiest way of achieving

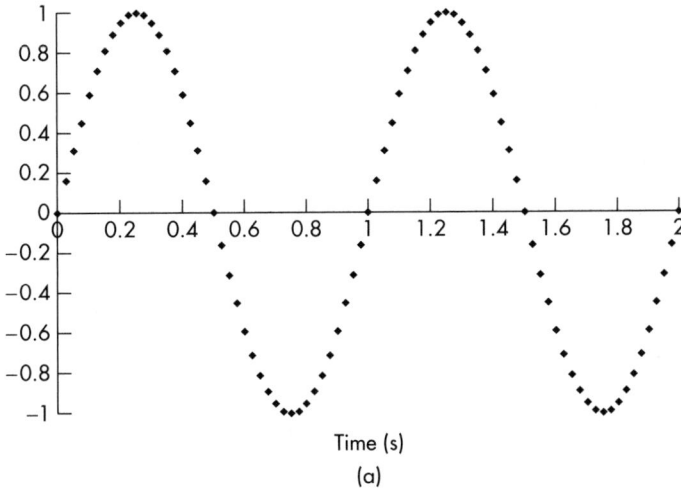

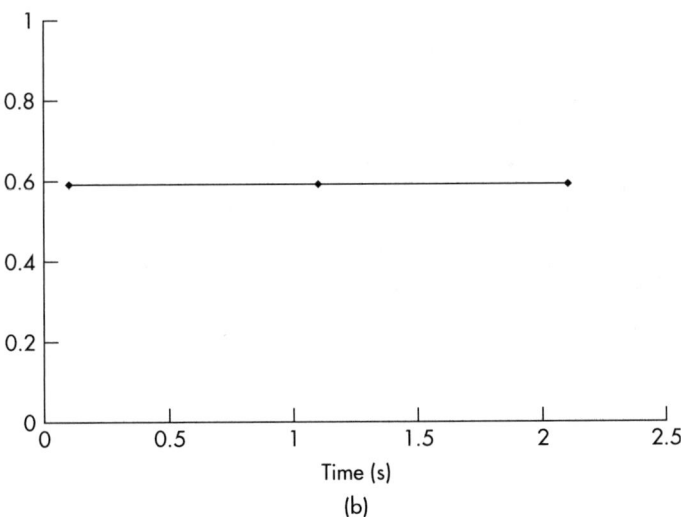

Figure 8.1 The effect of sampling a sinusoidal curve at a range of sampling rates. (a) Original sine curve with a frequency of 1 Hz; (b–f) sampling frequencies of 1, 2, 4, 10 and 20 samples s^{-1}, respectively.

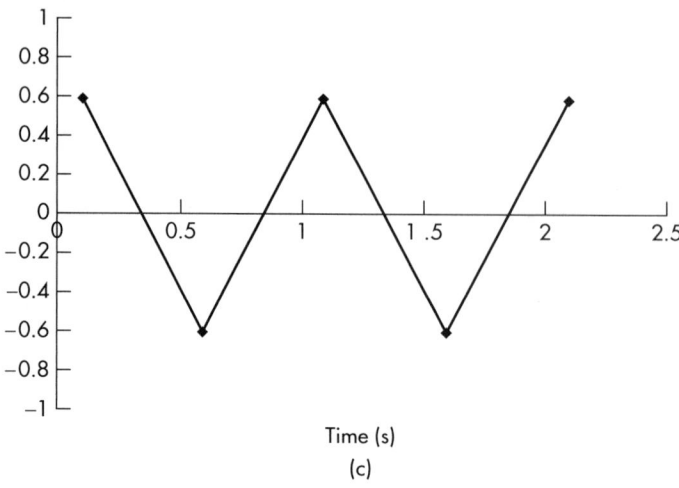

(c)

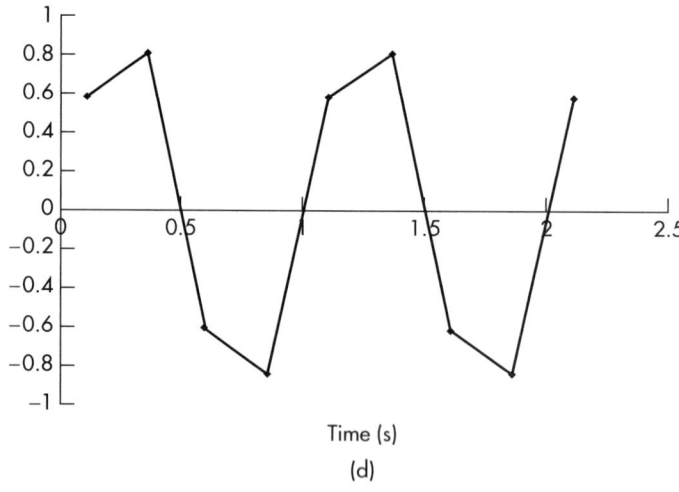

(d)

Figure 8.1 Continued

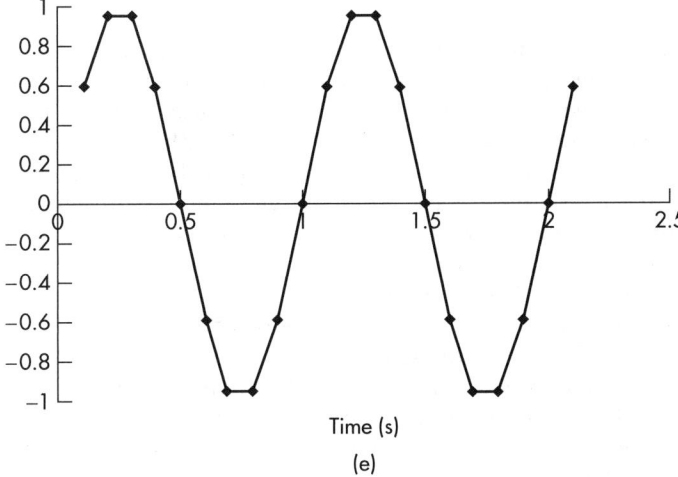

(e)

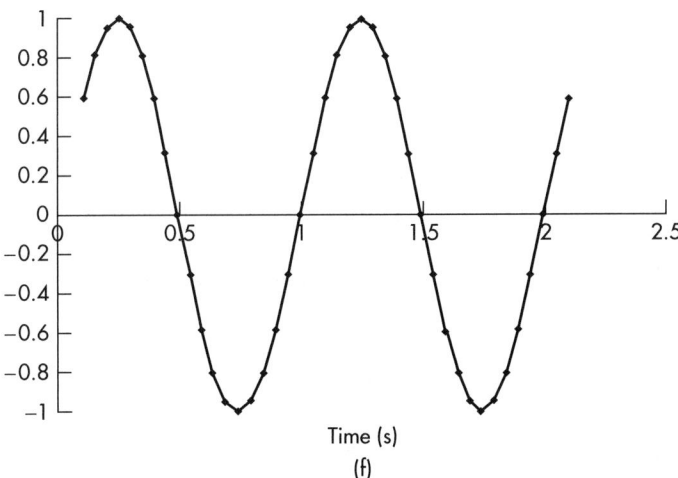

(f)

Figure 8.1 Continued

this is to raise the sampling rate again, as will be seen below.

If, however, we are going to build an embedded system that stores one or several cycles, a high sampling rate will require a correspondingly large memory, so a compromise must be reached between accuracy and size. One effect of low-pass filtering is that a moving average function of the data is obtained. This is also shown in Figure 8.2 and clearly demonstrates that a sampling rate of one every 10 ms is far too low, as the peak reading has fallen by 10%.

One way to determine the optimum sample rate is to digitise the sensor's waveform on an instrument of much higher specification than might otherwise be thought necessary. The digitised information can then be fed into a Fourier transform that generates either an amplitude–frequency or power–frequency bar chart. An amplitude–frequency chart has both real and imaginary levels, so phase differences may be recovered. As phase is not of interest, the two levels are converted into power. Imaginary levels are often denoted by i or j. An Excel macro, = IMABS (column)^2, performs the conversion function.

Often the terms fast Fourier transform (FFT), discrete Fourier transform and discrete cosine transform are used in this context. If the input sample buffer length is 2^N, where N is an integer, then the FFT algorithm may be used. The FFT removes a large number of duplicated calculations; hence it is called 'fast'.

Resolution

The analogue-to-digital converter (ADC) converts an analogue input voltage to a number. There is a range of analogue input voltages that output the same code number, as a result of quantisation, and this number represents the width of the LSB. Assuming a full-scale input range of 5 V, Table 8.1 lists the LSB width of various size ADCs and output number range.

An N-bit converter outputs numbers from 0 to $2^N - 1$, so an 8-bit converter outputs a range of binary numbers from 0 to 255, and resolution

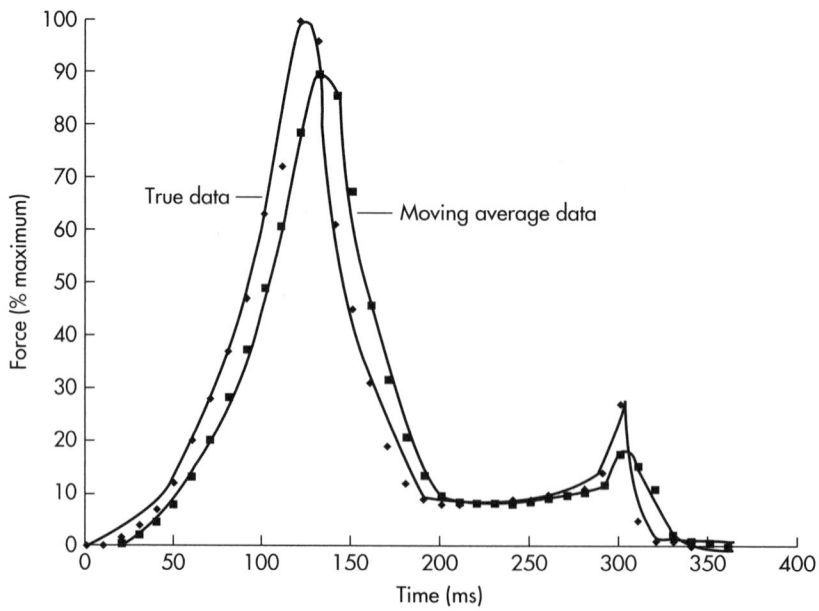

Figure 8.2 Force–time waveform from an eccentric tablet press, showing the original waveform and the averaging effect of low-pass filtering.

Table 8.1 Analogue-to-digital converters, their 'least significant bit' (LSB) width and output number range

Size (bit)	LSB width (µV)	Output number range
8	20 000	0–255
12	1200	0–4095
14	300	0–16 383
16	76	0–65 535

is better than 1%, but the use of 8-bit converters is not recommended. For one thing, input signals are unlikely to use the full analogue input voltage or dynamic range, and ADCs are not themselves perfect: they may exhibit non-linearity, missing codes and noise jitter. However, if the ADC chosen is better than actually required, these factors may be ignored.

Earlier chapters have suggested that a strain gauge is able to offer a resolution of 1 part in 10 000. If the gauge is only to be operated over the lower half of its range, a 12-bit converter would be adequate. At the other end of the scale, it is possible to use 16-bit converters, but apart from the increased cost, it is very difficult to ensure that the preceding electronics are capable of operating to that level of performance.

Microprocessors and microcontrollers usually handle data in 8-bit bytes and 16-bit words; therefore 2 bytes are required to store each sample. As far as data handling is concerned, it does not matter whether the ADC is of 12-, 14- or 16-bit resolution.

Electronics sub-systems

Ultimately, data from all sensors will be fed into a microcomputer for subsequent analysis. To achieve this, the sensor's analogue output voltage must be amplified, filtered and digitised. There are good reasons to perform all these tasks as near to the sensor as possible, including lower noise and higher accuracy, but it may be more convenient to fit a commercially available analogue-to-digital card in a PC and feed this card with differential analogue signals. The following items are all required for instrumentation in general. As the range of components is now so vast, we will concentrate on those components that generate at least 1000 readings per second with 12 to 16 bits of resolution.

Amplification

The input voltage range of an analogue-to-digital converter is likely to be 0–5 V, so the few millivolts of sensor output have to be amplified by approximately 1000 (60 dB). This may be achieved with a commercial amplifier module such as the SGA 100 (a generic name for amplifiers made by a range of manufacturers), but at 20 mm × 33 mm × 13 mm, this might be physically too large for some configurations. A surface-mount op-amp may be used instead, but this op-amp must have a suitable gain–bandwidth product. The LM358, for example, has an open loop frequency of only 1 kHz at a gain of 60 dB (1000), which means that any sub-millisecond change in force will be attenuated. As gain goes up, frequency response goes down. The specifications of any strain gauge amplifier must be carefully scrutinised to find the frequency response at the chosen gain. If that response is considered too low, then it is acceptable to connect two amplifiers of lower gain in series.

The frequency response of any data gathering system must be high enough to log the highest expected frequency component of a signal to a given accuracy. If the input waveform were an ideal square wave, then the amplifier would require a flat response up to the 9th harmonic for less than 10% error.

Some amplifiers are designed for use with a single supply; others require both positive and negative supplies. Furthermore, some amplifiers have what is known as 'rail-to-rail' inputs, the input voltage range including negative or ground, and rail-to-rail outputs. In fact, a rail-to-rail amplifier should not be used within 100 mV of the rails for highest accuracy.

If the ADC is more than a few metres away from the amplifier, then it is sensible to use a differential output device. As long as these two signals are within a common-mode input voltage range, external influences on the signal will tend to cancel out.

At this point it is useful to consider feeding any bridge reference voltages of the sensor to another ADC, again differentially.

Low-pass filter

The low-pass filter is also known as a baseband anti-aliasing filter. It should be fitted as near to the ADC's input as possible. This filter has two roles: removing any high-frequency noise picked up and, more importantly, attenuating any signal component above the Nyquist frequency.

Because a low-pass filter's attenuation above its cut-off frequency rises at 6 dB per octave per pole, and it is difficult to design extensive multi-pole filters, an engineering compromise must be made between circuit complexity, maximum useful frequency and sampling rate. The simplest design, the one-pole RC filter, is just a series input resistor and a capacitor to ground. More complicated optimised filter designs exist; some utilise operational amplifier systems (op-amps), capacitors and resistors, in which the component values are chosen to optimise parameters such as pass-band ripple, cut-off slope and final attenuation; others are switched-capacitor filters.

The problem with op-amp filters is that as more poles are added, the components need to have greater accuracy and temperature stability, otherwise the filter is likely to become unstable and may oscillate.

Switched-capacitor filters are available in surface-mount or dual-in-line integrated circuit packages, in various optimised filter configurations and need clock signals at typically 100 times the required cut-off frequency. These optimised filter designs include those described as Butterworth, elliptic and Chebychev, according to their particular characteristics.

In the Butterworth design, also known as 'maximally flat', the gain–frequency response curve is flattest at frequencies below the cut-off frequency; in other words, the filter exhibits less gain variation as a function of frequency than do the other configurations. This will produce the most accurate results from dynamic signals whose frequency components lie within the pass band.

The elliptic configuration exhibits a better transient response at the expense of more gradual attenuation above the cut-off frequency, while the Chebychev design (also spelt Tchebychev) has the steepest attenuation above the cut-off frequency.

It is preferable to choose a sampling frequency between 10 and 100 times the highest significant frequency in order to simplify the filter design. The choice of too high a frequency leads not only to greater ADC cost but also involves far more data to transfer, store and analyse.

Texas Instruments offers a free op-amp active filter design programme called FilterPro™. As a starting point, the user should select a gain of 1, a cut-off frequency of 1 kHz and a Butterworth-type response. If the sampling frequency is 100 kilosamples every second (100 kSa s^{-1}), then the Nyquist frequency will be 50 kHz; FilterPro™ will then show the design's attenuation at that frequency and will indicate that a simple single-stage circuit is adequate to give 20–40 dB of attenuation.

The attenuation required is dependent on the maximum expected frequency components from the sensor, the electrical noise level and the ADC resolution. If a commercial PC-compatible analogue-to-digital card is used, then it is important to check if anti-aliasing filters are supplied on the card, and if so, what their characteristics are.

Analogue multiplexer

It is possible to multiplex several analogue channels into a single ADC so that sequential conversion can be achieved. The multiplexer is an analogue switch that is controlled by the system. Sufficient time for a sample/hold circuit to charge up to a given accuracy must be allowed before a conversion can start. This reduces the time the converter can be producing valid data. It is difficult to design a multiplexed system capable of much better than 12-bit performance. Signals from one channel can influence other channels, the multiplexer's 'on' resistance varies with input voltage, and this makes fault-finding difficult. Designing such a sub-system from scratch is not recommended.

Some ADCs already incorporate an input multiplexer; these have been specifically matched to the ADC section and are capable of good performance.

Sample and hold

The filtered signal is now ready for the ADC sub-system. The ADC typically chosen for an instrumentation application requires that its input voltage does not vary by more than 1 bit equivalent voltage during the conversion process. If a 12-bit converter has an input voltage range of 0–5 V, then a change of input voltage of 5/4096 (roughly 1.0 mV) will change the output signal by one bit. A sinusoidal wave of more than 10 Hz will exceed this if the conversion time is more than 8 μs, and so for our applications, a 'sample and hold' must be used. The term 'track and hold' is more accurate but is not used as often.

A sample and hold comprises a CMOS (complimentary metal oxide semiconductor) analogue switch and a capacitor. The capacitor is charged to the input voltage when the switch is closed. The switch is opened by a logic signal for the duration of the conversion process; the capacitor then holds its original voltage to allow the conversion process to generate the correct output. The capacitor must be large enough that its voltage will not drop by the 1-bit equivalent voltage as a result of the current passing into the ADC. However, the value must not be too great, otherwise it would not charge up fully because of the series resistance of the analogue switch and the previous circuit's current drive capability. In fact, the series resistance and capacitor act as a single-pole RC filter.

It is possible to combine the anti-alias filter and sample-and-hold sections by using a switched-capacitor low-pass filter such as the MF10 (a generic device made by a number of manufacturers) and arranging the clock to stop during conversion time. Most modern ADCs have the sample-and-hold circuitry built in.

Analogue to digital converters

ADCs may be grouped into three simplified groups that convert an analogue voltage to a digital output by completely different methods. They are described in order of conversion speed.

Dual-slope integrating devices

Dual-slope ADCs are ideal for converting a strain gauge signal in devices such as weighing scales; their advantages including very low cost and a resolution of up to 24 bits. The fact that they can only convert at around $10\,\text{Sa s}^{-1}$ rules out their use in the applications discussed here, for which at least $1000\,\text{Sa s}^{-1}$ is needed.

Successive approximation register devices

Successive approximation register devices depend on an algorithm first suggested by Tartaglia, a mathematician, in 1556. The device takes one clock pulse per bit of resolution, so a 12-bit converter requires 12 clock pulses to perform one conversion. This limits their effective speed to a range of $100\,\text{kSa s}^{-1}$ to $3\,\text{MSa s}^{-1}$. They have two types of output: serial and parallel. Serial outputs are synchronous, that is they consist of a clock signal and serial data, with possibly a couple of signals to facilitate devices to be daisy-chained. The protocols used vary slightly; hence we have inter-integrated circuit (I^2C) and serial peripheral interface (SPI) as two examples. The SPI clock frequency is typically 10 MHz, so 16 bits of data may be transferred in less than 2 μs. Many microcontrollers have a built-in serial interface and so are easy to connect with what is referred to as 'glue-less' logic.

Parallel-output ADCs have eight or more data pins. Their disadvantage is that they have a higher pin count, and hence are larger, and their cost is greater. If they have only eight data pins and higher resolution, they will have high- and low-byte-enable pins as well.

The first device to achieve industry-standard status was the 574AJD (Figure 8.3c), made by Analog Devices Inc. and which was introduced in 1978. It had a resolution of 12 bits and a 35 μs conversion time: however, it required three separate power supply voltages and an external sample-and-hold circuit.

Figure 8.3 shows an interesting example of the progressive miniaturisation of electronic devices. An example of a modern device, the AD7680 has a 100 kSa s^{-1}, 16-bit ADC, and a size of 3 mm × 5 mm is shown in Figure 8.3b and can be compared with the early 574AJD.

Flash or parallel devices

Flash ADCs are the fastest, and most expensive, group of converter topologies and they have typical conversion rates of 10 MSa s^{-1} to 1 GSa s^{-1}. They can contain 2^{N-1} comparators so they are large and expensive, and they require plenty of power, though the latest designs feature lower power, pipelined architectures. Resolutions are typically 8 bit, but higher resolutions are available for applications such as video camcorders, radar and software-defined radios. The first monolithic flash ADCs were built in 1979.

Summary of analogue-to-digital converters

The successive approximation register device is the best choice for the uses covered in this volume. As it can produce more samples per second than needed, it gives the choice of either operating the device at a lower clock frequency to reduce power consumption or using a simpler anti-alias filter and averaging readings for less noise.

To minimise noisy readings, the circuit board must have a copper ground plane. This formerly involved making four-layer printed circuit boards; the ADC manufacturer Analog Devices Inc. offered the advice that if the printed circuit board facility could not make a multi-layer board then the purchaser should change to another company who can! However, if care is taken to choose devices with a high level of integration, it is possible to make double-sided printed circuit boards with all the tracks on one side, leaving the other as a 0 V ground-plane.

Figure 8.3 Analogue-to-digital converters. (a) MF10 switched capacity filter; (b) AD7680, a modern analogue-to-digital converter; (c) 574AJD, the first industry-standard analogue-to-digital converter dating from 1978.

Some two-channel ADCs are designed for stereo audiofrequency use. They are not recommended for instrumentation use, as they tend not to have good DC specifications.

It may be necessary to supply a voltage reference, though the supply voltage can often be used. If the sensor's excitation voltage is fed into another ADC channel, any change in the reference voltage will affect both signals, and subsequent data processing can cancel out these problems. This technique allows the use of non-regulated voltage on a Wheatstone bridge.

Embedded systems

As electronics become smaller, signal-processing circuitry can be positioned closer to the sensor; in this case the computer interface may be digital and the outgoing data may be calibrated and corrected for temperature before transmission. Clearly, this requires some local processing power so that analogue-to-digital conversions are scheduled at the sampling rate, while the data may be pre-processed to reduce noise by averaging and filtered, scaled and transmitted as a part of a packet.

An embedded system is a microcontroller with its own memory, program and input/output ports. The program itself is often quite simple: it will initiate ports and interfaces then loop round continually reading the inputs and either storing the data or forwarding it to another device. The input/output ports may or may not be physically in the microcontroller package.

To apply digital filtering to tabletting processes, a digital signal processor chip can be used. This chip is similar to a microcontroller but also has a multiplier, accumulator register and two data address generators. Texas Instruments and Analog Devices are the world leaders in the design of digital signal processors.

Microchip Technology's dsPIC series of digital signal microcontrollers (programmable input–output controllers (PIC)) are ideal for the purposes described in this volume. A surface-mounted package, 12 mm × 11 mm in dimensions, provides an 8-channel analogue input multiplexer, 12-bit 200 kSa s^{-1} ADC, digital signal processor core, 16-bit central processor unit, programme memory, 2 kbyte data memory, clock generator, timers, serial interface and debug port. The digital signal processor core may be used for post-conversion filtering, decimation and averaging, while the microcontroller section can assemble data into a packet for transmission either immediately or later.

The dsPIC series operate with a supply voltage between 2.7 and 5 V at currents ranging from 20 to 40 mA. However, if a particular application is intended to run on small coin cells, another approach is to use a micropower PIC chip and an external ADC of very low power. Current consumption would be reduced to the order of 100 μA. The PIC has an 8-bit internal data bus and only one register and, therefore, requires more effort to write programs.

Many of the dsPIC series processors are available in both surface mount as well as plastic DIP packages, so that prototype circuits may be easily constructed on 0.1 inch pitch Veroboard. Most other manufacturers no longer offer DIP components. The dsPIC contains electrically erasable memory, so the program may be altered as required during the debugging process, and even subsequently changed for a completely different application. Microchip Technology supply a free assembler and simulator called MPLAB®: a 'C' compiler and USB ICD2 programmer are available at reasonable cost.

Computer interfacing

There are many interfaces available as standard in the modern PC. This section will discuss each interface's relevance to instrumentation. The choice of interface depends on many factors. If the data-logging system is built into a punch and relies on an infrared window to pass data to the outside, then it makes sense to ensure it follows the Infrared Data Association (IrDA) standard so it may simply be held near a laptop, or IrDA-equipped PC, and interrogated. Conversely if a commercial PCI ADC card is used, then differential analogue interfacing would be employed. Many multi-channel cards offer the option of differential inputs.

Built-in interfaces

Recommended standard 232

Recommended Standard (RS) 232 provides for an asynchronous serial data stream of between ±3 V and ±12 V, and it is typically used to connect to a local dial-up modem. The data rate is comparatively slow owing to slew-rate limiting and cable capacitance, but over short distances of a few metres, common baud (Bd) rates are 9600 Bd, 38.4 kBd and 115 kBd. The baud rate is the maximum number of signalling transitions per second and is named after the French inventor J. M. E. Baudot (died 1903).

RS232 communications parameters are often specified by a series of numbers, such as 9600, N, 8, 1. In this example, 9600 represents the baud rate: at 9600 Bd, approximately 1000 characters per second may be transmitted. N defines the parity bits, which can be selected from N (none), O (odd) or E (even). The next numeral indicates the number of data bits, and can be 5, 7 or 8, and the last digit is the number of stop bits to be transmitted and can be 1, 1.5 or 2.

As an aside, the international telex network uses 5 data bits and 1.5 stop bits, at 50 Bd. This gives an equivalent speed of 66 words per minute and can be decoded mechanically. Special codes for letter shift and figure shift are used to select upper case letters or figures and punctuation.

Codes of 7-bit size are used to send text messages. Punctuation, figures, upper and lower case letters may be represented by a 7-bit code referred to as ASCII (American Standard Code for Information Interchange). In order that 8-bit data (in fact computer programs) could be sent over a communications link, a system called UU Encode was invented. It fits three 8-bit bytes into four 7-bit bytes. UU Decode is performed at the receiver.

A 9-way 'D' connector (Figure 8.4) uses pins 2, 3 and 5; the other pins are supplied for handshaking, where the other device can stop transmission if it gets overloaded with data. Computers are now sufficiently fast that handshaking should not be required. Sometimes the software driving the serial port insists on hardware handshaking. If this is the case, it is sometimes possible to convince the port to transmit by connecting the clear-to-send pin to a positive voltage of 5 to 12 V.

Two PCs may be connected together with a null-modem cable. This cable swaps 2 to 3, 3 to 2 and connects 5 to 5, so that one transmitter connects to the other receiver. Because both PCs are connected together, ground loops will occur. This can cause problems. Maximum possible baud rate depends on many factors: distance, cable capacitance and quality of electrical earthing. At 9600 Bd, a sensible limit would be 30 m.

Parallel printer

The parallel printer port is usually bidirectional and has eight data wires plus a number of control signals, such as Ack, Busy and Paper Out. Over

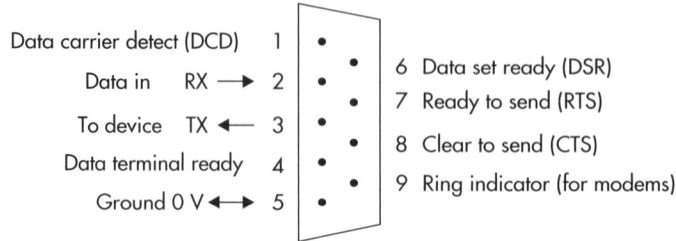

Figure 8.4 A 9-way 'D' connector.

short distances, say 5 m, the data rate can be fairly fast as eight bits are transferred simultaneously. A 25-way 'D' connector is used (Figure 8.5), and signalling is at TTL (transistor–transistor logic) levels of typically 0.5 V for logic 0 and 3.5 V for logic 1. Cable capacitance is a limiting factor.

The universal serial bus interface

The universal serial bus (USB) interface is perhaps the most useful as the latest USB2.0 standard allows the transfer of up to 480 megabits per second (Mbps) as well as compatibility with old standards of 1 and 12 Mbps.

Reasonably inexpensive commercially available data converters now use USB and tend to be self-powered from the supplied 5 V on the USB connector. Some units feature isolation between input and output to prevent earth loops; if differential signals are used, then this feature is not so important. These units will generally need external anti-alias filters. Some units have an analogue multiplexer feeding one converter; faster units have multiple converters that allow full speed on all channels and simultaneous sampling. USB is not intended to go much further than 5 m.

Ethernet

Ethernet is a serial data stream designed for computer networking, and runs at either 10 or 100 Mbps using CAT5 UTP unshielded twisted pair (UTP) cable. A transformer electrically isolates each node, so ground loops cannot exist. Even though cable lengths are limited to approximately 100 m, the use of hubs, network switches and even the Internet increases the distance. A unique 'media access controller' address and an 'Internet protocol' address identify each node on the network if TCP/IP or UDP protocols are used. The TCP/IP protocol specifies that all messages that are sent get an acknowledgement reply if received correctly; otherwise the message must be retransmitted. This guarantees a reliable service. By comparison, the UDP protocol just sends messages, and, if the link is reliable, throughput is higher. Lost packets are ignored.

PCI interfaces

Other ways of sending and receiving data at high speed over longer distances involve plugging adapter cards into PCI slots on the motherboard. Internally, the computer is likely to have a number of PCI slots on the motherboard into which circuit boards, such as ADCs, may be plugged. Each PCI slot is individually addressable by software so multiple cards may be used. The data throughput on the PCI bus depends on the motherboard but is over 33 Mbytes s^{-1}. Laptop computers do not have PCI slots but instead they have one or two PCMCIA slots, into which compatible adapters may be plugged.

The recommended standard 422 card

The RS422 standard defines a way that serial data may be sent differentially up to 1000 m, at up to 10 Mbps, on 4-core cables consisting of two pairs. Each pair is twisted together and cables often have characteristic impedance of approximately 100 Ω. Any electrical interference picked up will be cancelled out. One pair of wires transmits from a master device and will often send commands. The slave devices transmit on the master's receive pair and will only transmit when

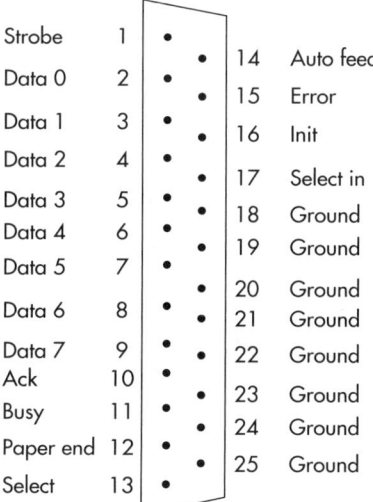

Figure 8.5 A 25-way 'D' connector.

asked. If there is more than one slave device, each slave transmitter must have tri-state control. Tri-state control effectively disconnects the transmitter from the cable.

Receivers have a common-mode input voltage range of at least ±7 V; provided that each unit's earth is within the 7 V range, a direct low-impedance connection between the two units is not necessary. Therefore ground loops do not cause problems.

Signalling voltages are similar to TTL level, but currents are higher, so each pair can be resistively terminated with matching resistors to absorb reflections. A 9-way D connector is often used, but there is no single wiring standard, so a data analyser is the most efficient tool for successful RS422 interface work. To further confuse the engineer, each pin's nomenclature varies from device to device: terms such as RX+, RX−, TX+, TX−, A, B, X and Y are often found (Figure 8.6).

The recommended standard 485 card

RS485 is a two-wire version of RS422. Again, it is differential, with signalling states of 0 V, 5 V and tri-state control. Up to around 32 devices may be connected to the single data pair; software must control each transmitter if data collisions are to be avoided. Data rates of 10 Mbps are achievable to 10 m, at 9600 Bd, over 1 km.

HP-IB

HP-IB is a parallel bus able to connect to plotters, data-loggers, etc. It is also known as GP-IB and IEEE-488. Connectors are 24-pin Amphenol. Units may be daisy-chained together so a PC may control many instruments. Distances are limited to 20 m.

IrDA

Most laptops have an infrared port, called IrDA, which is a serial data interface using infrared pulses. It is possible to purchase IrDA adapters for desktop PCs. Common IrDA data rates are up to 115 kbps, though UFIR (Ultra Fast IR) operates at 16 Mbps: 100 Mbps is under development. The range is 1 m. Ground loops do not exist, as the IrDA interface is electrically isolated.

Radiofrequency interfaces

Several licence-free bands have been allocated for ISM (Industrial, Scientific, and Medical) purposes: modules, integrated circuits, antennae, etc. are commercially available. Each band has its own characteristics with regard to bandwidth, range and duty cycle. The standards are constantly evolving, and manufacturers are adding proprietary modulation methods to raise data rates, although this can limit interoperability.

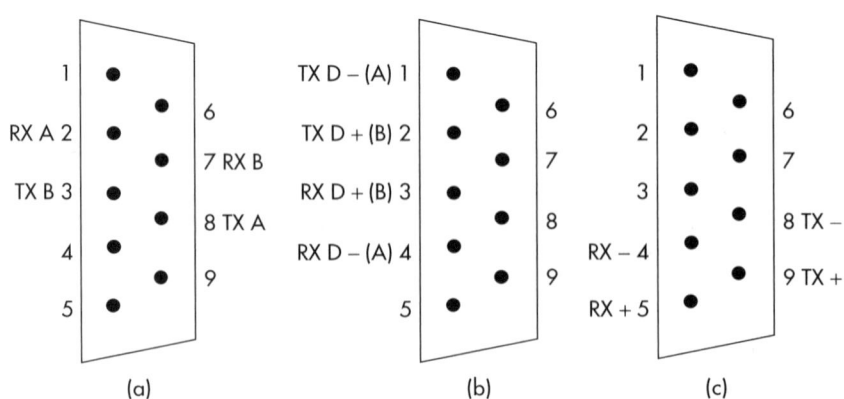

Figure 8.6 Wiring standards for RS422 cards using 9-way D connectors: (a) Sony wiring standard, (b) Moxa wiring standard, (c) Radamec wiring standard.

433 MHz. Modules are available for this band, but the possibility of interference from other users, including wireless car key-fobs and 70 cm amateur radio, rules out serious use of this frequency. Both AM (amplitude modulation) and FM (frequency modulation) are common. AM is sometimes known as 'on–off keying'.

868 MHz. A range of up to 300 m (outdoors) may be expected, and the data rate is up to 9600 bps. A segment of this small band allows low duty-cycle operation only. Addlink (part of Adcon Telemetry GmbH) make a one-channel transceiver module with 10-bit analogue channels and serial data interface. The frequency is fixed, so if more than one pair is within range of another, they should be organised to operate in separate time slots in order to avoid interference. Mantracourt Electronics offer a similar module measuring 35 mm × 42 mm × 14 mm with a built-in 24-bit resolution converter and strain gauge interface. It has 36 channels, but at only 10 Sa s^{-1}, it is far too slow for our application.

915 MHz. This is the American equivalent of the European 868 MHz band.

ZigBee™/IEEE 802.15.4. This standard operates on the 2.4 GHz band and offers data rates of 250 kbps, at a range of up to 20 m. ZigBee is designed for very-low-power operation, with a battery life of at least a year. Because the 2.4 GHz band is quite wide, it is possible to replace the modulation method by orthogonal frequency-division multiplexing (OFDM), in which many carrier frequencies are utilised. Some of these may be cancelled out if multi-path reflections from various objects arrive at the receiver's antenna exactly out of phase. OFDM sends extra information to enable the receiver to correct for this condition. Chipcon design integrated circuits for this standard such as the 7 mm × 7 mm CC2430. Their latest device, the CC2500, is only 4 mm × 4 mm. The voltage controlled oscillator components are integrated into the package. These devices are available from Texas Instruments.

Wireless Ethernet 802.11. This standard is typically used to network together computers such as laptops, and is ideal for sending large quantities of data in short bursts. Data rates vary dynamically to enable reliable throughput and the standard is being improved every year with a higher maximum data rate denoted by the suffix letter. Currently, data rates are 11, 54 and 125 Mbps: the band is at 2.4 GHz. Two relatively large integrated circuits are required for this implementation. Wireless Ethernet is not designed for low-power operation.

Commercially available analogue-to-digital cards

A vast range of commercially available cards exists, so just three will be covered here to give an idea of what is current practice.

USB analogue converter

One of the advantages of USB devices is that they may be hot-swapped, that is they may be connected and disconnected while the PC is running. A typical USB ADC is Data Translation's DT9800. It is available in both 12- and 16-bit versions and has 16 channels that may be used as eight differential channels. One converter operates at 100 kSa s^{-1}; therefore, the sample rate per channel is this rate divided by the numbers of channels being used. A 512-entry 'first in, first out' (FIFO) memory is used to reduce the likelihood of data loss by the host computer. To prevent earth loops, this unit has built-in isolation.

PCI analogue card

Many vendors sell PCI plug-in ADC cards. The PCI bus has superseded the old PC-ISA bus. The PCI bus is faster and also independently addressable, so each card does not have to be individually set up with DIP switches. An example of this type of card is the PCI 260+. It can sample at 500 kSa s^{-1} with 16-bit resolution and has 16 single-ended inputs, which may be reconfigured as eight differential inputs. Because the inputs are multiplexed into a single ADC chip, the sampling rate per channel reduces as more channels are selected. Differential inputs should be used if possible for highest accuracy if the signals have to travel from one board to another. Some form of anti-alias filtering will have to be used unless there is no energy in the input signals above the Nyquist frequency.

The card has a FIFO memory, so the host processor can temporarily ignore the card without loss of data; an internal timer fills the FIFO with data at a constant sample rate.

Plug-in cards have their total data throughput limited by the amount of time the PC's central processor unit devotes to copying the FIFO's contents on to a hard drive. The operating system can absorb a large percentage of the available processing time, often when least convenient for incoming data streams. PCI cards are not hot-swappable!

PCI analogue card (intelligent)

One particular card that deserves a mention is the MicroStar iDSC 1816 real-time anti-alias filter card, available from Amplicon Liveline Ltd. Resolution is 16 bits, and the card has 16 Mbytes of memory. This card accepts up to 16 single-ended or eight differential analogue inputs from optional strain gauge amplifiers and includes everything required for data acquisition at up to 153 kSa s^{-1} simultaneously. This performance is a consequence of an on-board 100 MHz 486 processor and two 100 MHz digital signal processors. 'Brick wall' anti-alias filters are included, which means they have a very steep, almost vertical, attenuation curve.

At these sampling rates, data are very likely to be lost if the host computer is not running an efficient real-time operating system, so either a large FIFO memory or an on-board processor and memory must be used. FIFO memory depth is of the order of a few thousand bytes in size, which is not large enough to store, say, one second's worth of data. The design of this card uses 16 Mbytes of random access memory (RAM) that easily stores that quantity of data. Fastest sample time is 7 μs, which is one or two orders of magnitude faster than is needed for the hypothetical tablet press under consideration; frequencies of up to 50 kHz should be digitised accurately.

Of course, this powerful card is not inexpensive; it comes at a price that is approximately 10 times the cost of the PCI 260+.

Software

Commercial ADC boards intended for computer interfacing are normally provided with a certain amount of software. The simplest examples have the basic ability to store incoming data as a text file, while more elaborate versions may have full analysis and display capability. Once data have been stored in the computer's memory, it is then possible to process the information, and one of the easiest ways is to use a spreadsheet program such as Excel.

Excel is reasonably 'intelligent' in that if one tries to open a text file, it will ask if the text file is space, comma or tab delimited. If the interface software is able to store the incoming information in delimited format, Excel will fill in the spreadsheet's cells automatically.

The next step is to normalise bridge readings. Each reading can be divided by the bridge's excitation voltage, provided that the advice given above has been taken.

Temperature readings can also be used to make corrections if any sensors are temperature dependent. Excel allows a mathematical equation to apply over an entire column, with the results generating a new column as in normal spreadsheet practice.

Once temperature and bridge excitation voltage compensations have been applied, variables such as force may be calculated in absolute units if each sensor can be calibrated. A column, or a part column, can be highlighted by two clicks of the computer's mouse. The highlighted region can then be displayed as a graph and expanded as necessary to show areas of interest in more detail. One click on the bar chart icon brings up the graph menu. If the first column is arranged to be an incrementing sample number, then the xy scatter graph should be chosen.

A data-analysis menu provides a selection of commands including the Fourier transform. This command uses the FFT algorithm because it will indicate if the sample length is not 2^N, or over

4096. Output is a column of half the input column's size where each cell is split into real and imaginary floating-point numbers. A macro (a small program) converts these two numbers into a single, positive power.

It is recommended that a brief search on the Internet is the best way of finding out how to perform various calculations in Excel.

Data backup

The final results will be a set of graphs displaying force against time and possibly the variation of peak force over a set of tablets. These graphs may be printed out and possibly used to adjust the press for consistent results. It may also be necessary to store the raw sampled data in order that the performance of a press can be checked from time to time for gradual wear and tear. It is often said that the data stored in a computer are worth far more than the computer itself. We shall list some methods of data storage.

Hard drive

Hard drives are fairly reliable but when they do fail, there is either a catastrophic failure or a gradual degradation of the surface. Hard drives are inexpensive on a cost per gigabyte basis, so it is common practice to have removable hard drives for archival purposes. One disadvantage is that the hardware interface is a moving target. Hard drives have evolved through different standards: ST506, IDE, SCSI (several standards) and now SATA; therefore, finding a compatible computer in years to come may prove difficult.

Different ways of organising the file allocation tables such as FAT12, FAT16, FAT32 and NTFS etc. may also cause complications for retrieval in the future.

Floppy disk

Historically, there have been three successful standards, 8, 5.25 and 3.5 inch, but only the 3.5 inch 1.44 Mb disk remains in use. Floppy disks are not very reliable, and most current PCs no longer provide a 3.5 inch drive as standard.

Zip drive

The Zip drive is a more robust version of the floppy disk and also has a considerably greater capacity. The first Zip disks held 100 Mb; nowadays this has been increased to 250 Mb, and also 750 Mb. This storage medium is relatively expensive. Even though the case is substantial and should protect the disk inside, some users have reported mechanical failure.

Tape drives

Magnetic tape has been used as a storage medium for computer data for a long time. For PC use, the tape is held in a cartridge of some description, so manual threading is not necessary. Recording density has increased over the years through improvements in the magnetic properties of the tape and the use of smaller tape heads. Tape width and cartridge size have changed with these improvements. Figure 8.7 shows a selection of different cartridges, roughly in date order. It was considered important to re-tension tapes from time to time to prevent magnetic feed-through if tapes are stored in an archive, as opposed to being used in a cyclic backup regime. As with the hard disk, different index schemes are used, so not only must a compatible drive be used but also the correct software.

Compact disks

Compact disks CD-R, CD-RW etc. allow storage of up to 700 Mbytes at an economical cost. As this technology is relatively new, it is too early to predict long-term reliability; however, there have been reports of failures in mass-produced CDs caused by ink gradually seeping into the central layers, and sunlight also may erase data. CD-R and CD-RW disks generate smaller signal levels when being read; these levels may be too low for older CD-ROM drives.

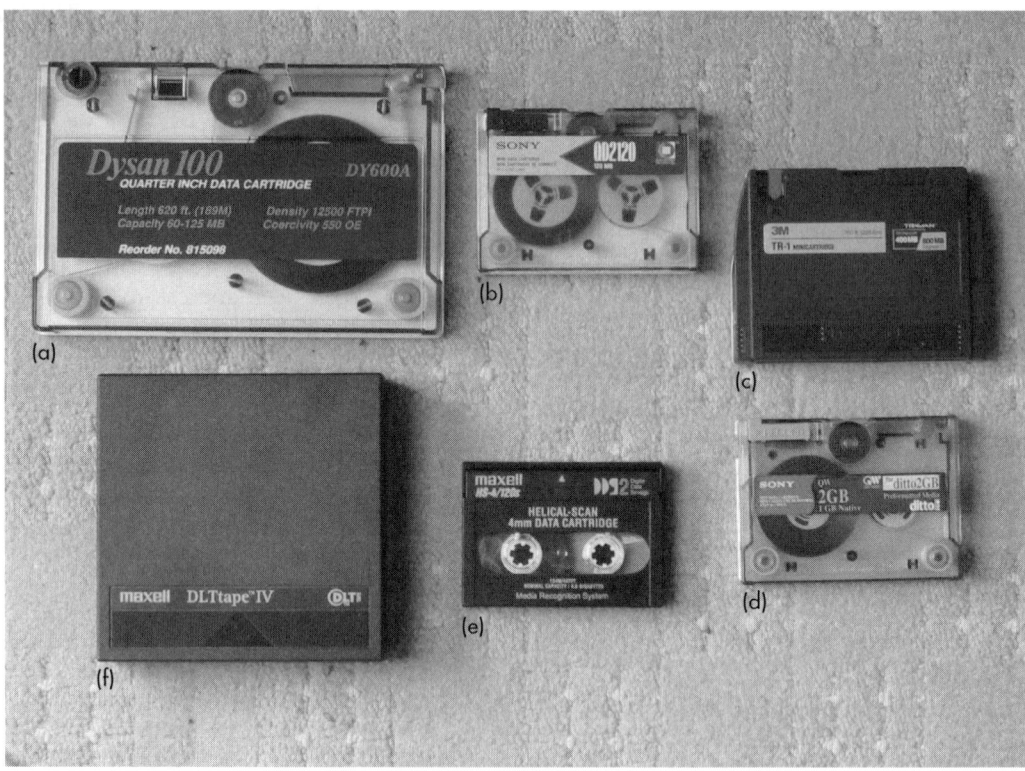

Figure 8.7 A selection of magnetic tape cartridges. (a) DY600A, 60 Mbytes quarter inch cartridge (QIC); (b) QD2120, 120 Mbytes, also quarter inch tape, but the cartridge is much smaller; (c) Travan TR-1, 400 Mbytes, 8 mm tape width, (d) QW2GB, 1 Gbyte, 8 mm tape width; (e) HS-4/120S, 4 Gbytes, DDS2 for rotating digital audio tape; (f) DLT tape IV, 40 Gbytes, 12.5 mm tape width.

Digital versatile disks (DVDs) may also be written to in the same way, but storage capacity is 4.7 Gbytes for a single-sided disk. DVD-RAM disks may be rewritten up to 100 000 times but are not compatible with all drives; data are organised in 2048 byte sectors similar to those of hard disks. An 'Archive Gold' CD-R designed specifically for long-term storage, when carefully stored, has an expected life of 100–300 years.

Magneto optical (MO) disks

Magneto optical (MO) disks were designed for long-term archival storage, where a read/write CD disk is protected by a plastic cartridge. When a data bit is to be written, a spot on the surface is heated above its Curie point by a small laser beam. Then, a 'write' pulse is fed into the relatively large magnetic write head. The magnetic field only affects the state of the hot material. In this way, the recording density is far higher than could be achieved by conventional magnetism alone. MiniDisks operate on the same principle.

Write speed is slow, because every sector is compared against the buffer on the next rotation of the disk. The cartridge is twice as thick as a 3.5 inch floppy disk, but otherwise is of the same length and width. Lack of compatibility, expensive media and drives, and slow write speeds have so far prevented widespread acceptance of what may be the most reliable backup device. Disks of 128, 230 and 640 Mbyte capacity are available. All of these are the same size.

Flash solid-state memory

Flash memory relies on storing electrical charge on the gate of a transistor. The gate to source resistance is so high that it would take many years for the charge to leak away. Flash memory integrated circuits are the basis of Compact Flash, and other size memory cards. Some memory cards plug into PCMCIA sockets and others USB. Sizes range from 8 Mbytes to 2 Gbytes. There is a limit on how many times each block of memory can be erased: from 10 000 to 1 000 000 times, depending on the technology used.

Further reading

Kosler W (ed) (2004). *Analog–Digital Conversion*. Norwood, NJ: Analog Devices Inc.

Williams T (2007). *EMC for Product Designers*, 4th edn. Oxford: Newnes.

9

Applications of tablet press instrumentation

N Anthony Armstrong

Introduction

There are three principal areas in which tablet press instrumentation is used. The first of these is the automatic control of tablet weight. This depends on the relationship between the compression force and the tablet weight. The force used to compress material into a tablet is proportional to the density of the tablet. Hence, if it can be assumed that the volume of the tablet is constant, then force is proportional to weight. Consequently, by linking the transducer used to measure force to the device for adjusting tablet weight, automatic control of tablet weight in a production environment becomes feasible. Many presses are fitted with automatic weight-control equipment, and this topic is dealt with more fully in Chapter 11.

The second main application of press instrumentation is to establish the relationship between compression force and tablet properties. Virtually all tablet properties, such as crushing strength, tensile strength, friability, disintegration time and dissolution time, are dependent on the force that was applied during the formation of the tablet. It follows, therefore, that knowledge of that force is essential if these properties are to be studied in a meaningful manner. The instrumentation that is needed for this is relatively simple, since only the maximum punch force needs to be measured. If the transducer output is linked to some form of visual recorder, then measurement of the peak height of the upper punch trace followed by reference to a calibration curve is perfectly adequate.

Figure 9.1 shows a series of outputs of transducers fitted to the upper punch of an eccentric press. The output is linked to a recorder, and the height of the trace is a function of both the applied force and the degree of amplification of the transducer output, the former being controlled, in turn, by the setting of the eccentric cam of the press. Alternatively if the transducer output is interfaced with a computer, then the maximum output is detected, a calibration factor applied and the maximum force calculated.

There are numerous applications described in the literature in which compression force measurements have been used to evaluate the effects of changes in formulation and processing variables on tablet properties. For example, Du and Hoag (2003) characterised the properties of multi-vitamin tablets with respect to force of compression. The influence of compression force on the properties of tablets made from new direct compression excipients were investigated by Armstrong (1998) and Olmo and Ghaly (1999). Information such as that shown in Figure 9.1 can also be used as an indication of the uniformity with which the formulation flows into the die. Since the compression force is proportional to the weight of material in the die, any variation in the latter will be reflected by an irregular trace height. Billardon *et al.* (1987) used an instrumented tablet press to devise optimal tablet formulations, employing the principles of factorial design and response surface methodology.

The third major application of instrumented tablet presses is to try to establish the underlying principles whereby a particulate mass is turned into a tablet. This, in turn, can be used to characterise the behaviour of solids under compression and perhaps predict the ease with which they can be transformed into tablets.

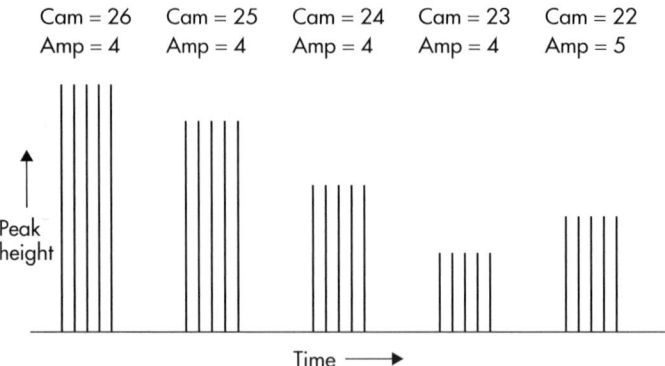

Figure 9.1 Upper punch force transducer output at several force settings on an eccentric press. Cam, setting of the eccentric cam of the press; Amp, amplifier setting.

Punch displacement–time profiles

The output of transducers fitted to tablet presses are usually obtained as a function of time. If the instrumentation is linked to some form of chart recorder, then a trace is obtained with the transducer output as the ordinate and time as the abscissa. The detail shown on the trace will be dependent on the chart speed. A very slow chart speed will give the output as a vertical line, the height of which is proportional to the maximum value of that output, as shown in Figure 9.1. An increase in chart speed will stretch out the time axis, thereby showing in more detail how the transducer output changes with time.

If the transducer output is interfaced with a computer, then the sampling speed of the latter will govern the amount of information collected. It should be noted that a measurement of transducer output might not coincide with the maximum value of the output, and therefore the maximum value would not be recorded. A rapid sampling rate will avoid this possibility (see Chapter 8).

The two parameters of principal interest in tablet presses are force and punch displacement, and both of these are obtained as a function of time. Displacement–time profiles will be considered first. In theory at least, these can be obtained in two ways. They can be predicted by knowledge of the dimensions of some components of the press or they can be obtained by recording the variation with time of the output of a displacement transducer that records punch movement. The shape of the force–time profile, by comparison, is strongly influenced by the properties of the substance being compressed and cannot be reliably predicted.

The patterns of punch movement with respect to time differ markedly between eccentric presses and rotary presses.

Displacement–time profiles of eccentric presses

Figure 9.2 represents the drive shaft, eccentric sheave, eccentric strap, lower bearing and upper punch holder of an eccentric press. The position of any component below the lower bearing, for example the tip of the upper punch, is given by Equation (9.1), which was derived by Armstrong et al. in 1983.

$$y = a + r \sin(90 + \omega t) + \sqrt{l^2 - r^2 \cos^2(90 + \omega t)} \quad (9.1)$$

where y is the position of the punch tip, a is the distance between the centre of the lower bearing and the punch tip (a constant), l is the length of the eccentric strap, r is the radius of the eccentric sheave, t is time and ω is the angular velocity of the shaft of the press. As the eccentric rotates, the angle θ is made with the horizontal, as shown in Figure 9.2.

When θ is 270°, the tip of the upper punch is at its highest level (point A in Figure 9.3), equal to $a + (l - r)$. As θ increases from 270°, the punch begins its descent, and reaches its lowest point B, equal to $a + (l + r)$ when θ is 90°. At point C, the punch tip enters the die, and if there is a solid present in the die, a force will be exerted by the upper punch on that solid. At point B, the punch changes direction from a downward movement to an upward movement, and it leaves the die at point D.

The depth of penetration of the upper punch tip into the die is altered by means of an eccentric cam. This is usually connected to an arbitrary scale and the higher the scale setting, the deeper the punch penetration. Deeper punch penetration means that point C occurs earlier in time and point D later. It must be stressed that the scale on the cam acts purely as a reference point and is neither a measure of punch position nor of the force that is being exerted. The latter depends, in addition to the cam setting, on the position of the lower punch and on the solid being compressed. Therefore, though a higher cam setting means that a higher force is being exerted, force cannot be predicted from the cam setting unless the lower punch setting is kept constant and a solid of identical composition is used on each occasion.

For a Manesty F3 eccentric press, l is 204 mm. The value of r is governed by the position of the eccentric cam, but if a mid-range value of 25 mm is selected, then changes in the position of the upper punch at a range of values of θ can be calculated. The distance of the upper punch position from its maximum displacement is shown in Table 9.1. The highest position of the punch tip ($a + 179$ mm) occurs when θ is 270° (point A) and the deepest penetration ($a + 229$ mm) occurs when θ is 90° (point B). There is, therefore, a difference of $2r$ (50 mm) between the highest and the lowest positions of the punch tip. The distance of the punch tip from its point of deepest penetration as θ changes is shown in Table 9.1.

Figure 9.3 shows the almost sinusoidal pathway that the upper punch tip describes. It should

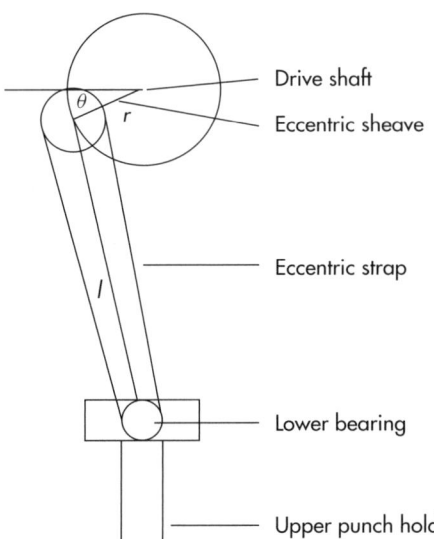

Figure 9.2 Drive shaft, eccentric strap, eccentric sheave and upper punch assembly of an eccentric tablet press. θ, angle made by eccentric with the horizontal.

Table 9.1 The upper punch tip position of a Manesty F3 eccentric press

Angle of eccentric with the horizontal, θ (°)	Distance from maximum punch displacement (mm)
270	50.0
290	48.7
310	44.8
330	38.7
350	30.8
0	26.5
10	22.1
30	13.7
50	6.5
55.1	5.0
70	1.7
90	0.0
110	1.7
130	6.5
150	13.7
170	22.1
180	26.5
190	30.8
210	38.7
230	44.9
250	48.7
270	50.0

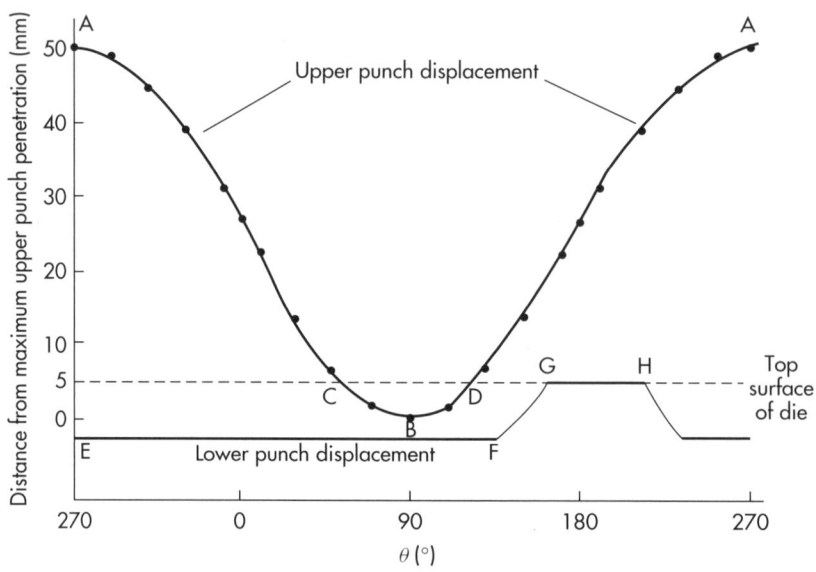

Figure 9.3 The change in the positions of the upper and lower punches over time in an eccentric tablet press. θ is the angle the eccentric makes with the horizontal. Upper punch: A, the tip of the upper punch is at its highest level; B, lowest point of the punch where it changes direction from a downward movement to an upward movement; C, punch tip enters the die; D, punch leaves the die. Lower punch: E, lowest position; F, initiation of rise in lower punch; G, tip of lower punch level with the top surface of the die: H, recommencement of punch descent.

be noted that the displacement trace is symmetrical about its axis at θ of 90°. The graph also has a smooth curve in that as soon as the punch stops descending, it commences its ascent. Consequently, punch speed is zero only for that instant when reversal of direction takes place. In the above discussion, the position of the punches is defined in terms of the angle θ and so is independent of the rate at which the press operates (i.e. its output).

If the rate of rotation of the press is known, then the units of the abscissa in Figure 9.3 can be changed from 'degrees' to 'time', since 360° equals the time taken for the press to complete one cycle. Figure 9.3 now becomes a plot of distance against time, and hence the slope of the line is a measure of punch speed at that point. This aspect will be considered in more detail later in this chapter.

In an eccentric press, compression takes place entirely via the downward movement of the upper punch, the lower punch remaining stationary as compression occurs. When the upper punch is at its highest level, the lower punch has already fallen to its lowest position (point E). This is defined by the volume in the die above the lower punch tip that is needed to accommodate the amount of solid that will in turn give a tablet of the required weight. After maximum upper punch penetration has occurred, the upper punch rises, and shortly afterwards, the lower punch also begins to rise (point F) until its tip is level with the top surface of the die (point G). This movement ejects the tablet from the die. When the tablet has been swept away by the leading edge of the feed shoe, the lower punch commences its descent (point H) to its former low level.

The duration of the several stages of punch movement is controlled by the rate at which the press operates (i.e. its output). Thus at an output of 60 tablets min^{-1}, the whole cycle takes 1 s. If the output is reduced to 30 tablets min^{-1}, then the duration of each stage is doubled. The time interval between points C and D depends both on the speed of rotation of the press and on the required depth of penetration of the upper punch into the die. If a value for the latter of

5.0 mm is arbitrarily chosen, this is achieved when θ is 55.1°. Therefore C to D represents a rotation of $2 \times (90 - 55.1) = 69.8°$, which is approximately 19% of the whole cycle. Hence at a rate of 60 tablets min^{-1}, the upper punch tip is within the die for a period of approximately 0.19 s per cycle.

Charlton and Newton (1984) also derived an equation describing the movement of the upper punch tip of an eccentric press. Their mathematical treatment differed from that used by Armstrong et al. (1983), notably in their definition of θ. Hence their equation differs from Equation (9.1) but gives identical results.

Displacement–time profiles of rotary presses

The movement of a punch tip in a rotary press is more complicated than that of an eccentric press, and it differs from the latter in two important respects. Firstly, compression in an eccentric press is exerted solely by the downward movement of the upper punch, the passive lower punch remaining stationary until after the compressive event has been completed. In a rotary press, compression is achieved by the simultaneous downward movement of the upper punch and the upward movement of the lower punch, so that the tablet is compressed from both faces. Secondly, the punches of an eccentric press move in only a vertical dimension, whereas in a rotary press, the punches move in three dimensions – vertically but also in two horizontal dimensions as the die table rotates.

Equations describing the movement of the punches of a rotary press as they pass between the pressure rolls have been described by Rippie and Danielson (1981) and Charlton and Newton (1984). Though the equations derived by these two sets of workers appear dissimilar, the methods used in their derivation are almost identical. Charlton and Newton pointed out that the analysis of the compression process in a rotary press is complicated by the relative movements of the two punches. Calculations were simplified by the assumption that both punches move equal amounts in opposite directions. However, since the position of the upper pressure roll is fixed, and that of the lower roll can be changed, since it forms the mechanism for changing the compression pressure, this assumption is not necessarily true in practice.

Figures 9.4 and 9.5 show the spatial relationships between punch head, pressure roll and die table of a rotary tablet press from above and from the side, respectively. The radius of the pressure roll is r_1, and that of the circle on the die table in which the dies rotate is r_3. The punch head is initially considered to be a hemisphere of radius of curvature r_2.

The horizontal movement of the punches is straightforward, in that they describe a circular motion, the radius of which is r_3. Movement in a vertical dimension is more complex. The upper punch is lowered by coming into contact with a cam track. If a force is exerted at all on the die contents at this stage, its magnitude is only that of the weight of the punch and can be ignored. Then when angle θ becomes small enough, the punch head comes into contact with the pressure roll and active compression begins. Punch movement is now controlled by the radii of curvature of the pressure roll and the punch head. A line joining the centres of curvature of these two components will always pass through the point of contact between them. The vertical displacement of the punch (z) is given by Equation (9.2).

$$z = \sqrt{[(r_1 + r_2)^2 - x^2]} \quad (9.2)$$

where x is the horizontal distance between the centre line of the pressure roll and the point of contact between the roll and the punch head. The distance between the centre line of the pressure roll and the centre line of the punch (x_1), can be calculated from Equation (9.3).

$$x_1 = 2r_3 \sin \varphi/2 \quad (9.3)$$

when φ is the angle between the punch axis and the vertical centreline of the pressure roll. When φ is small, Equation (9.3) reduces to

$$x_1 = r_3 \sin \varphi \quad (9.3a)$$

Combination of Equations (9.2) and (9.3) gives Equation (9.4)

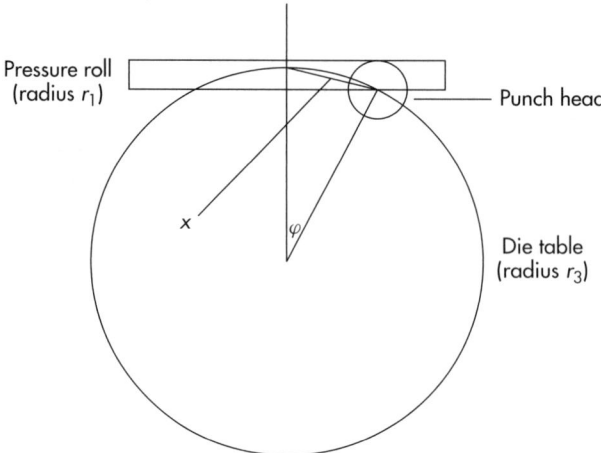

Figure 9.4 Spatial relationships between punch head, pressure roll and die table of a rotary tablet press, viewed from above.

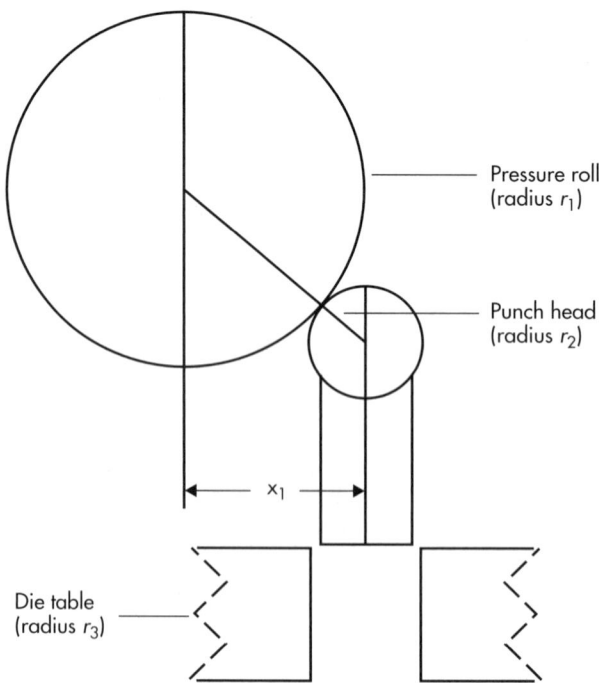

Figure 9.5 Spatial relationships between punch head, pressure roll and die table of a rotary tablet press, side view.

$$z = \sqrt{[(r_1 + r_2)^2 - (r_3 \sin \varphi - x_2)^2]} \quad (9.4)$$

Vertical punch movement is described by Equation (9.4) provided the punch head has a radius of curvature of r_2. However, the overwhelming majority of punches have a flat area of radius r_4 on their heads and a curved periphery of radius of curvature of r_2 as before. The pressure roll will come into contact with the flat area of the punch head when $r_3 \sin \varphi = r_4$ (Figure 9.6). Punch displacement is now constant while the pressure roll passes across a flat surface of diameter $2r_4$. During this period, the centre line of the punch passes beneath the centre line of the pressure roll.

There is no fixed connection between the pressure roll and the punch head, so after compression has ceased, the punch will remain in contact with the tablet until it comes into contact with another cam. This actively lifts the punch from the die. Thus unlike an eccentric press, the punches of a rotary press do not necessarily have displacement curves that are symmetrical about the point of maximum compression.

The pattern of upper punch movement can be generated for any rotary press provided the following dimensions are known: r_1, the radius of the pressure roll; r_2, the radius of curvature of the punch head; r_3, the radius of the circle in which the punches and dies rotate; r_4, the radius of the flat area on the top of the punch head. For a Manesty B3B 16-station press, the relevant dimensions are as follows: $r_1 = 102$ mm; $r_3 = 120$ mm. The punch head dimensions are assumed to be $r_2 = 12$ mm and $r_4 = 3$ mm. Punch movement using these dimensions is shown in Figure 9.7 and Table 9.2. The flat area of the punch head is reached when $r_3 \sin \varphi = 3$ mm, φ then being equal to 1.43°.

The depth in the die to which the lower punch of a rotary tablet press descends is controlled by the weight-adjusting cam. After the die has been filled, rotation of the die table causes the head of the lower punch to come into contact with the lower pressure roll. The punch rises, and there then follows a period when no further upward movement of the punch occurs as the flat portion of its head comes into contact with the pressure roll. Then the punch falls slightly under the influence of its own weight until it is moved upwards by the ejection cam. This upward movement continues until the lower punch tip is level with the top of the die.

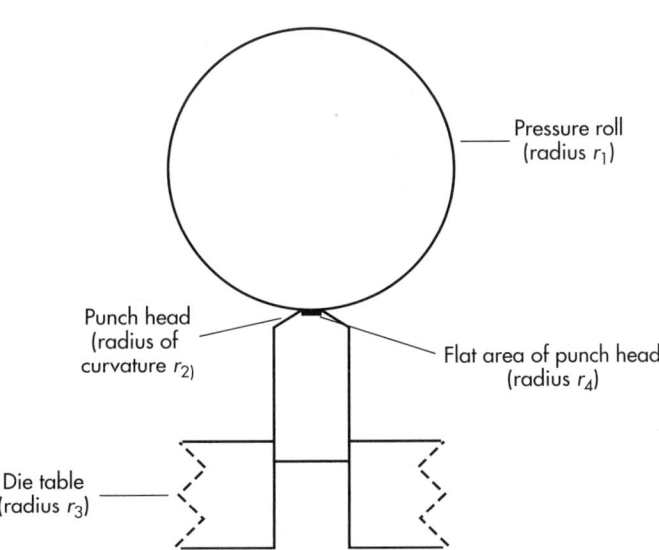

Figure 9.6 Spatial relationships between upper punch head, pressure roll and die table of a rotary tablet press; side view showing the flat area on the punch head.

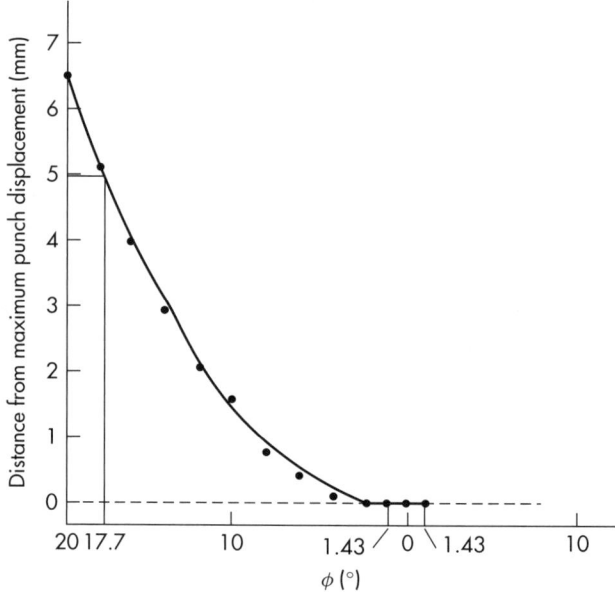

Figure 9.7 The pattern of upper punch movement of a Manesty B3B 16-station rotary tablet press.

The tablet is removed by the sweep-off arm and then the punch descends for the die to be refilled.

The factors that control the movement of the lower punch are the same as those that apply to the upper punch. The axes of the two pressure rolls may not be precisely vertically opposed to each other because, although the force-adjustment control of a rotary press raises or lowers the lower punch, in addition it slightly moves the pressure roll in a horizontal direction. Therefore, the start and end of the compression event may differ a little between the upper and lower punches. However, the duration will be the same provided the geometries of the upper and lower punch heads are identical.

With some rotary presses, the contents of each die are compressed twice since the press has two sets of compression rolls. The additional pair of rolls is called the pre-compression rolls, and the patterns of punch movement as the punches pass them are controlled by the same factors as discussed above.

The period over which the displacement of the punches in a rotary press remains constant is termed the dwell time. There is no dwell time in an eccentric press since displacement is only constant for the instant when punch movement changes from a downward to an upward direction. For any given rotary press operating at a specified frequency, the duration of the dwell time is directly proportional to the diameter of the flat area on the punch head. This can differ between different punch manufacturers. Also with some punch manufacturers, the punch length is brought into specification by grinding down the punch head, so in theory at least, the dwell time can vary within a set of punches made by the same manufacturer. For a Manesty B3B rotary press rotating at its maximum frequency of 44 rpm, and with a punch head flat of radius 3 mm, the duration of the dwell time is 10.8 ms.

Though the overall patterns of punch movement in a given rotary press is independent of the frequency of rotation of the turret of the press, the latter controls the duration of each event in the compression cycle.

Table 9.2 Punch tip position of a Manesty B3B 16 station rotary tablet press

$\varphi(°)$	Distance from maximum punch displacement (mm)
20.0	6.50
19.0	5.82
18.0	5.17
17.7	5.00
17.0	4.57
16.0	4.00
15.0	3.47
14.0	2.97
13.0	2.51
12.0	2.09
11.0	1.71
10.0	1.36
9.0	1.06
8.0	0.79
7.0	0.55
6.0	0.36
5.0	0.20
4.0	0.09
3.0	0.01
2.0	0.01
1.43	0.00
1.00	0.00
0.00	0.00
1.00	0.00
1.43	0.00

Punch shortening and press distortion during compression

It must be stressed that the calculations of punch movement described above are based solely on machine dimensions. They are not obtained by actual displacement measurements and are valid only when the dies are empty. If a solid is present in the die, then compression forces will be generated. These will have the effect of shortening the punches and also stretching the machine frame; both these will alter theoretically derived patterns of punch displacement. Twisting of the press can also occur in those presses, such as the Manesty F3, where there is only one connection between upper and lower punches via the frame of the press. Presses with symmetrical connections, including some eccentric presses and all rotary presses, are less prone to this type of distortion.

It is instructive to calculate the extent to which punches are subjected to shortening during compression. Figure 9.8 represents the lower punch of a rotary press and its dimensions. For convenience, the punch can be divided into two cylinders, representing the punch body and the punch tip, and the deformation of each part considered separately. A maximum load of 4 tonnes (40 kN) is assumed, and a realistic value of Young's modulus for the steel from which the punch is made is $21 \times 10^{10}\,\text{N}\,\text{m}^{-2}$.

Stress is force per unit area and strain is elongation per unit length. Young's modulus is defined as stress/strain, and therefore strain is stress divided by Young's modulus.

The diameter of the punch body is 19 mm and, therefore, the cross-sectional area is 283.5 mm^2 or $283.5 \times 10^{-6}\,\text{m}^2$. Therefore the strain is $4 \times 10^4 / [(283.5 \times 10^{-6}) \times (21 \times 10^{10})]$ or 6.72×10^{-4} strain (672 microstrain). Multiplying this by the length of the punch body (101.4 mm) gives a resultant change in length of approximately 68.1 μm. A similar calculation for the punch tip gives a deformation of approximately 83.1 μm. Therefore the total deformation of the whole punch is $(68.1 + 83.1) = 151.2$ μm, an amount that is of a similar order of magnitude to some of the punch displacements shown in Table 9.2.

A deformation of this size will involve a corresponding error in the measurement of punch tip position if this were to be estimated from geometrical considerations alone. There will also be a similar error in estimating the position of the tip of the lower punch, and a further error caused by deformation of components of the press itself. However, the extent of all these deformations is proportional to the applied force, and so a knowledge of the latter can be combined with the appropriate value of Young's modulus to apply a correction factor to the punch tip displacement measurements. Altaf and Hoag (1995) carried out a comprehensive review of deformation occurring in a Stokes B2 rotary press. All deformations were found to be elastic, and their magnitudes could be predicted by use of the appropriate Young's modulus.

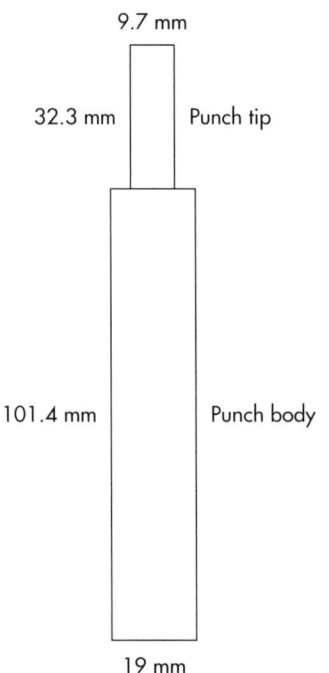

Figure 9.8 Diagrammatic representation of the lower punch of a rotary tablet press.

Oates and Mitchell (1990) showed that the predicted patterns of punch movement based on punch and press geometry differed considerably from those obtained experimentally using a press fitted with displacement transducers. However, provided allowance was made for punch and press distortion, excellent agreement was obtained between predicted and experimental results. These workers suggested that, when the difficulties associated with the practical measurement of punch displacement were taken into account, reliance on appropriately corrected displacements derived from punch and press geometry was, in fact, preferable. Similar conclusions were reached by Munoz-Ruiz *et al.* (1992) and Schmidt and Vogel (1994).

Force–time profiles

The force that is exerted on a particulate mass constrained in a die is a vital determinant of the properties of the resultant tablet. Hence the ability to measure the applied force is essential for a meaningful consideration of tablet properties to take place. Furthermore the shape of the force–time relationship can provide useful information about the mechanisms by which powder consolidation is taking place.

Mechanisms of powder consolidation

The process of tablet formation is essentially one of forcing an aggregate of particles ever closer together so that short-range interparticulate attractive forces can have a progressively greater effect. Thus the compression force transforms a quite highly porous mass into one of much reduced porosity.

The initial reduction of porosity is brought about by particle rearrangement, a process that requires only a very low force. The punch then encounters a resistance to its motion because further consolidation by rearrangement becomes impossible. Subsequent consolidation by application of a higher force is brought about by several possible mechanisms. Figure 9.9 shows how an individual particle may behave if it is subject to an ever-increasing force.

Initially elastic deformation occurs, and the relative change in length is proportional to the stress. This is reversible, in that the particle will immediately return to its original shape and size when the force is removed. Hence any interparticulate bonds that have formed as a result of elastic deformation are likely to be broken when the applied force returns to zero. A further increase in force will cause the elastic limit of the particle to be exceeded. Plastic or viscous flow will occur, the velocity of which is proportional to the force. Since velocity is distance divided by time, it follows that viscous movement is not instantaneous. On removal of the load, deformation by viscous flow is not reversed, and hence interparticulate bonds formed at this stage would be expected to remain intact.

A subsequent increase in force will cause the particle to fragment. This too is an irreversible process. The fragmented particles will then find new positions; this will further decrease porosity, and in consequence will favour the formation of

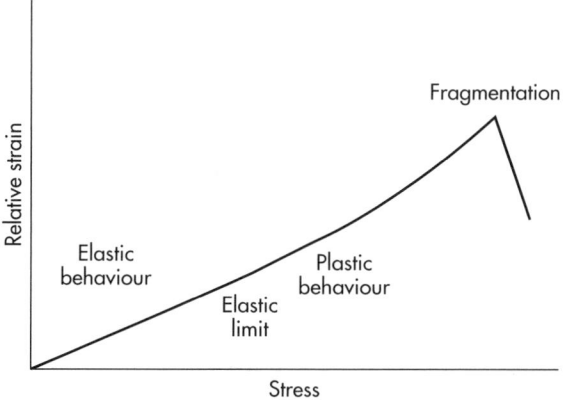

Figure 9.9 The effect of applying an increasing stress to a single particle.

interparticulate bonds because these smaller particles will have a greater number of points of contact. Most solids undergo elastic deformation, plastic deformation and fragmentation, but one consolidative mechanism will usually predominate. These mechanisms are summarised for a collection of particles in Figure 9.10.

All materials undergo elastic or plastic deformation, but the force at which some of them fragment is so low that consolidation by brittle fracture is by far the most important consolidation mechanism. Typically, inorganic materials such as calcium phosphate consolidate almost exclusively by fragmentation, whereas microcrystalline cellulose consolidates primarily by plastic deformation.

Force–time profiles derived from eccentric presses

Consider an eccentric press in which both the upper and the lower punch are equipped with force transducers. The die is filled with particulate material and the compression process begins. The change in force with respect to time for both punches is shown in Figure 9.11.

In an eccentric press, force is applied by the downward motion of the upper punch and transmitted via the contents of the die to the face of the lower punch, the latter remaining stationary. Hence the phrase 'lower punch force' refers to the force received by the lower punch rather than the force applied by it.

The first stage of consolidation is caused by particle rearrangement. This requires a very low force that is probably not detectable by the force transducers, the outputs of which remain zero. The upper punch then encounters a resistance and the output of its transducer begins to rise (point A), slowly at first and then more rapidly.

Force is transmitted through the powder bed to the lower punch and a similar rise in force is detected by the transducer located there. As maximum upper punch penetration is achieved, so force maxima are detected on both punches. That detected at the upper punch (point B) is larger than that detected by the lower punch (point C). The reasons for this will be discussed later.

Once force maxima have occurred, the upper punch begins an upward movement, and the force detected on both punches falls. That on the upper punch returns to zero when contact is lost between the ascending punch and the top surface of the tablet (point D). The force on the lower punch is still present after that detected by the upper punch returns to zero. This is because the tablet is still held in the die, from which it is ejected by upward movement of the lower punch. This results in a peak in the lower punch force profile (point E), after which the latter returns to zero (point F).

Though all eccentric presses will give the same basic force–time profile, the duration of each

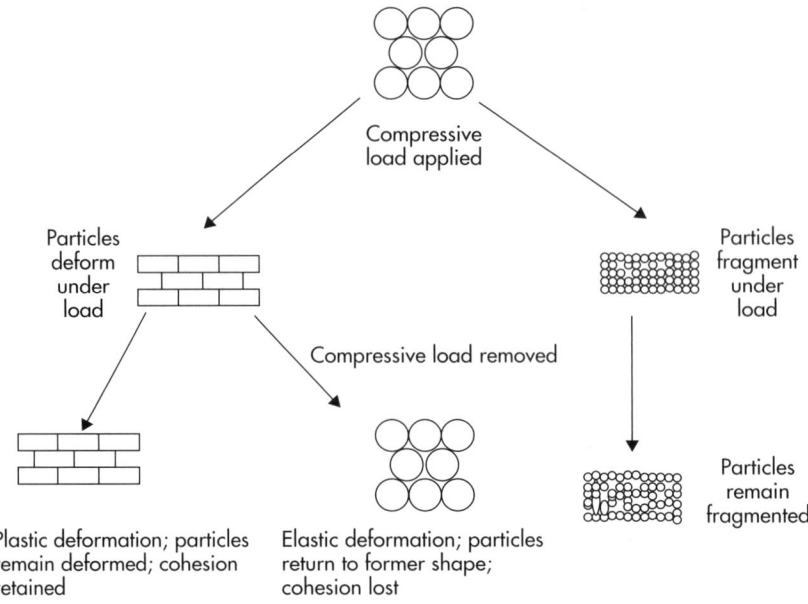

Figure 9.10 Mechanisms of consolidation of a collection of particles contained in a die and exposed to a compressive force.

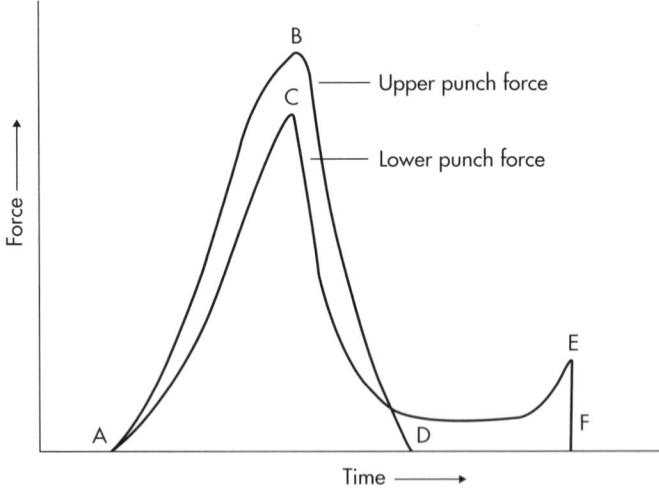

Figure 9.11 Force–time profiles for upper and lower punches of an eccentric tablet press. See text for details.

stage will depend on the speed of rotation of the press. The magnitude of the peak force is governed by the depth of penetration of the upper punch into the die.

Force–time profiles derived from rotary presses

The force transducer outputs from a rotary press are somewhat different to those from an eccentric press since both upper and lower punches are in motion during the compression stage. This means that the particles in the die are being compressed from both top and bottom at the same time, rather than the lower punch having the passive role that it does in an eccentric press. Force adjustment in a rotary press is achieved by moving the lower pressure roll. Upward movement of this causes a consequent deeper upward penetration by the lower punch within the die, and an increased force is exerted in this direction. Hence the force detected by the lower punch transducer is not necessarily lower than that detected by transducers on the upper punch.

During the dwell time, when there is a period of no punch movement, the compressive force would also be expected to be constant. This is not necessarily the case, and this point will be discussed further below. Forces generated by the tablet being retained in the die, followed by ejection owing to upward movement of the lower punch, are essentially the same as in an eccentric press.

The shape of the force–time profile

The upper punch profile shown in Figure 9.11 is not symmetrical about its peak. It might intuitively be expected that after the maximum punch penetration has been achieved, the upper punch would begin to withdraw from the die and contact between the tablet and the punch face would immediately be lost. This would result in the 'decompression' curve being a line extending vertically downwards from point B. This behaviour would be expected from a solid that had undergone fragmentation or purely plastic deformation. In both these cases, the tablet would retain a constant thickness on removal of the compressive force. However, if there has been any elastic deformation, then as the punch begins to move upwards, the tablet will expand in a vertical direction and will push against the punch face as this ascends in the die. A force will be detected for as long as the tablet remains in contact with the face of the punch, and a perfectly elastic body such as a rubber plug would be expected to give a force–time profile that was symmetrical about the maximum force axis.

When elastic expansion is complete, or when the punch is ascending quicker than the tablet can expand, contact will be lost and so the force detected on the upper punch will fall to zero (point D). This lack of symmetry about the maximum force suggests a means of distinguishing between plastic and elastic behaviour. This will be further explored below.

In an eccentric press, the maximum force applied by the active upper punch is always greater than that detected by the passive lower punch. As a force F is applied to particles in a powder bed, the topmost layer of particles will receive the force coming down in a vertical direction (Figure 9.12).

Consider the two particles labelled A and B in the figure. Particle A is not vertically above particle B and particle B is in contact with the die wall. Particle A receives the applied force and transmits it to B, the angle between the centres of A and B with the vertical being θ. The transmitted force can be resolved into vertical and horizontal components, which are $F\cos\theta$ and $F\sin\theta$, respectively.

If the angle of contact (θ) had been zero (i.e. the two particles had been vertically opposed), then $F\cos\theta$ would equal F, and $F\sin\theta$ would be zero. If however, θ exceeds zero, then $F\cos\theta$ will be less than F, and $F\sin\theta$ will be greater than zero. Therefore, the force diminishes as it is transmitted downwards through the powder bed, and, consequently, the force received by the lower punch would be less than that applied by the upper punch, accounting for the differences in peak heights B and C in Figure 9.11. Furthermore, if $F\sin\theta$ exceeds zero, a force will be exerted horizontally on the die wall. This gives rise to a frictional force between the die wall and the edge of the tablet.

The treatment shown in Figure 9.12 has been considerably simplified. However, a rigorous mathematical treatment by Train (1956) has shown that differences in vertical and horizontal force transmission can have a profound effect on the porosity of tablets and their resultant physical strength. The data also conform to the Shaxby–Evans equation (Shaxby and Evans, 1923), which predicts that the force applied by

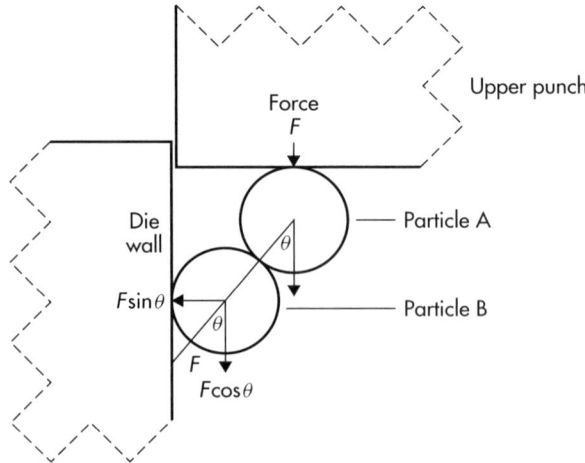

Figure 9.12 Force transmission between two particles in a powder bed.

the upper punch decays exponentially towards the lower punch at a rate that is dependent on tablet dimensions and on a constant that is substance specific.

It might be assumed that the point at which peak force is achieved during tablet compression would coincide with maximum punch penetration into the die. However, Müller and Casper (1984) found that this was not the case. Peak force was reached approximately 20 µm before maximum punch penetration with microcrystalline cellulose and approximately 10 µm with lactose. Similar results were obtained by Ho and Jones (1988a), using a compaction simulator. They found that only with totally elastic bodies such as steel or rubber plugs did coincidence between maximum force and maximum punch penetration occur. Konkel and Mielck (1997), using an eccentric press, found that the time difference between maximum force and maximum punch displacement was of the order of 5–10 µs, depending on the applied force and the substance being tabletted. The ability to measure such small differences in distance and time depends on the sensitivity of the instrumentation and, where a computer is used, on the sampling speed of the computer.

Ylirussi and Antikainen (1997) derived a number of parameters in an attempt to define the shape of the force–time profile of an eccentric press. These are listed in Table 9.3, and for full definitions, the reader is referred to the original paper. In an attempt to compensate for the differences between upper and lower punch forces, they calculated what they termed the effective force F_{eff}, which is the geometric mean of the upper and lower punch forces. They also com-

Table 9.3 Parameters used to characterise the force–time profile derived from an eccentric press (Ylirussi and Antikainen, 1997)

Upper punch force (F_{up})
Lower punch force (F_{low})
Effective force ($F_{eff} = (F_{up} \cdot F_{low})^{1/2}$)
Reflected force
Maximum of the effective force
Time of maximum punch displacement (t_{max})
Time when the maximum effective force is reached
Width of F_{eff}–time curve at 50% F_{eff} (left side)
Width of F_{eff}–time curve at 50% F_{eff} (right side)
Elasticity factor
Relative elasticity factor
Area of F_{eff} curve before t_{max}
Area of F_{eff} curve after t_{max}
Area difference of an ideal elastic and real material after time t_{max}

pared the compression and decompression force–time profiles and surmised that the decompression curve of a perfectly elastic substance would be the mirror image of the compression curve. They then derived a list of parameters similar to those listed in Table 9.3 based on the differences between this hypothetical decompression curve and that actually obtained.

There have been several attempts to characterise the shape of the force–time curve of a rotary tablet press. Probably the most comprehensive have been that carried out by Schmidt and co-workers at the University of Tübingen and Mielck and his colleagues at the University of Hamburg.

Vogel and Schmidt (1993) considered that the force–time curve could be divided into three phases: the compression phase in which the punches approach each other, the dwell time when the flat portions of the punch heads are in contact with the pressure rolls, and the relaxation phase in which the punches move away from each other.

With reference to Figure 9.13, the compression phase can be characterised both by the area A_1, the area under the curve during the compression phase, and the slope Sl_C. This is the slope of the straight line joining the points at which 20% and 80% of the maximum force is achieved. Using these limits of 20% and 80% avoids the curved regions at the extremities of the force–time profile. This is a similar parameter to the rise time originally proposed by Ho and Jones (1988b), who defined the rise time as the time interval between initiation of compression and the point of maximum compression, using a standard force of 15 kN at a constant rate of force application.

It would intuitively be expected that punch displacement and applied force would remain constant during the dwell time, but this is not necessarily true. If the material being compressed undergoes stress relaxation, then plastic flow will lead to a thinner tablet. Hence the force will fall and, in turn, the punches will penetrate further into the die and some of the elastic stress within the press will be recovered. Schmidt and Vogel (1994) quantified this behaviour by dividing the dwell time into two equal parts, the areas under the force–time profile for these two halves being A_2 and A_3, respectively. A horizontal line was then drawn through the minimum force obtained during the dwell time. The area between this horizontal and the force–time profile up to the midpoint of the dwell time was A_5 and that after the midpoint was A_6; A_5 and A_6 formed parts of A_2 and A_3, respectively. A low value of the area ratio A_6/A_5 indicated a high degree of stress relaxation. This ratio is similar in concept to a suggestion by Dwivedi *et al.* (1991) that particle behaviour under load could be defined by the peak offset time, which is the difference in time between the achievement of maximum force and the midpoint of the dwell time. Finally the relaxation phase could be characterised by the area under the decompression force–time profile (A_4) and the decompression slope Sl_D. The latter is the slope of the straight

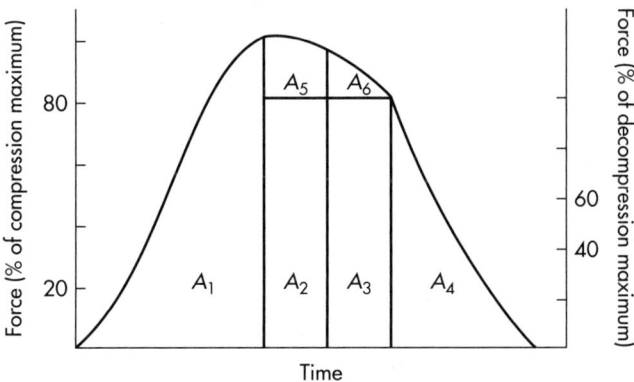

Figure 9.13 Force–time diagram from a rotary tablet press. A, area under the curve.

line connecting the points where 60% and 40% of the minimum force achieved during the dwell time are reached. These workers found that the compaction load decreased during the dwell time by plastic flow, and that brittle behaviour could be easily distinguished from plastic deformation (Vogel and Schmidt, 1993; Schmidt and Leitritz, 1997).

Dietrich and Mielck (1984, 1985) attempted to characterise the variation of applied pressure with time during the compression event. They pointed out that the asymmetry of the pressure–time curve characterises the ability of a material to undergo fragmentation under load, and they quantified this asymmetry by means of a modified Weibull function (Equation 9.5):

$$P(t) = \frac{\gamma}{\beta} \cdot \frac{t^{(y-x)}}{\beta} \cdot e^{-\left(\frac{t}{\beta}\right)^{\gamma}} \qquad (9.5)$$

This is a mathematical model in which the value of the constant γ describes the skewness of a curve. If $\gamma = 1$, then an exponential curve is described, and $\gamma = 3.3$ describes a perfectly symmetrical curve. Dietrich and Mielck found that a lower value of γ correlated with a greater extent of irreversible deformation, and they concluded that use of the Weibull function could characterise the pressure–time function using only the measurement of the applied pressure, rather than both this and the more difficult measurement of punch displacement. Konkel and Mielck (1997) later extended the use of the Weibull function to force–time relationships obtained from an eccentric tablet press.

Force–porosity relationships

Introduction

When a force is applied to particulate matter constrained in a die, consolidation occurs and porosity decreases. As described above, only a low force is needed to cause particle rearrangement, but further densification, via particle fragmentation and/or deformation, requires a much higher force. Particles are thrust into ever-closer contact with each other, interparticulate bonds form and a coherent structure results. Therefore, an essential stage in the formation of a tablet is the progressive reduction in the volume occupied by the powder particles in the die, resulting in a decrease in porosity.

Tablet porosity or pore fraction can be measured in two ways. The first is based on the dimensions of the tablet after it has been ejected from the die. Consider a solid of apparent particle density ρ_T, this usually being measured by gas pycnometry. This is sometimes called the 'true' density of the solid, but this is incorrect, since if there are any closed pores, these cannot be accessed by the gas. This solid is compressed into a cylindrical tablet of weight w, diameter d and thickness t. The volume of this tablet is $\pi d^2 t/4$ and its apparent density ρ_A is $4w/\pi d^2 t$. The relative density (ρ_{rel}) is the fraction of the volume of the tablet that is occupied by solid, and is equal to ρ_A/ρ_T. The pore fraction ε is that part of the tablet volume that is not occupied by solid. Hence $\varepsilon = (1 - \rho_{rel})$ and is given by Equation (9.6).

$$\varepsilon = 1 - \frac{4w}{\pi d^2 t \rho_T} \qquad (9.6)$$

This is called the 'out-of-die' or 'zero pressure' porosity. However, unless the tablet is made from totally non-elastic materials, the tablet will have expanded when the compressive force was removed. This expansion, by increasing t and perhaps d, will have led to an increase in porosity.

The second method by which porosity can be measured is to employ press instrumentation. If displacement transducers are fitted to both upper and lower punches, then the distance separating the faces of the punches can be calculated (see Figure 9.3). If both punches are in contact with the tablet, then the distance between punch faces is equal to the thickness of the tablet. Hence porosity can be measured while the force is still being applied and before any elastic expansion can occur. This is termed the 'in situ', 'in die' or 'at pressure' porosity. The in situ porosity includes densification caused by both elastic and non-elastic deformation. By the time the 'out-of-die' porosity is measured, elastic recovery will be complete, and only densification caused by plastic deformation and fragmentation will be assessed. The difference between the 'in situ' porosity and the 'out-of-die' porosity is thus a measure of the elastic expansion that the tablet undergoes after the compressing force is removed.

The relationship between force and porosity

A typical relationship between force and porosity is shown in Figure 9.14.

This is not an easy relationship to represent mathematically. If it is assumed that the limiting porosity is zero, then the curve will approach the abscissa asymptotically, zero porosity being attained with an infinitely high force. If the limiting porosity is other than zero, then that value too will be reached at infinite force. In other words, the abscissa will not be crossed. However, the curve will definitely cross the ordinate, since when the force is zero, the porosity will have a specific value related to the bulk density of the solid by Equation (9.7).

$$(1-\varepsilon)\rho_T = \frac{4w}{\pi d^2} \qquad (9.7)$$

The upper limiting value of the pore fraction must be less than one.

Compression equations

Notwithstanding the difficulty of representing Figure 9.14 in mathematical form, several compression equations have been devised that attempted to represent the relationship between force and porosity. It is important to realise that these equations are attempts to find a satisfactory model of experimental data and that they may not in themselves have any underlying scientific foundation.

The Heckel equation

Among the equations that have been devised to describe the consolidation process, that of Heckel (1961) has received most attention in the pharmaceutical field. In the *International Pharmaceutical Abstracts* database up to the end of 2006, there were 100 references to the Heckel equation, more than 10 times the number of references to all other compression equations put together.

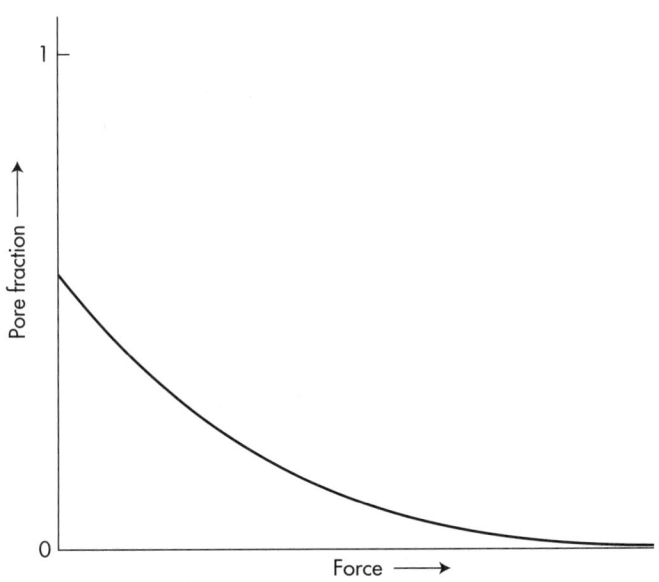

Figure 9.14 Generalised relationship between force and pore fraction.

Heckel considered the compaction of powders to be analogous to a first-order chemical reaction in which the pores were the reactant and densification was the product. Therefore, the 'kinetics' of the process could be described by a proportionality between the change in density with pressure and the pore fraction, giving Equation (9.8).

$$\frac{dD}{dP} = k(1-D) \qquad (9.8)$$

where D is the relative density of the powder mass (hence $\varepsilon = (1 - D)$); P is the pressure and k is a constant.

Integration of Equation (9.8) gives Equation (9.9), the Heckel equation.

$$\ln \frac{1}{(1-D)} = kP + A \qquad (9.9)$$

where A is another constant.

If this equation is valid, then a plot of $\ln 1/(1 - D)$ against P should give a straight line of slope k and with an intercept of A on the ordinate (Figure 9.15). In his original work, Heckel (1961) found that these plots were curved at low pressures, and he suggested that constant A was a consequence of consolidation that had occurred during die filling and particle rearrangement before fragmentation and/or deformation had taken place. Slope k, he suggested, gave a measure of the ease with which the solid underwent plastic deformation. The greater the slope, the greater the plasticity. Hersey and Rees (1971) suggested that the reciprocal of k was numerically equal to the yield stress of the powder, and this suggestion was taken further by Rowe and Roberts (1996). The latter workers pointed out that the reciprocal of k was the mean deformation stress whether it was a plastic deformation stress for materials that deform plastically, a fracture stress for fragmenting materials or a combination of both. Rowe and Roberts quoted the yield stresses of a number of solids often used in tablet formulation.

The Kawakita equation

The Kawakita equation relates the reduction in volume of a column of powder to the applied pressure. It can be represented as Equation (9.10) (Kawakita and Ludde, 1970).

$$\frac{(V_0 - V)}{V_0} = \frac{abP}{(1+bP)} \qquad (9.10)$$

where V_0 is the initial volume of the powder column and V is the volume at pressure P. Hence the degree of volume reduction C is $(V_0 - V)/V_0$, and a and b are constants.

C can be expressed in terms of porosity as Equation (9.11)

$$C = \frac{(\varepsilon_0 - \varepsilon_P)}{(1 - \varepsilon_P)} \qquad (9.11)$$

where ε_0 and ε_P are the initial porosity and the porosity at pressure P, respectively. Equations (9.10) and (9.11) can be rearranged to give Equation (9.12).

$$\frac{P}{C} = \frac{P}{a} + \frac{1}{ab} \qquad (9.12)$$

If P/C is plotted as the ordinate and P as the abscissa, then a straight line of slope $1/a$ should result, with an intercept of $1/ab$ on the ordinate.

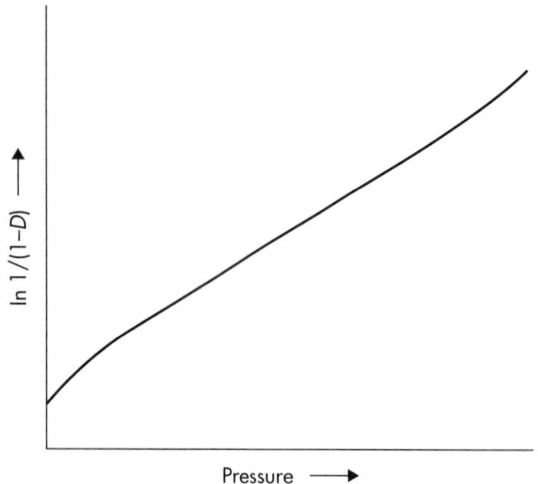

Figure 9.15 Pressure–pore fraction diagram represented by the Heckel equation (Equation 9.9).

Kawakita considered that constant a described the maximum volume reduction that could be achieved (i.e. the compressibility of the powder), and b was an indication of how easily the volume reduction was achieved. Sheikh-Salem and Fell (1981) found that the value of the constant C was very dependent on the way the die was filled. Ramberger and Burger (1985) found that the Kawakita equation proved to be a good representation of the densification of sulfathiazole provided porosity data obtained after ejection (i.e. 'zero pressure') was used.

The Cooper-Eaton equation

Cooper and Eaton (1966) considered that compaction of powders takes place in two stages. The first is the filling of voids in the particle bed by a process of rearrangement in which the voids are larger than the particles themselves. When this stage is complete, voids that are smaller than the particles are then filled by particle fragmentation or deformation. They derived Equation (9.13)

$$\frac{(1/D_0 - 1/D)}{(1/D_0 - 1)} = a_1^{-k_1/P} + a_2^{-k_2/P} \quad (9.13)$$

where D_o is the relative density at zero pressure, D is the relative density at pressure P, and a_1 and a_2 are constants that indicate the fraction of densification that is achieved, respectively, by the two mechanisms described above. Hence $(a_1 + a_2)$ should equal unity.

Therefore if $\ln[(1/D_o - 1/D)/(1/D_o - 1)]$ is plotted against $1/P$, then the values of the constants a_1, a_2, k_1 and k_2 can be calculated from the intercepts and slopes of the two linear regions of the graph. This procedure is dependent on there being two distinct linear portions of the curve, and several workers, for example Armstrong and Morton (1979), have pointed out that two such parts of the curve are sometimes not readily discernable. Chowhan and Chow (1980) compared the ability of the Heckel and Cooper–Eaton equations to describe the densification of lactose and found both were equally appropriate.

Critical consideration of compression equations

Compression equations have been widely used in attempts to characterise the behaviour of a collection of particles when a compressive force is applied; see, for example, Roberts and Rowe (1985) and Morris et al. (1996). However, several workers have drawn attention to the fact that the coefficients of the equations are very dependent on experimental conditions. For example, York (1979) collated data from several studies and showed that the coefficients of the Heckel equation were dependent on the speed at which the compression pressure was applied, the die diameter and the state of lubrication. Rue and Rees (1978) found that variations in particle size, die diameter and the rate of compaction all resulted in changes to the Heckel coefficients, and different coefficients were obtained depending on whether the porosity was measured in situ or after ejection from the die. Kiekens et al. (2004) used a compaction simulator to study the effects of compression pressure, punch diameter, punch tip shape and filling depth of the die on the yield pressure derived from the Heckel equation, analysing their results by means of response surface methodology. They found that the compression pressure was the most important factor, and that filling depth and die diameter were minor factors, though their effect was more pronounced when concave punches were employed.

Sonnergaard (1999) drew attention to the wide variation in yield pressures quoted by different authors for the same material. He pointed out that no proof exists in the literature that the linear portion of the Heckel plot describes plastic deformation or that the apparent yield pressure is the pressure at which plastic deformation commenced. He showed that for 'in situ' measurements of porosity, an error of $\pm 10\,\mu m$ in the measurement of punch displacement could give an error of over 5% in the value of the yield pressure. The slope of the Heckel plot was very susceptible to variation in experimental conditions.

Muller and Augsburger (1994) have noted how little attention has been paid to the shape of the displacement–time relationship when data derived from the Heckel equation have been generated. Using a series of differing displacement–time waveforms generated on a tablet press simulator, they obtained Heckel constants and compared them with those obtained from an instrumented tablet press. They found that for all the materials they studied, the relationship between pressure and volume was affected by the shape of the waveform. The found that a sawtooth profile (i.e. a constant punch speed) was better for Heckel analysis, but pointed out that this is not representative of punch movement in a rotary press.

It must also be borne in mind that the originators of the Heckel, Kawakita and Cooper–Eaton equations were using powders that are very much harder than those used in pharmaceuticals, and there may be a threshold material hardness below which these equations are not applicable. Sun and Grant (2001) were of the view that the Heckel equation does not accurately describe the compaction of pharmaceutical powders at high pressure, and data obtained when the pore fraction is less than 0.05 must be interpreted with caution.

Virtually all compression equations endeavour, by mathematical manipulation, to transform a curved line into one that is straight, at least over part of its length. Identifying this straight section is not easy and is often subjective. In earlier studies, in which perhaps less than 10 combinations of pressure and porosity formed the graph, it was relatively easy to identify a straight portion of the graph. However, if the instrumented tablet press is linked to a computer, a curve can be obtained with many more points, the number being determined by the sampling frequency of the computer. It is the author's experience that the more points there are on the compression curve, the more difficult it is to identify the straight-line portion.

It may be that as more data become available, more appropriate equations will be derived. An alternative approach might be to measure the change in porosity brought about by a specific change in pressure with no underlying assumptions as to the physical significance of its constants. For example Equation (9.14) is a relatively simple equation that describes data that give a line that is asymptotic to both axes or lines parallel to them.

$$(F + a)(\varepsilon - b) = C \tag{9.14}$$

where F is the force, ε is the pore fraction, and a, b and C are constants. A graphical representation of this equation is shown in Figure 9.16. The equation does not demand that porosity approaches zero at very high pressures. It accepts that a limiting porosity greater than zero can

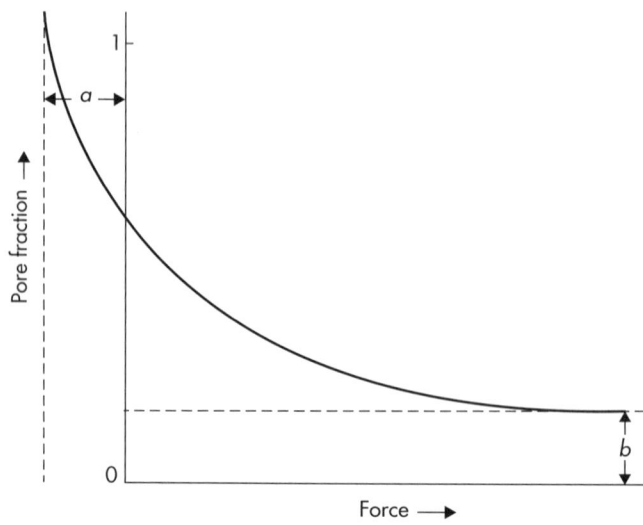

Figure 9.16 Force–pore fraction diagram representing Equation (9.14).

occur, and even a porosity less than zero can be accommodated, such situations arising when hollow particles are compressed.

The force–displacement curve

Introduction

The formation of a coherent tablet from a particulate mass involves an input of energy, provided as mechanical energy by the moving punch(es). This energy is used in a variety of ways, the relative importance of which depends on the nature of the solid.

1. Moving particles within the die to fill voids during the consolidation process, when interparticulate frictional forces and friction between particles and the die wall must be overcome
2. Deforming particles, either elastically or plastically, and also deformation of the punches and other components of the press
3. Breaking particles when the energy is transformed into surface energy as the surface area increases
4. Formation of bonds between particles.

After compression is complete, then energy used for elastic expansion will be released, and part of that will be delivered back to the punches. Other energy will be released as heat. Immediately after being compressed, tablets are warm to the touch.

Because the energy needed to compress a given solid will depend on its propensity to undergo elastic deformation, plastic deformation and fragmentation, it might be thought that the tabletting properties of a solid could be characterised by measuring the amount of energy expended in making tablets from it. This is the basis of the force–displacement curve.

Figures 9.3 and 9.11 show, respectively, the changes in upper punch displacement and upper punch force in an eccentric press with time. If the force is plotted against the displacement, then a force–displacement curve is obtained (Figure 9.17). Conventionally, force is plotted as the ordinate and punch displacement as the abscissa, with downward movement of the

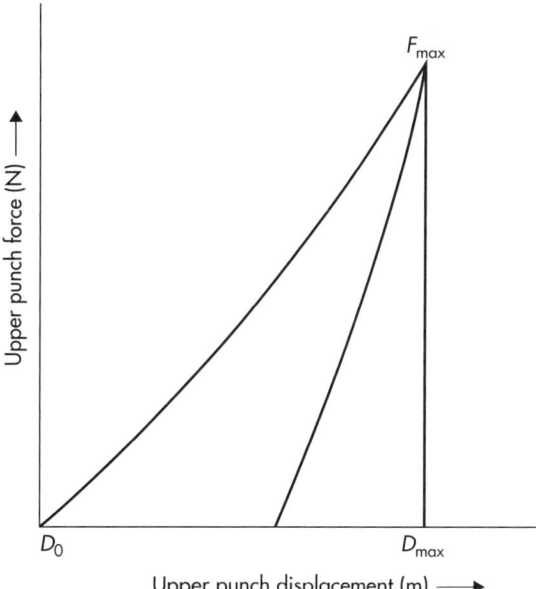

Figure 9.17 A diagram of force plotted against punch displacement. D, displacement; F, force.

punch represented by movement to the right along the abscissa.

At a displacement of D_0, the upper punch, having previously entered the die, first exerts a detectable force on the particles. Force increases to a maximum (F_{max}) at displacement D_{max}. If the powder were totally non-elastic in behaviour, then as soon as the punch begins to rise in the die after maximum penetration, the force would immediately fall to zero, and the force–displacement curve would be completed by a vertical line to the abscissa at D_{max}. A totally elastic material would retrace the whole curve back to its origin. In practice, there will always be some elastic recovery, and so the punch retreats to a point somewhere between D_{max} and D_0 before contact with the tablet is lost and the force then becomes zero.

A significant feature of the force–displacement curve is that the area under it (i.e. the area bounded by D_0, F_{max} and D_{max}) has the units of force (newtons) multiplied by distance (metres) and this is dimensionally equal to work or energy (newton.metres or joules). Therefore, the area under the force–displacement curve can be given by Equation (9.15):

$$\text{Area} = \int_{D_0}^{D_{\max}} F dD \qquad (9.15)$$

In other words, the area under the force–displacement curve will be equivalent to the work expended to produce the tablet, and as such has a physical significance. However, there has been some controversy regarding how the force and displacement data are to be treated, and precise measurements, particularly of punch movement, are especially important if meaningful information is to be obtained.

Calculation of the area under the force–displacement curve

Nelson et al. (1955) were the first workers to describe the construction and significance of the force–displacement curve. However, it was not until the 1970s that a group led by Polderman at the University of Leiden in the Netherlands made a detailed study. Their work is of added significance in that it is the first published report of an instrumented tablet press being directly linked to a computer (de Blaey and Polderman, 1971). Like most studies in this area, they used an eccentric press fitted with force transducers on both punches and displacement transducers that enabled the distance between upper and lower punch tips to be measured.

De Blaey and Polderman recognised that though Equation (9.15) represented the input of work into the particulate mass, it included work needed to overcome friction and also work recovered from the tablet after elastic expansion. To estimate frictional work, de Blaey (1972) adopted the suggestion that, at any point in the compression cycle, the difference between the force applied by the upper punch and that received by the lower punch is attributable to friction and he calculated the work required to overcome friction using Equation (9.16).

$$FW = \int_{D_0}^{D_{\max}} (UPF - LPF) dD \qquad (9.16)$$

where FW is the work required to overcome friction; UPF and LPF are the upper and lower punch forces, respectively; D_0 is the displacement of the upper punch when the force rises above zero; and $D_{\max}$ is the maximum displacement of the upper punch.

Jarvinen and Juslin (1971) tried to derive an estimate for the work of friction that was based on the movement of particles in contact with the die wall rather than the force and displacement of the upper punch. They derived Equation (9.17), which gives a value of FW of about half that predicted by de Blaey and colleagues.

$$FW = \int_{D_0}^{D_{\max}} \left\{ UPF - \left(\frac{(UPF - LPF)}{\ln(UPF/LPF)} \right) \right\} dD \qquad (9.17)$$

Experimental support for the suggestion that Equation (9.17) was valid was provided by Ragnarsson and Sjögren (1985). A consequence of this approach is that rather than calculating FW separately, the mean of the upper and lower punch forces can be used as a measure of the compression force that is independent of friction. To be precise, the geometric mean rather than the arithmetic mean should be used, but in practice, this change has little effect.

The force transmitted to the upper punch by the tablet during the decompression phase is a consequence of elastic expansion of the tablet in an axial direction. However, de Blaey and Polderman (1970) recognised that this force falls to zero when the punch face and the tablet cease to be in contact with each other and elastic recovery might not be completed at this point. They, therefore, modified their press so that each tablet was compressed twice without it being ejected from the die. Their reasoning was that by the time the second compression occurred, elastic expansion would be complete, and the work expended on the second compression would be that required to re-impose elastic deformation. Thus the net work of compression would be equal to the area under the compression curve of the first compression minus the area under the compression curve of the second compression.

If the work of the second compression did only represent that needed to reverse the elastic expansion of the tablet, then logically it would follow that work expenditure of the third and subsequent compressions all ought to be identical to the second. Armstrong et al. (1982) compressed tablets up to 13 times without ejection and found that the maximum force achieved at each compression changed, the magnitude of the change being dependent on the force applied, the time interval between successive compressions and also on the substance being compressed. For example, dicalcium phosphate dihydrate showed a progressive decrease in maximum force, indicative of further fragmentation of the particles. Microcrystalline cellulose, when compressed repeatedly at short time intervals, showed a rise in maximum force, believed to be the result of the development of a structure that was progressively more difficult to consolidate. Armstrong et al. (1982) concluded that energy was being expended at the second and subsequent compressions in ways in addition to overcoming elasticity, probably by further fragmentation of the particles.

Ragnarsson (1996) considered that since the true work of expansion is itself small, then the error introduced by ignoring all expansion other than that detected on the first compression is negligible. He asserted that more important sources of error are associated with measurement and data registration. This is particularly true of displacement. Already in this text, reference has been made to errors caused by inappropriate siting of displacement transducers, and also uncertainties caused by punch and press deformation and distortion (Chapter 6). Such errors are particularly important in calculating the area under a force–displacement curve owing to the shape of the curve. The total area under the curve is the sum of a series of 'slices' of thickness dD, and the largest contribution made to the area will be by those 'slices' obtained at the highest force when punch and press deformation are also at a maximum. If any error in the measurement of displacement occurs with these 'slices', it will have a major impact on the overall work measurement.

A further complication in the evaluation of a force–displacement curve is that it has been shown, as described above, that peak force is not necessarily achieved at the same time as maximum punch penetration into the die. Ho and Jones (1988a) have shown that peak force is recorded before the punch has fully penetrated into the die for many solids of pharmaceutical interest, and they attributed this to viscoelastic behaviour. This gives a force–displacement curve as shown in Figure 9.18, in which the sharp peak is replaced by a more rounded shape. If a computer is connected to the instrumented tablet press, then the observed shape of the force–displacement curve near the maximum force is very dependent on the sampling speed of the computer. When this is high, more detail is obtained and greater deviation from the expected shape can be detected.

All the work on force–displacement curves described up to now has been carried out on eccentric presses. The use of a rotary press is more complex, since both punches are in motion and, therefore, imparting energy to the tablet. An interesting approach was adopted by Oates and Mitchell (1989), who found that by considering the dimensions of certain components of a rotary press, and by making allowances for punch and press distortion, they could predict punch displacement more accurately than they could measure it. They then

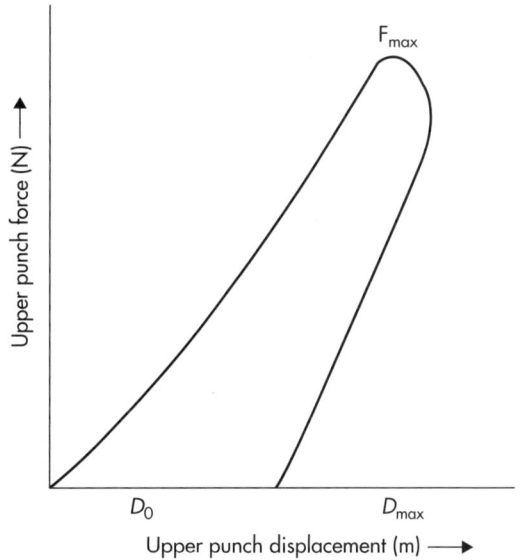

Figure 9.18 A force–displacement curve showing the effect of non-coincidence of maximum punch displacement and maximum force.

went on to use these predicted values to calculate work of compaction.

Antikainen and Yliruusi (1997) have suggested that a closer consideration of parts of the force–displacement curve may give rise to a greater understanding of the compression event. For example, they characterised the shape of the curve by measuring the displacement at 25, 50, 75 and 100% of the maximum force, both for the compression and the decompression curves, and related this information to tablet properties such as crushing strength. While such studies may be of value, they, and all others using force–displacement curves, are dependent on accurate measurement of force and displacement, especially the latter.

Attempts have been made to compare the energy used to compress a tablet as measured with a force–displacement curve with the heat content of the tablet. Coffin-Beach and Hollenbeck (1983) used calorimetric techniques. They found that the tablet after compression was in a lower energy state and this was attributed to a dissipation of surface energy. A smaller reduction in energy occurred if weak tablets were formed. Similar findings were made by Rowlings *et al.* (1995), who pointed out that a loss in surface energy ought to be reflected in a reduced internal surface area of the tablet, but they found that this was not the case. Both these sets of workers used a hydraulic press rather than a tablet press, and so compression was slow and no realistic examination of the decompression stage could be made. Hoag and Rippie (1994) carried out a thermodynamic analysis of tablets during the decompression stage using a Colton 216 rotary press. They concluded that the work represented by the area under the force–displacement curve was only a small fraction of the total energy exchange.

Punch velocity

Introduction

The outputs from force and displacement transducers are usually recorded as a function of time. Therefore, if the basis of the time measurement is known (the chart speed of a chart recorder, the sampling rate of a computer), then the duration of various events in the tabletting or encapsulation process can be calculated. Also the rate of change of force and/or displacement can be measured. The latter is particularly important, as it enables the punch speed to be calculated, a property that can have a major influence on tablet properties, since some formulations are sensitive to changes in punch speed.

It is important when dealing with tablet presses or capsule-filling equipment to distinguish between 'speed' and 'speed of production'. The latter is the same as 'output' and is measured as the number of tablets or capsules produced in unit time. This might range from a few thousands of tablets per hour produced on a simple eccentric press to a million or more per hour from a modern multi-punch rotary tablet press. 'Speed' is the distance moved in unit time and refers to components of the equipment, for example. the punches of a tablet press. As the direction of movement of a particular component at a given time is known, strictly speaking it is the vector 'velocity' rather than the scalar 'speed' that is being considered here.

For a given piece of equipment, there is a direct proportionality between 'output' and 'velocity'. If the output of a press is doubled, then all its components move twice as fast and twice as many tablets will be produced in a given time. However, this proportionality does not occur when two different types of press are considered or when presses from different manufacturers are compared. A change of press may well occur during scale-up from a trial formulation to production scale, or when formulations developed at one site in a multinational company are transferred to other sites. If the formulation is speed sensitive, a change in press can have profound effects on tablet quality.

Scientific studies on the effect that punch velocity can have on the compaction process only began in the 1980s, but there was an earlier awareness that it may affect product quality. For example, Smith (1949) commented that 'with some intractable tablets, the trouble may be overcome by reducing the speed of the machine'. In the first textbook on tabletting, written by Little and Mitchell in 1963, there is the statement 'increased speed of production may necessitate modification to formulae' though it was accepted that 'many formulae can be compressed at high speeds without any change in the formulation being required'. So it was already recognised that some formulations were speed sensitive and some were not.

The underlying reason for such differences is the mechanism by which consolidation occurs. If the predominating mechanisms are fragmentation or elastic deformation, on the one hand, then because these are virtually instantaneous processes, they should be independent of the rate at which the compression force is applied. On the other hand, the particles can behave plastically and plastic deformation is not instantaneous but takes a finite time. If the speed at which the force is applied does not make sufficient time for this, then consolidation will be incomplete and product quality will suffer.

Punch velocity in an eccentric press

Punch velocities can be measured in two ways. Firstly, they can be measured by the rate of change of the output of a displacement transducer connected to the punch. Secondly, they can be estimated if the dimensions of certain components of the press are known plus its rate of rotation.

The movement of the upper punch in an eccentric tablet press is described by Equation (9.1). Differentiation of this equation gives Equation (9.18) (Armstrong et al., 1983).

$$\frac{dy}{dt} = \omega r \cos \theta \left[1 + \frac{r \sin \theta}{\sqrt{(l^2 - r^2 \cos^2 \theta)}}\right] \quad (9.18)$$

where y is the position of the punch tip, l is the length of the eccentric strap, r is the radius of the eccentric sheave, t is time, ω is the angular velocity of the shaft of the press and θ is the angle made with the horizontal, as shown in Figure 9.2.

Therefore, provided the necessary dimensions of the press and its rate of rotation are known, then the velocity of the tip of the punch at any point in its up and down path can be calculated. Taking the Manesty F3 press as an example, as the eccentric rotates (i.e. θ changes), the distance of the punch tip from its point of maximum penetration into the die varies as shown in Table 9.1; if it is assumed that the press is rotating at its maximum rate (85 tablets min^{-1}), the corresponding velocities can be calculated. These are shown in Table 9.4, downward motion being designated as positive.

To facilitate comparison with other presses, the punch velocity when the punch is 5.0 mm from its lowest point is shown. With the Manesty F3 press, this occurs when θ equals 55.1°, at which point the punch tip is moving downwards at approximately 140 mm s^{-1}. Figure 9.19 shows the variation in punch tip velocity with changes in θ.

Punch velocity in a rotary press

The horizontal velocity (V_H) of a punch in a rotary press is given by Equation (9.19).

$$V_H = 2r_3 f \quad (9.19)$$

where f is the frequency of rotation of the turret, r_3 is the radius of the circle on the die table in which the dies rotate and φ is the angle between the axis of the punch and the vertical centre line of the pressure roll; a change of 1° in φ is, therefore, $1/360f$.

The vertical velocity V_V at any particular value of φ can be calculated from Equation (9.4). This equation gives vertical displacement as a function of angle φ. However, the time interval represented by φ is also a function of f in that the turret of the press will rotate through φ° in $\varphi/360f$ seconds. As both displacement and time are known, the vertical velocity can be calculated.

For a Manesty B3B 16 station press with the dimensions quoted earlier, vertical punch velocities are shown in Figure 9.20 and Table 9.5, the turret rotating at its maximum rate of 44 rpm (equivalent to approximately 700 tablets min^{-1}) The flat area of the punch head is reached when $r_3 \sin \varphi = 3$ mm, φ then being equal to 1.43°. At this point, the vertical velocity of the punch is zero.

When the punch is 5.0 mm from its deepest penetration into the die, the vertical punch velocity is approximately 160 mm s^{-1}. This is not

Table 9.4 The punch tip position and the punch velocity for a Manesty F3 eccentric press when rotating at its maximum rate of 85 revolutions minute^{-1}

Angle of eccentric with the horizontal, θ (°)	Distance from maximum punch displacement (mm)	Punch tip velocity (mms^{-1})
270	50.0	0.0
290	48.7	67.3
310	44.8	129.6
330	38.7	180.9
350	30.8	214.5
0	26.5	222.6
10	22.1	223.9
30	13.7	204.6
50	6.5	156.5
55.1	5.0	140.2
70	1.7	84.9
90	0.0	0.0
110	1.7	−84.9
130	6.5	−156.5
150	13.7	−204.6
170	22.1	−223.9
180	26.5	−222.6
190	30.8	−214.5
210	38.7	−180.9
230	44.9	−129.6
250	48.7	−67.3
270	50.0	0.0

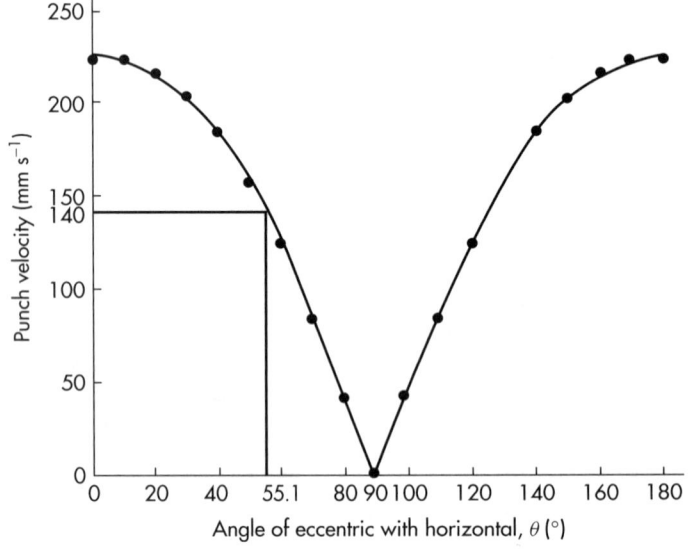

Figure 9.19 Velocity profile of the upper punch tip of a Manesty F3 eccentric press producing its maximum output of 85 tablets min^{-1}.

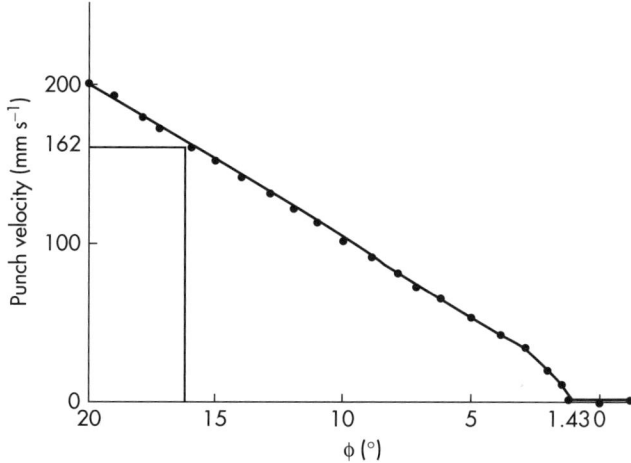

Figure 9.20 Velocity profile of the upper punch of a Manesty B3B rotary press rotating at its maximum rate of 44 rpm; φ is the angle between the axis of the punch and the vertical centre line of the pressure roll.

Table 9.5 Punch tip position and punch tip velocity for a Manesty B3B 16 station rotary tablet press operating at its maximum rate of rotation of 44 rpm

Angle between axis of punch and vertical centre line of pressure roll, (°)	Distance from maximum punch displacement (mm)	Punch tip velocity (mm s^{-1})
20.0	6.50	183.9
19.0	5.82	174.3
18.0	5.17	164.7
17.7	5.00	161.9
17.0	4.57	155.1
16.0	4.00	145.4
15.0	3.47	135.6
14.0	2.97	125.8
13.0	2.51	116.0
12.0	2.09	106.1
11.0	1.71	96.2
10.0	1.36	86.3
9.0	1.06	76.3
8.0	0.79	66.3
7.0	0.55	56.3
6.0	0.36	46.2
5.0	0.20	36.1
4.0	0.09	26.0
3.0	0.01	15.9
2.0	0.01	5.8
1.43	0.00	0.0
1.00	0.00	0.0
0.00	0.00	0.0
1.00	0.00	0.0
1.43	0.00	0.0

too dissimilar to the punch speed at this point in a Manesty F3 eccentric press. Hence for speed-sensitive formulations, a change from an F3 eccentric press to a B3B rotary press would not be expected to cause major problems. Furthermore it should be noted that, although the output of a rotary press is much greater than that of an eccentric press, it does not necessarily follow that punch speeds are significantly different.

Provided that the appropriate press and punch dimensions are known, punch velocity at any given point and the duration of events such as the dwell time can be calculated for any rotary tablet press. Using the same assumptions of press head geometry ($r_2 = 12$ mm, $r_3 = 3$ mm), Table 9.6 shows such data for a series of Manesty tablet presses, using the actual press dimensions for r_1 and r_3 for the rotary presses.

A graphical representation of these data is shown in Figure 9.21.

The Manesty B3B is a relatively 'slow' press, and if velocity data from this are compared with those from a press such as the Manesty Novapress, the differing environments are striking. When the upper punch of the latter is 5 mm from its point of deepest penetration into the die, it is travelling at approximately 700 mm s^{-1} and dwell time lasts only approximately 2 ms. Schmidt et al. (1986) have performed similar calculations for the Fette F2, Fette P1000, Kilian LX28 and Korsch PG343 rotary presses. For any given press, punch velocity is directly proportional to the press rotation rate. Therefore, if the presses listed in Table 9.6 are operating at half their maximum rate, velocities are also halved and events last twice as long. The radius of curvature of the punch has only a minor effect on punch velocity. For example, using punch heads with radii of curvature of 6, 12 and 24 mm on a Manesty Novapress, punch velocities of 727, 720 and 680 mm s^{-1}, respectively, are obtained 5 mm from the point of deepest die penetration.

Measurement of the speed dependency of tablet formulations

Since the quality of tablets produced by some formulations depends on the punch velocity, it is obviously of interest to ascertain if such dependency can be predicted.

Using a compaction simulator in which a wide range of punch velocities could be achieved, Roberts and Rowe (1985) studied the relationship between applied pressure and tablet porosity at two punch velocities, 0.033 and 300 mm s^{-1}. The calculation of tablet porosity contained appropriate corrections for punch and instrument distortion. They based their study on the Heckel equation (9.9), which they plotted for each substance at the two velocities. They measured the slopes of these plots, using a mathematically rigorous method for identifying the linear portion of the Heckel plot, and then calculated yield stress as the reciprocal of this slope. They derived a parameter that they termed 'strain rate sensitivity' (SRS) using Equation (9.20).

$$\text{SRS} = \frac{P_{y2} - P_{y1}}{P_{y2}} \times 100 \qquad (9.20)$$

where P_{y1} and P_{y2} are the yield pressures at 0.033 and 300 mm s^{-1}, respectively.

Their results are shown in Table 9.7. The materials that were more strain-rate sensitive were those whose main mechanism of consolidation

Table 9.6 Velocity-related data for some Manesty tablet presses, all operating at their maximum production rates

Press	Production rate per die (tablets min^{-1})	Time for punch to descend last 5 mm (ms)	Punch speed at first contact (mm s^{-1})	Dwell time duration (ms)
Manesty F3	85	68.6	140	0.00
Manesty B3B	44	61.4	162	10.84
Manesty Express	100	26.7	416	3.94
Manesty Unipress	121	19.1	485	3.16
Manesty Novapress	100	10.0	720	2.14

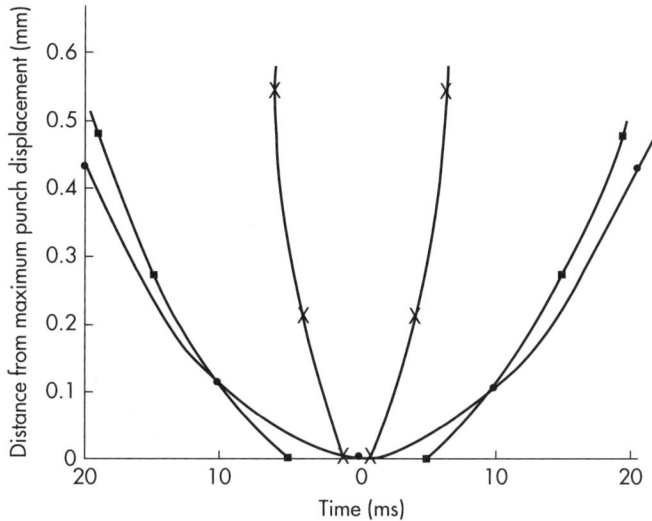

Figure 9.21 Punch movement as a function of time for three types of press. Manesty F3 eccentric press (●); Manesty B3B rotary press (■); Manesty Novapress (×).

was plastic deformation. Materials that are known to consolidate by fragmentation showed little change in yield pressure when punch velocity was increased.

A further attempt to quantify velocity dependency was the introduction by Armstrong *et al.* (1983) of the concept of power of compaction. The unit of power, the watt, is defined as the power dissipated when one joule is expended for one second. For the tabletting process, this can be defined in two ways that are dimensionally equivalent.

Firstly the area under the force–displacement curve can be divided by the time over which the force is applied as shown in Equation (9.21).

$$\text{Power} = \left[\int_{D1}^{D2} F.dD - \int_{D2}^{D3} F.dD \right] t^{-1} \quad (9.21)$$

where F is the applied force, t is time, D_1 is the displacement when the force deviates from zero, D_2 is the maximum punch displacement and D_3 is punch displacement when the force returns to zero.

Secondly, the power expended equals the force applied by the punch multiplied by the punch velocity at the time that force is being applied. Punch velocity is calculated as described earlier. Thus

Work = force × distance

Power = work/time = (force × distance)/time

or

Power = force × (distance/time)

which is equivalent to force × punch velocity.

Table 9.7 Values of strain-rate sensitivity of some materials used in tabletting (Roberts and Rowe, 1985)

Material	Strain-rate sensitivity (%)
Calcium phosphate	–
Calcium carbonate	–
Heavy magnesium carbonate	–
Paracetamol DC	1.8
Paracetamol	10.6
Lactose	16.2
Tablettose	19.2
Anhydrous lactose	20.3
Avicel PH101	38.9
Sodium chloride	39.9
Mannitol	46.4
Maize starch	49.3
Corvic	54.1

The practical importance of the sensitivity of some materials to changes in punch velocity was shown by Armstrong and Palfrey (1989). They tabletted four direct compression diluents at two press outputs (24 and 160 tablets min^{-1}) using the same eccentric press in each case. They then compared the crushing strength of the resultant tablets. Their results are shown in Table 9.8. Tablets made from an inorganic material showed little change in strength as output was increased, whereas tablets made from directly compressible starch lost almost 40% of their strength.

These workers also examined a range of particle sizes of lactose and found that, although crushing strength was reduced as punch velocity was increased, the degree of reduction was not dependent on particle size; that is, the effect of change of velocity was dependent on the substance but not its physical state.

Die wall stress

Introduction

Probably one of the least well-characterised measurements in the area of tablet compression studies has been that of die wall stress. It is usually possible to assess such variables as punch force and displacement without much modification to the normal mechanical operations of the tablet press, but it has proved much more difficult to estimate die wall stress without such alterations. The main reason for this is that the stress levels at the outer edge of the die are too low to give adequate signals from foil strain gauges, and so it is necessary to produce modified dies with local concentrations of stress. Removing some of the metal from the die is one way of achieving this, and strain gauges applied to the thinned regions are then able to provide larger signals.

The first published work in this area, like so much in tablet press instrumentation, was carried out at the University of Wisconsin and was reported by Nelson in 1955. Nelson's apparatus was more a simulator than a tablet press. It contained a stationary punch and die assembly and used a slow acting hydraulic press to apply the necessary compressive force to produce single tablets. The time required to make one compression was between 25 and 45 min.

The die had a rectangular aperture in the side, and this was closed by a third punch, the tip of which was shaped to match the curvature of the die wall. A commercial load cell was mounted at the rear end of the third punch and registered any forces exerted on it by the compressed granulation. A more or less linear relationship was obtained between punch pressure and die wall pressure, with approximately 30% of the punch pressure appearing as die wall pressure. Addition of magnesium stearate increased this percentage. The technique could be criticised on the grounds that the measured signal at the die wall was, in part, dependent on the extrusion properties of the material under compression.

Table 9.8 The effect of press output on the tensile strength of tablets made from directly compressible diluents (Armstrong and Palfrey, 1989)

	Diluents			
	Dicalcium phosphate dihydrate	Microcrystalline cellulose	Spray-dried lactose	Pregelatinised starch
Tensile strength (MPa) for press output (tablets min^{-1})				
24	0.59	6.6	0.98	0.42
160	0.56	5.9	0.82	0.26
Reduction in tensile strength (%)	5.1	10.6	16.3	38.1

The segmented die

The segmented die has proved to be the most popular method of investigating die wall stress. It was introduced by Windheuser *et al.* (1963), also of the University of Wisconsin, who avoided the problems of extrusion by leaving the bore of the die intact, but cutting away a segment of the outer wall to provide a site for strain gauges. Two strain gauges were fitted to the modified die. One was bonded to the thinned segment and had its axis at right angles to the die bore, and the other was bonded to the unaltered portion of the die, parallel to the bore, as a temperature-compensating element (Figure 9.22). They too used a hydraulic press rather than a tablet machine.

Windheuser and colleagues used plugs of soft rubber to calibrate the die, since this material was believed to have the properties of a liquid when under pressure. The upper punch force was measured with a load cell and was plotted against the strain gauge bridge output at pressures up to approximately 130 MPa. They found that the calibration was linear over the range tested, but that it was also dependent on the thickness of the rubber 'tablet' in the die. In all subsequent measurements, Windheuser and colleagues used a fixed lower punch position and a final tablet thickness of 5 mm, each charge of material being weighed out individually in order to produce a tablet of this thickness.

When tested on various types of granulation, the die gave response curves that differed appreciably in shape from material to material, being most non-linear for the alkali halides and closest to a straight line for potassium carbonate. From these curves, Windheuser and colleagues concluded that materials which achieved a good conversion of normal pressure to lateral pressure tended to form good tablets. The tablet height and its position in the die were both critical to the measurement, which gave average values over a considerable area of the die wall pressure.

Dies from which one or more segments of the outer wall have been removed have formed the basis of most experimentation on die wall stress. The principle was used by Leigh *et al.* (1967), working with a Manesty E2 tablet press, who thinned a 12.7 mm die at two diametrically opposed regions to a wall thickness of 3.17 mm. Semiconductor strain gauges, forming opposite arms of the measuring bridge, were attached to the two thinned portions of the die, and the cut-away segments were then filled with silicone rubber. There was no temperature compensation in this arrangement, since the two gauges formed opposite rather than adjacent arms of the Wheatstone bridge circuit. Data were viewed on an ultraviolet recorder. Calibration was effected by sealing the ends of the die cavity and pressurising the interior space with water. As before, the strain gauge signal was dependent on the height of the pressurised section, and it was thus necessary to keep to a particular tablet thickness in all subsequent work.

Drawing on an earlier mathematical treatment by Long (1960), Leigh *et al.* (1967) identified three types of relationship between axial and radial forces, as shown in Figure 9.23. These are called compression cycles.

Figure 9.23a represents the behaviour of a perfectly elastic body or a body exposed to a force below its elastic limit. As the axial force is increased, particles are rearranged and then begin to undergo elastic deformation. A radial force is generated, the magnitude of which is the product of the axial force and the Poisson ratio. The compression and decompression curves are superimposable, and when the axial force is totally removed, the radial force also becomes zero.

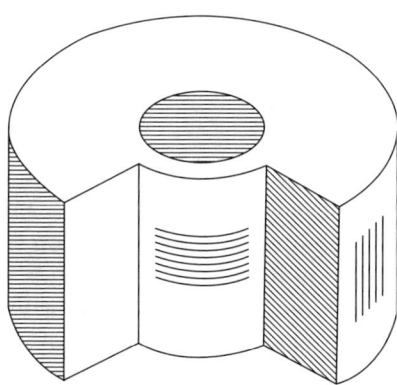

Figure 9.22 The segmented die. (After Windheuser *et al.*, 1963.)

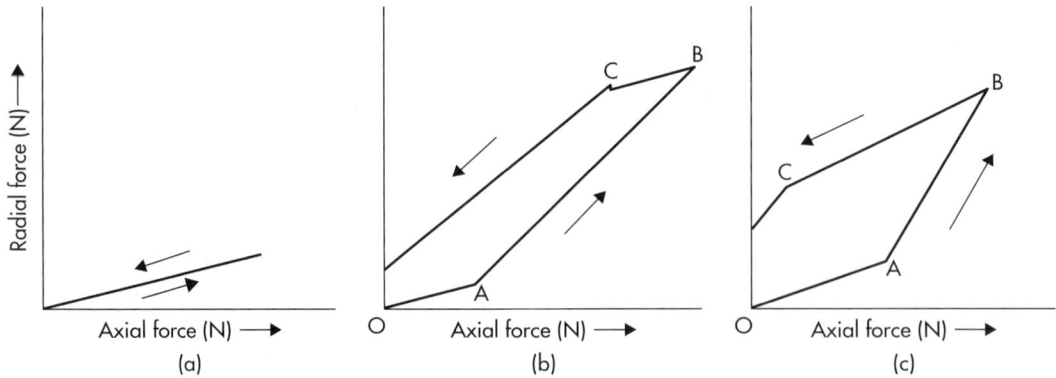

Figure 9.23 Compression cycles from a completely elastic solid (a), a solid that exhibits constant yield stress in shear (b) and a Mohr's body (c). See text for details. (After Leigh et al., 1967.)

If the elastic limit is exceeded, then one of two types of compression cycle can be obtained, dependent on the behaviour of the material under load. Firstly, it can exhibit constant yield in stress, in which the relationship between axial force and radial force (i.e. the slope of AB) is unity (Figure 9.23b). Secondly, it can behave as a Mohr's body, in which the yield stress is related to the normal stress across the shear plane; here the slope of AB is not necessarily unity (Figure 9.23c). Constant yield materials are associated with plastic deformation and Mohr's bodies tend to undergo brittle fracture.

After maximum force has been applied, both axial and radial forces now fall. For constant yield materials, changes are now again linked to the Poisson ratio, and slope BC is equal to the slope of OA. Yield then takes place and continues until the axial force returns to zero. At this point, there is still a residual radial stress, which is large in the case of Mohr's bodies. Therefore, for both constant yield materials and for Mohr's bodies, the compression and decompression curves cannot be superimposed, and a hysteresis loop or compression cycle is obtained. It has been suggested that analysis of these cycles can predict whether or not satisfactory tablets will be obtained (Leigh et al., 1967; Obiorah, 1978).

The major problem with the segmented die technique is that the calibration may often only be relevant to tablets of a fixed thickness. This was investigated in 1979 by Hölzer and Sjögren. Using a symmetrically segmented die, the authors fitted pairs of foil gauges (Kyowa type KFC-03–C1, 120 Ω) to each side. In each case, the compensating gauges were aligned vertically and were positioned some 13 mm from the upper end of the die. The active gauges were aligned horizontally. Several rubber-like materials were tested as calibration standards, but many of these were found to show marked hysteresis between increasing and decreasing forces. The most satisfactory was a blend of polyvinyl chloride and nitrile rubber, sold as Breon Polyblend 503 (BP Chemicals). Molybdenum disulphide was used as a lubricant.

The relationship between punch force and strain gauge output was then checked for different compact heights, at seven positions of the lower punch. In all, around 650 observations were made. It was found that for a given gauge arrangement, the output signal was reasonably proportional to the height of the compact. If the lower punch was kept at a fixed height of 10.2 mm from the upper end of the die, then the ratio of signal to the cross-sectional area of the compact was practically independent of the compact height. It was, therefore, possible to estimate the die wall pressure from the regression line of the signal per unit cross-sectional area versus applied punch pressure. This relationship has been used since then by other groups including Krycer et al. (1982) and Carstensen et al. (1981).

Huckle and Summers (1985) took the problem of calibration even further, working with both full and half-bridges on two similar dies.

Paradoxically these authors observed that the die wall signal from a full bridge was at a minimum when the site of compression was close to the gauges, but it increased as the site was moved away. If a half-bridge was used, the situation was completely reversed. It appeared that the vertically orientated temperature-compensating gauges, used in the full bridge, were being strained rather unexpectedly at the level of the compact. Hence the bridge signal was reduced to well below its predicted value. The authors commented that this effect had not been reported by previous workers, but there may be a logical explanation for this. The die used by Hölzer and Sjögren had been cut away to a wall thickness of 5.0 mm over a fairly narrow angle. However, Huckle and Summers had reduced the wall thickness to only 3.0 mm over what, from their illustration, seems to be an angle of 90°. The differences in the wall thickness and width might be important. Clearly, the internal pressure of the Polyblend plug would tend to distort the wall of the die towards a spherical form, since that represents the shape with a minimum surface area. The deflection of a uniformly loaded plate is inversely proportional to the cube of its thickness and to the square of its width. So doubling the thickness and halving the width reduces the deflection to approximately 3% of its original value. The thinner and wider the die wall, the more likely it is that the area around the plug will bulge outwards in three dimensions, hence affecting the compensating gauges.

Apart from this, Huckle and Summers (1985) found that the best gauge configuration was that in which several horizontally aligned gauges were fitted side by side over the whole surface of the thinned area to form a large assembly. The upper and lower gauges on each side were connected as half-bridges, and no vertically aligned gauges were used. The signals from the system were almost independent of the lower punch position.

The original cutaway die described by Windheuser and colleagues had only one segment removed, but most segmented dies used since then have been cut away on both sides to give a symmetrical form. It has been suggested that the effect of compressing a powder within such a die will be to produce a non-circular cross-section, because the thinned sections are free to expand outwards. The parts that have not been thinned will be drawn in and might in some circumstances bind on the punch. Ho and Jones (1979) studied the distortion both of punches and of dies under load and found that a die wall signal could be produced even in an empty die if the die itself was secured rigidly in the press. The signal disappeared when the die was allowed to 'float' freely around the punch and it was concluded that it had been produced by the punch pressing on the die wall. The maximum signal detected in this way represented nearly 4000 N, and this could introduce appreciable error into a measurement.

In 1981, Rippie and Danielson reported the use of an instrumented die on a Colton 216 rotary tablet machine. Their die comprised three layers. The upper and lower layers were in conventional circular forms, and the cutaway portion formed the centre of the 'sandwich'. The centre layer was only 1.5 mm in thickness and had been reduced to a wall thickness of 3.2 mm for half its circumference. It was gauged with one active element (Micro Measurement type CEA-06-125WT-350) and a second compensating gauge was mounted on the flat surface of the lower section. Plots of the die wall signal versus punch signal were linear.

A similar die was used by Khossravi and Morehead (1997), who repeatedly compressed the same tablet without ejecting it between compressions. They found that materials that consolidated by plastic deformation have similar compression cycles for the first and subsequent compression cycles. For brittle materials, brittle fracture occurs in first cycle and subsequent cycles gave no further yield. However, this work was carried out on an Instron tester with a punch speed of only 5 mm min^{-1}.

Yeh et al. (1997) also used a die with a narrow middle section fitted to a Stokes B2 press. Using finite element analysis, they calculated the cutting angle of the segments of the die wall that were to be removed and also the optimum die wall thickness. The ejection cam of the press was removed to allow measurement of residual die wall stress, and lower punch stress was measured by means of an instrumented compression roller-pin. Rubber plugs were used for calibration. They found that the measured die wall stress was linear and independent of the height and position of the tablet. They constructed compression cycles for sodium chloride and microcrystalline cellulose.

Other die constructions

One of the objections raised against Nelson's original transverse piston design was that it gave readings that were at least partly dependent on the extrusion properties of the granulation. In 1972, Lindberg made a further attempt to use this method of measurement, but unlike Nelson he left the inner surface of the die intact. He then milled a 'flat' across the outer portion of the die so that the wall was reduced to 2.0 mm at its thinnest part. A steel piston was fitted with one end against the thinned wall of the die and its other end against a piezoelectric load washer (Kistler 901A), so that the expansion of the die wall was transmitted to the load washer. In practice, the alignment of the components proved difficult.

A somewhat similar arrangement was used by Hiestand et al. (1977). Because the studies involved some rather slow changes of the applied pressure, a piezoelectric load cell would have been unsuitable. They, therefore, fitted a diaphragm type strain gauge load cell (Entran model EPS-122S).

In a subsequent development, Amidon et al. (1981) took the basic idea of the flexible die wall element several stages further. Instead of constructing the die purely as a measuring device, this group proposed to use the flexible die wall as a novel production tool in a rotary press. In essence, the die was made of a thin-walled tube, positioned inside a conventional cylindrical housing. The space between the die and the housing was completely filled with a rubber compound that could transmit pressure changes in the manner of a fluid, and a horizontal piston was arranged to communicate with the rubber-filled space. During compression of a powder, the piston was forced into the die space, thereby contracting the bore of the die by a small amount. After compression the piston was released, allowing the die to expand away from the completed tablet. The 'triaxial compression' system was intended to produce improved tablets from formulations that were prone to sticking or capping on conventional presses.

Cocolas and Lordi (1993) devised a novel die for use with a compaction simulator. In this, a spiral groove was cut around the outside of the die, into which were fitted four piezoelectric transducers at approximately 60° intervals. The centre lines of these transducers were roughly 5, 8, 11 and 14 mm from the top of the die, as such an arrangement gave an output that was not dependent on lower punch position nor tablet height. A further advance was the use of high-density polyethylene granules for calibration, since these formed a zero porosity compact at pressures of approximately 100 MPa. Data were collected by computer and viewed on an oscilloscope.

Like all experimental techniques, the predictive use of compression cycles is governed by the quality of the data that are collected. Earlier in this chapter, the problems of accurately measuring die wall force have been discussed, particularly with regard to siting the transducers in the die wall. A further problem arises with the quantity of data that is generated. In earlier studies, only a relatively small number of pairs of axial and radial force data was generated, and points on the compression cycle could be joined up by straight lines to produce the convincing parallelograms shown in Figure 9.23b,c. However, if transducer outputs are linked to a computer, then many more pairs of data points are available, and the successive linear segments of the cycle cannot be discerned so readily. There is an analogy here to the apparent lack of linearity in graphs showing the relationship between force and tablet porosity that have many data points.

Nevertheless, compression cycles can be used to investigate mechanisms of consolidation of solids. For example, Cocolas and Lordi (1993) constructed compression cycles for six pharmaceutical excipients. They found for all of them that the slope of that segment of the compression cycle approached unity after the yield point had been exceeded but was not linear.

Khossravi (1999) measured the manner in which the area of the hysteresis loop changed as the axial force was increased. He found that for plastically deforming materials, the relationship between the axial force and the hysteresis loop area was rectilinear, whereas for brittle materials, the relationship was quadratic. This implies that with increasing axial force, not only does the loop get longer, but it also gets wider at roughly the same rate (see Figure 9.23b,c). A further point of interest in Khossravi's work was that he compressed each tablet five times and found that for wholly plastic materials such as polyethylene glycol, the five compression cycles were virtually superimposible, but those of fragmenting materials changed significantly with every successive compression.

It is interesting to note that Carstensen and Touré (1980) found similar effects when they plotted the change in force received by the lower punch against that transmitted by the upper punch of an eccentric tablet press. Here, too, compression cycles were obtained whose shape was governed by the consolidation mechanism of the solid.

Applications of press instrumentation to lubrication studies

Introduction

A lubricant is an essential component of almost all tablet formulations, and of many formulations that are to be packed into hard gelatin capsules. A lubricant has two functions that are really quite distinct, but they are often included in the overall term 'lubrication'. The true lubricant function is to facilitate movement between two surfaces that are in close contact with each other (i.e. friction must be overcome). This occurs when the tablet is being ejected from the die, the two surfaces being the die wall and the edge of the tablet, and relative movement is resisted by die wall friction. In contrast, the lubricant function that should properly be called 'anti-adherence' applies when the tablet sticks to the face of one or both punches. Since the tablet face and the punch do not move relative to one another while contact between them is maintained, overcoming friction is not involved in this case.

As most effective lubricants are water repellent, tablet lubrication is usually a compromise situation: enough lubricant is needed to overcome friction but not so much as to seriously inhibit water penetration into the tablet and hence affect the release of the active ingredient. Weakening of the tablet structure is also often a concomitant of lubricant action. The presence of a lubricant film around each particle weakens interparticulate bonds, especially if the particles consolidate by deformation rather than fragmentation.

Instrumentation has facilitated studies into both qualitative and quantitative aspects of lubrication.

Measurement of lubricant action

As with so many aspects of tabletting research, studies on lubricant activity were initiated by Higuchi's group at the University of Wisconsin. Using their instrumented eccentric tablet press, they noted that the force applied by the upper punch was higher than that transmitted to the lower punch (Strickland et al., 1960). They explained this by postulating that some of the applied force was transferred to the die wall. This, in turn, generated die wall friction that had to be overcome before the tablet could be ejected from the die (Figure 9.12). Strickland and colleagues termed the ratio between the lower punch force and the upper punch force the 'R' value. This ratio can never exceed unity, and the nearer the R value is to one, the better the state of lubrication. The R value has been widely used in the assessment of lubricants (e.g. Miller and York, 1988) but suffers from some limitations. The ratio between lower punch force and upper punch force is not constant throughout the compression cycle (Figure 9.11), and so the actual value of R depends on the force applied by the upper punch. Furthermore, the loss of transmitted force and hence the generation of die wall friction is not solely a function of the lubricant but also depends on the Poisson ratio of the overall formulation. Müller et al. (1982) have shown that if the formulation is already adequately lubricated, then slight improvements in lubrication cannot be detected by changes in the

R value. Nevertheless, for a given formulation compressed at a uniform maximum force, the R value has proved a useful measure of lubricant efficacy. Alkali metal stearates usually have R values between 0.9 and 0.95.

A more direct method of measuring lubricant action is to measure the force required to eject the tablet from the die. With an eccentric press, this is achieved by fitting force transducers to the lower punch assembly; for a rotary press, transducers are usually attached to the ejection ramp (Chapter 6). Hölzer and Sjögren (1977) reviewed four methods for evaluating friction during tabletting: the ratio of lower punch force to upper punch force (the R value), the difference between upper and lower punch forces, the force detected by the lower punch before ejection and the force needed to eject the tablet. They found that the compression force had an influence on all four techniques, and so a correction for the area of contact between the tablet and the die wall was applied. When the ejection force was corrected in this way, a straight-line relationship was obtained between it and the compression force. Lubricants reduced the slope of this line, the degree of reduction depending on the nature of the lubricant and its concentration.

Using an eccentric press fitted with force transducers on upper and lower punches, Delacourte *et al.* (1993) attempted to measure the value of the upper punch force that caused the press to jam. They gradually increased the upper punch force until tablet production for 3 min was not possible without ejection problems such as a grinding noise, scratches on the tablet edge or disturbance of the lower punch force signal. They used a standard mixture of lactose and dicalcium phosphate dihydrate mixed with a range of lubricants.

Measurement of adhesion to punch faces

If a tablet sticks to the lower punch of a press, then the force required to sweep off the tablet after ejection is increased. Sticking to the upper punch can even cause rupture of the tablet if the adhesive force between the tablet and the punch face is greater than the cohesive forces within the tablet. Usually only a film forms on the tablet face, but even this can be difficult and expensive to eradicate.

The first workers to study sweep-off forces were Mitrevej and Augsburger (1980). They fitted a strain-gauged cantilever beam to the feed frame of a Stokes B2 rotary press, the beam being in front of the sweep-off blade. They reasoned that the adhesion force was the difference between the total force measured at the beam and that owing to the momentum of the tablet. The latter could be calculated from its mass and its velocity as it travelled around the die table. They noted that the adhesion force increased with a rise in compression pressure or a decrease in lubricant concentration. Mitrevej and Augsburger also measured the force needed to eject the tablet from the die and found that differences in lubricant efficiency did not necessarily reflect differences in adhesion. A similar instrumented beam was used by Wang *et al.* (2004), who were able to relate adhesion forces to the intermolecular attraction between ingredients of the tablet formulation and the metal surface of the punch face.

Waimer *et al.* (1999a) adopted a different approach in that, rather than determining the sweep-off force, they measured the force required to separate the punch face from the tablet. As the punch begins to rise in the die, adhesive forces cause the punch to be in tension before contact is lost between punch face and tablet. However, these forces are extremely low. For a compression force of 40 kN, the adhesion force is only 30–40 N, so a sensitive and wide-ranging system of measurement is needed. In a subsequent paper (Waimer *et al.* 1999b), the same workers studied the influence of punch engravings on adhesion by screwing small cones into the punch face. They found that the angle of the cone had a major influence on the amount of adhesion.

References

Altaf SA, Hoag SW (1995). Deformation of the Stokes B2 rotary tablet press: quantitative influence on tablet compaction. *J Pharm Sci* 84: 337–343.

Amidon GE, Smith DP, Hiestand EN (1981). Rotary press utilizing a flexible die wall. *J Pharm Sci* 70: 613–617.

Antikainen OK, Yliruusi JK (1997). New parameters derived from tablet compression curves. 2: Force–displacement curve. *Drug Dev Ind Pharm* 23: 81–93.

Armstrong NA (1998). Direct compression characteristics of granulated lactitol. *Pharm Tech Europe* 10: 42–45.

Armstrong NA, Morton FSS (1979). An evaluation of the compression characteristics of some magnesium carbonate granulations. *Pharm Weekblad Sci Ed* 114: 1450–1459.

Armstrong NA, Palfrey LP (1989). The effect of machine speed on the consolidation of four directly compressible tablet diluents. *J Pharm Pharmacol* 41: 149 – 151.

Armstrong NA, Abourida NMAH, Gough AM (1983). A proposed consolidation parameter for powders. *J Pharm Pharmacol* 35: 320–321.

Armstrong NA, Abourida NMAH, Krijgsman L (1982). Multiple compression of powders in a tablet press. *J Pharm Pharmacol* 34: 9–13.

Billardon P, Ozil P, Guyot JC (1987). Planning experiments using an instrumented tablet machine in formulation. *Drug Dev Ind Pharm* 13: 2477–2493.

Carstensen JT, Touré P (1980). Compression cycles in tabletting. *Powder Technol* 26: 199–204.

Carstensen JT, Touré P, Duchêne D, Puisieux F (1981). Correlation between hysteresis loop areas of lower punch and of die pressures versus upper punch pressures. *Drug Dev Ind Pharm* 7: 645–648.

Charlton B, Newton JM (1984). Theoretical estimation of punch velocities and displacements of single punch and rotary tablet machines. *J Pharm Pharmacol* 36: 645–651.

Chowhan ZT, Chow YP (1980). Compression behavior of pharmaceutical powders. *Int J Pharm* 5: 139–148.

Cocolas HG, Lordi NG (1993). Axial to radial pressure transmission of tablet excipients using a novel instrumented die. *Drug Dev Ind Pharm* 19: 2473–2497.

Coffin-Beach DP, Hollenbeck RG (1983). Determination of the energy of tablet formation during compression of selected pharmaceutical powders. *Int J Pharm* 17: 313–324.

Cooper AR, Eaton LE (1980). Compaction behavior of several ceramic powders. *J Am Ceram Soc* 45: 97–101.

de Blaey CJ (1972). Compression of pharmaceuticals. 6: Quantitative determination of the friction during compression in terms of the work involved. *Pharm Weekblad* 107: 233–242.

de Blaey CJ, Polderman J (1970). Compression of pharmaceuticals. 1: The quantitative interpretation of force–displacement curves. *Pharm Weekblad* 105: 241–250.

de Blaey CJ, Polderman J (1971). Compression of pharmaceuticals. 2: Registration and determination of force–displacement curve using a small digital computer. *Pharm Weekblad* 106: 57–65.

Delacourte A, Predella P, Leterme P et al. (1993). A method for quantitative evaluation of the effectiveness of lubricants used in tablet technology. *Drug Dev Ind Pharm,* 19: 1047–1060.

Dietrich R, Mielck JB (1984). Parameterization of the time course of compression in tabletting by means of a modified Weibull function. 1: Idea and experimental design. *Pharm Ind* 46: 863–869.

Dietrich R, Mielck JB (1985). Parameterization of the time course of compression in tabletting by means of a modified Weibull function. 2: Statistical evaluation and interpretation of results. *Pharm Ind* 47: 216–220.

Du J, Hoag SW (2003). Characterisation of excipient and tabletting factors that influence folic acid dissolution, friability and breaking strength of oil and water-soluble multivitamin and mineral tablets. *Drug Dev Ind Pharm* 29: 1137–1147.

Dwivedi SK, Oates RJ, Mitchell AG (1991). Peak offset time as an indication of stress relaxation during tabletting on a rotary tablet press. *J Pharm Pharmacol* 43: 673–678.

Heckel RW (1961). Density-pressure relationships in powder compaction. *Trans Metal Soc AIME* 221: 671–675.

Hersey JA, Rees JE (1971). Deformation of particles during briquetting. *Nature* 230: 96–98.

Hiestand EN, Wells JE, Peot CB, Ochs JF (1977). Physical processes of tableting. *J Pharm Sci* 66: 510–519.

Ho AYK, Jones TM (1979). The effect of punch:die distortion on the measurement of radial die wall force during compaction. In *Proceedings of the International Conference of Pharmacy Technology and Production Manufacture*, Copenhagen.

Ho AYK, Jones TM (1988a). Punch travel beyond peak force during tablet compression. *J Pharm Pharmacol* 40: 75P.

Ho AYK, Jones TM (1988b). Rise time: a new index of tablet compression. *J Pharm Pharmacol* 40:74P.

Hoag SW, Rippie EG (1994). Thermodynamic analysis of energy dissipation by pharmaceutical tablets during stress unloading. *J Pharm Sci* 83: 903–908.

Hölzer AW, Sjögren J (1977). Comparison of methods for evaluation of friction during tabletting. *Drug Dev Ind Pharm* 3: 23–37.

Hölzer AW, Sjögren J (1979). Instrumentation and calibration of a single-punch press for measuring the radial force during tabletting. *Int J Pharm* 3: 221–230.

Huckle PD, Summers MP (1985). The effect of strain-gauge size and configuration on radial stress measurement during tabletting. *J Pharm Pharmacol* 37: 722–725.

Jarvinen MJ, Juslin MJ (1971). Studies on energy expenditure by compressing tablets with an eccentric tablet machine. 1: Determination of mechanical energy. *Farm Aikak.* 80: 242–248.

Kawakita K, Ludde K-H (1970). Some considerations on powder compression equations. *Powder Technol* 4: 61–68.

Khossravi D (1999). Compaction properties of powders: the relationship between compression cycle hysteresis areas and maximally applied punch pressures. *Drug Dev Ind Pharm* 25: 885–895.

Khossravi D, Morehead WT (1997). Consolidation mechanisms of pharmaceutical solids: a multicompression cycle approach. *Pharm Res* 14: 1039–1045.

Kiekens F, Debunne A, Vervaet C et al. (2004). Influence of punch diameter and curvature on yield pressure of microcrystalline cellulose compacts during Heckel analysis. *Eur J Pharm Sci* 22: 117–126.

Konkel P, Mielck JB (1997). Association of parameters characterising the time course of the tabletting process on a reciprocating and on a rotary tabletting machine for high speed production. *Eur J Pharm Biopharm* 44: 289–301.

Krycer I, Pope DG, Hersey JA (1982). An evaluation of the techniques employed to investigate powder compaction behaviour. *Int J Pharm* 12: 113–134.

Leigh S, Carless JE, Burt BW (1967). Compression characteristics of some pharmaceutical materials. *J Pharm Sci* 56: 888–893.

Lindberg NO (1972). Instrumentation of a single punch tablet machine. *Acta Pharm Suec* 9: 135–140.

Little A, Mitchell KA (1963). *Tablet Making*, 2nd edn. Liverpool, UK: Northern.

Long WM (1960). Radial pressures in powder compaction. *Powder Metal* 6: 73–86.

Miller TA, York P (1988). Pharmaceutical tablet lubrication. *Int J Pharm* 41: 1–9.

Mitrevej A, Augsburger LL (1980). Adhesion of tablets in a rotary tablet press. I: Instrumentation and preliminary study of variables affecting adhesion. *Drug Dev Ind Pharm* 6: 331–337.

Morris LE, Moore JC, Schwartz JB (1996). Characterisation and performance of a new direct compression excipient for chewable tablets: Xylitab. *Drug Dev Ind Pharm* 22: 925–932.

Müller BW, Steffens K-J, List PH (1982). Tribological laws and findings in tablet technology. 5: On methods for determining the tribological properties of solid lubricants in tablet manufacture. *Pharm Ind* 44: 636–640.

Müller F, Caspar U (1984). Viskoelastische Phänomene während der Tablettierung. *Pharm Ind* 46: 1049–1056.

Muller FX, Augsburger LL (1994). Role of the displacement-time waveform in the determination of Heckel behaviour under dynamic conditions in a compaction simulator and a fully instrumented tablet machine. *J Pharm Pharmacol* 46: 468–475.

Munoz-Ruiz A, Jiminez-Castellanos MR, Cunningham JC et al. (1992). Theoretical estimation of dwell and consolidation times in rotary tablet machines. *Drug Dev Ind Pharm* 18: 2011–2028.

Nelson E (1955). The physics of tablet compression. 8: Some preliminary measurements of die wall pressure during tablet compression. *J Am Pharm Assoc Sci Ed* 44: 484–497.

Nelson E, Busse LW, Higuchi T (1955). The physics of tablet compression. 7: Determination of energy expenditure in the tablet compression process. *J Am Pharm Assoc Sci Ed* 44: 223–226.

Oates RJ, Mitchell AG (1989). Calculation of punch displacement and work of powder compaction on a rotary tablet press. *J Pharm Pharmacol* 41: 517–523.

Oates RJ, Mitchell AG (1990). Comparison of calculated and experimentally determined punch displacement on a rotary tablet press using both Manesty and IPT punches. *J Pharm Pharmacol* 42: 388–396.

Obiorah BA (1978). Possible prediction of compression characteristics from pressure cycle plots. *Int J Pharm* 1: 249–255.

Olmo IG, Ghaly ES (1999). Compressional characterization of two dextrose-based directly compressible excipients using an instrumented tablet press. *Pharm Dev Technol* 4: 221–231.

Ragnarsson G (1996). Force–displacement and net work measurements. In Alderborn G, Nystrom C (eds), *Pharmaceutical Powder Compaction Technology*. New York: Marcel Dekker, pp 77–97.

Ragnarsson G, Sjögren J (1985). Force–displacement measurements in tabletting. *J Pharm Pharmacol* 37: 145–150.

Ramberger R, Burger A (1985). On the application of the Heckel and Kawakita equations to powder compaction. *Powder Technol* 43: 1–9.

Rippie EG, Danielson DW (1981). Viscoelastic stress–strain behaviour of pharmaceutical tablets:

analysis during unloading and postcompression periods. *J Pharm Sci* 70: 476–482.

Roberts RJ, Rowe RC (1985). The effect of punch velocity on the compaction of a variety of materials. *J Pharm Pharmacol* 37: 377–384.

Rowe RC, Roberts RJ (1996). Mechanical properties. In Alderborn G, Nystrom C (eds), *Pharmaceutical Powder Compaction Technology*, New York: Marcel Dekker, pp 283–322.

Rowlings CE, Wurster DE, Ramsey PJ (1995). Calorimetric analysis of powder compression. 2: The relationship between energy terms measured with a compression calorimeter and tabletting behaviour. *Int J Pharm* 116: 191–200.

Rue PJ, Rees JE (1978). Limitations of the Heckel equation for predicting powder compaction mechanisms. *J Pharm Pharmacol* 30: 642–643.

Schmidt PC, Leitritz M (1997). Compression force/time profiles of microcrystalline cellulose, dicalcium phosphate dihydrate and their binary mixtures – a critical consideration of experimental parameters. *Eur J Pharm Biopharm* 44: 303–313.

Schmidt PC, Vogel PJ (1994). Force–time curves of a modern rotary tablet machine. 1: Evaluation techniques and characterisation of deformation behavior of pharmaceutical substances. *Drug Dev Ind Pharm* 20: 921–934.

Schmidt PC, Tenter U, Hocke J (1986). Force–displacement characteristics of rotary tabletting machines. 1: Installation of a single punch for force measurements *Pharm Ind* 48: 1546–1553.

Shaxby JH, Evans JC (1923). The variation of pressure with depth in columns of powders. *Trans Faraday Soc* 19: 60–72.

Sheikh-Salem M, Fell JT (1981). The influence of initial packing on the compression of powders. *J Pharm Pharmacol* 33: 491–494.

Smith AN (1949). Compressed tablets. 1: resistance to wear and tear. *Pharm J* 163: 194–195.

Sonnergaard JM (1999). A critical evaluation of the Heckel equation. *Int J Pharm* 193: 63–71.

Strickland WA, Higuchi T, Busse LW (1960). The physics of tablet compression. 10: Mechanism of action and evaluation of lubricants. *J Am Pharm Assoc Sci Ed* 49: 35–40.

Sun C, Grant DJW (2001). Influence of elastic deformation of particles on Heckel analysis. *Pharm Dev Technol* 6: 193–200.

Train D (1956). An investigation into the compaction of powders. *J Pharm Pharmacol* 8: 745–761.

Vogel PJ, Schmidt PC (1993). Force–time curves of a modern rotary tablet machine. 2: Influence of compression force and tabletting speed on deformation mechanisms of pharmaceutical substances. *Drug Dev Ind Pharm* 19: 1917–1930.

Waimer F, Krumme M, Danz P et al. (1999a). A novel method for the detection of sticking of tablets. *Pharm Dev Technol* 4: 359–367.

Waimer F, Krumme M, Danz P et al. (1999b). The influence of engravings on the sticking of tablets. Investigations with an instrumented upper punch. *Pharm Dev Technol* 4: 369–375.

Windheuser JJ, Misra J, Eriksen SP, Higuchi T (1963). Physics of tablet compression. 13: Development of die-wall pressure during compression of various materials. *J Pharm Sci* 52: 767–770.

Yeh C, Altaf SA, Hoag SW (1997). Theory of force transducer design optimisation for die wall stress measurement during tablet compaction. Optimisation and validation of split web die using finite element analysis. *Pharm Res* 14: 1161–1170.

Ylirussi JK, Antikainen OK (1997). New parameters derived from tablet compression curves. 1: Force–time curve. *Drug Dev Ind Pharm* 23: 69–79.

York P (1979). A consideration of experimental variables in the analysis of powder compaction behaviour. *J Pharm Pharmacol* 31: 244–246.

Further reading

Armstrong NA (1989). Time-dependent factors involved in powder compression and tablet manufacture. *Int J Pharm* 49: 1–13.

Bolhuis GK, Hölzer AW (1996). Lubricant sensitivity. In Alderborn, G., Nystrom, C (eds), *Pharmaceutical Powder Compaction Technology*, New York: Marcel Dekker, pp 517–560.

Carstensen JT, Marty J-P, Puisieux F, Fessi H (1981). Bonding mechanisms and hysteresis areas in compression cycle plots. *J Pharm Sci* 70: 222–223.

Celik M (1992). Overview of compaction data analysis techniques. *Drug Dev Ind Pharm* 18: 767–810.

Kottke MK, Rudnic EM (2002). Tablet dosage forms. In Banker GS, Rhodes CT (eds), *Modern Pharmaceutics*, New York: Marcel Dekker, pp 333–391.

Doelker E, Massuelle D (2004). Benefits of die wall instrumentation for research and development in tabletting. *Eur J Pharm Biopharm* 58: 427–444.

Heckel RW (1961). An analysis of powder compaction phenomena. *Trans Metall Soc AIME* 221: 1001–1008.

Hoblitzell JR, Rhodes CT (1990). Instrumented tablet press studies on the effect of some formulation and processing variables on the compaction process. *Drug Dev Ind Pharm* 16: 469–507.

Morehead WT (1992). Viscoelastic behaviour of pharmaceutical materials during compaction. *Drug Dev Ind Pharm* 18: 659–675.

Müller F (1996). Viscoelastic models. In Alderborn G, Nystrom C (eds), *Pharmaceutical Powder Compaction Technology*. New York: Marcel Dekker, pp 99–132.

Paronen P, Ilkka J (1996). Porosity-pressure functions. In Alderborn G, Nystrom C (eds), *Pharmaceutical Powder Compaction Technology*, New York: Marcel Dekker, pp 55–75.

10

The instrumentation of capsule-filling machinery

N Anthony Armstrong

Introduction

The hard-shell capsule is a commonly used dosage form, and it has been estimated that approximately 60 billion capsule shells are used annually for pharmaceutical products (Podczeck, 2004). The product essentially consists of a hard shell, usually gelatin based, that is almost always filled with a particulate solid. Among the advantages claimed for the hard-shell capsule compared with tablets are enhanced bioavailability owing to the highly porous nature of the fill, less-demanding requirements for powder flow and the ability to fill formulations into a capsule shell that are not compressible to the extent needed for tablet manufacture. However, Jones (2001) has pointed out the fallacy of the belief that 'powder filled capsules are a very simple product that does not need much skill to prepare'. The requirements of formulations to be filled into hard shells can be quite complex, and understanding the filling process has been greatly facilitated by fitting appropriate instrumentation to the filling equipment.

There are several similarities between instrumentation for a tablet press and that of capsule-filling equipment. Tablet making and capsule shell filling involve compressing a particulate mass and so in both types of instrumentation, the parameters of most interest are force (pressure) and distance, both almost invariably recorded as a function of time. The transducers that have been used to measure these in tablet presses – various types of strain gauge and displacement transducer – have been used in capsule-filling equipment, and an instrumentation system for capsule-filling equipment requires the same components as for an instrumented tablet press, namely a power supply, suitably calibrated transducers, amplification circuitry and devices for recording and manipulating the amplified outputs of the transducers. Many of the approaches that have been used in the instrumentation of tablet presses have also been used to instrument capsule-filling equipment, and the use of simulators has been particularly successful.

However, there are important differences. The forces used in the compression of the contents of hard-shell capsules are much lower – of the order of tens of newtons – whereas in tablet manufacture, forces of tens of kilonewtons are needed. Hence a sensitive measuring system that is capable of distinguishing a signal of this magnitude from background noise is required to study capsule filling. Earlier chapters in this book have stressed the importance of the correct siting of transducers and this is just as important with capsule-filling machinery. Indeed there is a particular problem in this case. To give a stable and meaningful output, the transducer must be attached to a massive and secure component of the equipment. Suitable sites are readily available in a tablet press but are not so available in capsule-filling equipment.

The uses made of instrumented capsule-filling machines parallel those of instrumented tablet presses: the effect of compression force on plug properties such as physical strength, dispersion and dissolution have been studied, together with the lubricant requirements of the formulation.

Capsule-filling equipment

Though there are two types of tablet press – eccentric and rotary – the way in which they operate is essentially the same in that a particulate mass is constrained in a die and compressed between an upper and a lower punch. In contrast, there are several types of capsule-filling equipment, each with its own modus operandi. The fill material is treated in different ways in each case, and hence the challenges of fitting instrumentation also differ.

The two most popular types of capsule-filling equipment are those based on a tamping mechanism into a dosating disk and those that make use of a dosating tube. Both types have been fully described by Jones (2001) and Podzceck (2004), but a brief description of their operating principles is necessary here.

Dosating disk machines

The dosating disk machine, shown diagrammatically in Figure 10.1, in some ways resembles a tablet press. The dosating disk has a number of holes bored through it, all except one being closed off by a base plate. Powder flows into the first hole and is compressed by tamping pin 1. This hole is then moved to position two; further powder flows in and is compressed again, this time by tamping pin 2. This is repeated until the last hole is reached when, after excess powder has been scraped off, the dosating disk positions the plug of powder over a capsule body. The plug is then ejected by a piston, and the upper part of the shell is fitted. The overall arrangement is analogous to the die cavity of a tablet press that is progressively filled and the contents repeatedly consolidated. The several filling positions are usually arranged in a circle.

The tamping pins are spring loaded to avoid the application of an excessive force. The resultant plugs are fragile and of high porosity and hence dispersion and dissolution of the contents after ingestion are facilitated. This type of machine is exemplified by those made by Höfliger & Karg (now part of the Bosch Group) and Harro Höfliger.

Dosating nozzle machines

In dosating nozzle machines, the dosator consists of a tube, open at one end, within which is a moveable piston. The dosator is plunged into a bed of particulate solid contained in a hopper.

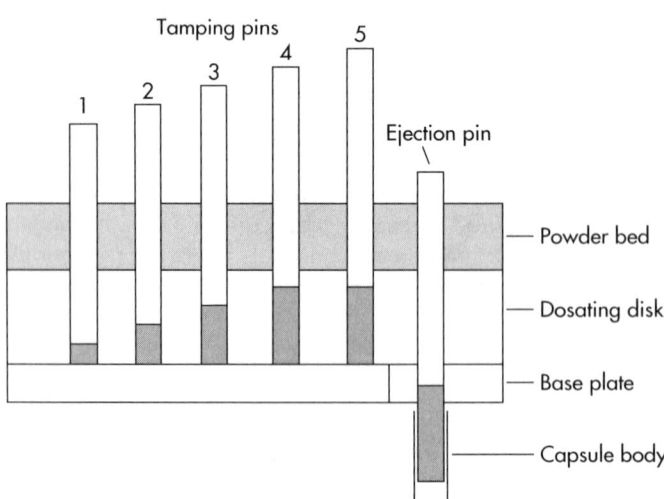

Figure 10.1 A dosating disk capsule-filling machine.

Powder enters the open end of the dosator and is then consolidated by a downward movement of the piston to form a plug. The dosator is then withdrawn from the hopper, taking the plug with it, and is positioned over the body of the capsule shell. The piston again moves downwards, ejects the plug into the capsule body and the upper part of the shell is then fitted. Since the plug must be retained inside the dosator tube while the latter is being moved into position over the capsule body, a free-flowing powder is not a prerequisite. However, the powder cannot be too cohesive since the powder bed must be maintained at a relatively uniform depth to facilitate reproducible filling. A lubricant such as magnesium stearate may be required. Examples of machines using this technique, shown diagrammatically in Figure 10.2, are the mG2, Zanasi and Macofar.

Instrumentation of dosating disk capsule-filling machines

The consolidation process on a dosating disk machine is more complex than that of a tablet press. In the latter, plugs are formed by a single compressive stroke, whereas in the former, the plugs are formed progressively by compression from a series of tamping pins, five in the case of Bosch equipment. The aim is to obtain a plug of powder of specific weight, and the increase in weight achieved at each tamping is dependent on the degree of volume reduction applied by the previous tamping pin. Each pin can be set to a different depth of penetration and so a large number of different combinations can be obtained. If two successive positions are adjusted to the same depth of penetration, then the fill weight gain at the second station depends on the volume of the void left after the preceding tamp.

The group headed by Professor Augsburger at the University of Maryland have been pioneers in instrumentation of capsule-filling equipment. In 1983, Shah *et al.* described the instrumentation of a Höfliger & Karg GKF330 machine, fitting strain gauges to the necks of two tamping pins. One instrumented pin was kept at the ejection station, and the other could be inserted at any of the five tamping positions. Data were stored on a data logger and viewed by oscilloscope. They observed that the fourth tamping position had the greatest influence on fill weight and compression force while the second position had the least.

The instrumentation was then developed so that each of the five tamping pins was equipped with strain gauges so that the whole compressive history of a single plug could be examined (Shah *et al.*, 1986). As the tamping pin penetrates the powder bed, it pushes particulate matter into the

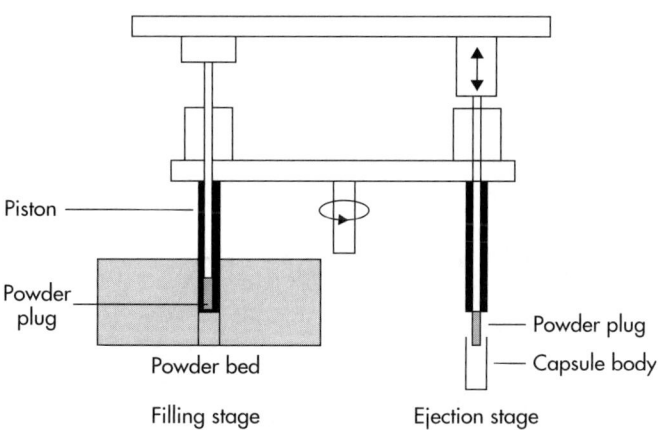

Figure 10.2 A dosating nozzle capsule-filling machine.

hole in the disk. Force rises smoothly to a maximum of approximately 200 N. Then the tamping pin begins to rise, and there is a consequent decrease in force until a plateau is reached. The height of the plateau is dependent on the maximum force, but its duration is constant at approximately 65 ms. The plateau is caused by a brief halt in the upward movement of the tamping pin, caused by the intermittent motion of a Höfliger & Karg machine, which brings on the next capsule shell to be filled. Thus the plateau is a feature introduced by the design of the machine and has nothing to do with the properties of the plug. Contact between the top surface of the plug and the end of the tamping pin is maintained by the partially relaxed relief spring, and hence a force continues to be detected. Once the plateau has been passed, decompression proceeds (Figure 10.3). The ejection force is approximately 50 N.

Shah *et al.* (1986) found that the target weight of the plug could be achieved after three tamps, but not after just two. They also found that effective compression began before the tamping pin enters the dosating disk, powder being pushed ahead of it. Therefore, the higher the tamping force, the heavier the plug. Shah *et al.* also measured the physical strength of the plugs using a three-point flexure test. As compression force was increased, not unexpectedly plug strength also increased. Because of repeated application of force during plug formation, it might be expected that the lowest part of the plug would show a progressive increase in strength as it was compressed for a second or third time. This was found not to be the case, provided that the subsequent compressions were at the same force. An increased force led to higher consolidation and hence greater plug strength.

Using this filling equipment, Shah *et al.* (1986) examined the plug-forming properties of anhydrous lactose, dicalcium phosphate dihydrate and microcrystalline cellulose. They found that, with the last named diluent, there was an optimum concentration of magnesium stearate of 0.1%, in that plugs with this level of lubricant were both stronger and heavier than unlubricated plugs for any combination of force and number of tamps. This they attributed to improved powder flow. Magnesium stearate concentrations in excess of 0.1% caused softening of the plugs, ultimately to a lower strength than those made with the unlubricated powder.

In a later paper (Shah *et al.*, 1987), the effects of multiple tamping on dissolution were studied, using the same instrumented equipment and hydrochlorothiazide as the active ingredient. In general, increasing the number of tamps resulted in slower dissolution. Higher compression forces accelerated drug release when anhydrous lactose was used as the filler, but the reverse was true with dicalcium phosphate dihydrate. They made the important observation that dissolution of the active ingredient was not changed by altering the tamping force or the number of tamps provided that sufficient disintegrating agent was present (4% croscarmellose was used).

The fitting of displacement transducers (LVDTs) to tablet presses in an attempt to quantify the compression process was pioneered by de

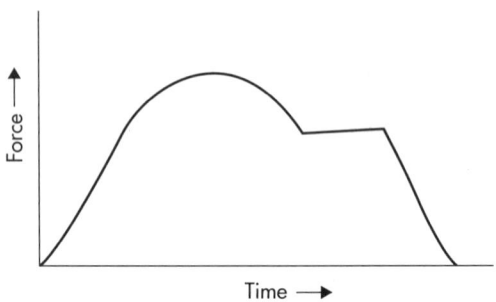

Figure 10.3 Force–time profile from instrumented Höfliger & Karg dosating disk capsule-filling equipment.

Blaey and Polderman at the University of Leiden (e.g. de Blaey and Polderman, 1970). By plotting force as a function of punch, the work expended in the compression event could be calculated since work is dimensionally equal to the area enclosed by the force–displacement curve.

Fitting displacement transducers in addition to strain gauges to capsule-filling equipment so that force and tamping pin position could be monitored simultaneously was reported by Cropp et al. in 1991. Two LVDTs were fitted to the instrumented Höfliger & Karg GLF330 equipment already described. The first of these monitored the movement of the brass ring to which the tamping pin holder assembly was anchored. The second was attached to a modified pin that rested on the head of the tamping pin and beneath the overload spring, and thus movement of the overload spring could be detected. This allowed measurement of the precise penetration of the dosating disk by the tamping pin using a combination of the output of the two LVDTs. Transducer outputs were stored and recorded by digital oscilloscope and computer.

Force–time relationships were obtained as described above (Shah et al, 1986), and both ring displacement and pin displacement showed a pause in upward movement shortly after maximum displacement had been achieved. This confirmed that the plateau in the force–time profile was caused by movement of the transport mechanism for the capsule shells. Peak maximum force occurred at the same time as maximum displacement. Some of the data obtained are shown in Table 10.1.

The authors pointed out that the choice of overload spring could affect the properties of the powder plug. An increase in the strength of the spring caused the applied force to rise with consequent changes in plug properties such as strength and dissolution, though a significant change in plug weight was not observed. Stronger springs are often used for reasons of durability but this approach may have negative consequences on product quality.

Plugs made from anhydrous lactose were subjected to a higher peak force during compression than microcrystalline cellulose plugs over a whole range of settings for the tamping pin. Microcrystalline cellulose has a lower bulk density than anhydrous lactose, and so pin displacement is greater for the former.

Cropp et al. (1991) combined force and displacement data to construct force–displacement curves for dosating disk capsule-filling equipment. A force was registered before any tamping displacement was detected. This was because the tamping pin detects a resistance to its movement while travelling through the powder prior to penetrating the dosating disk. After peak penetration is reached, displacement falls to a plateau and then the curve returns to the baseline. Calculation of the work of compaction after correction for elastic recovery gave much lower values (a fraction of a joule) than those needed to compress tablets. Though the displacement of the tamping pin of a few millimetres was similar to that of a punch in a tablet press, the applied force was only tens of newtons. The need for precise measuring systems in such circumstances is

Table 10.1 Data from Höfliger & Karg GKF330 capsule-filling equipment fitted with force and displacement transducers (Cropp et al., 1991)

Pin penetration setting (mm)	Peak force (N)	Force at plateau (N)	Peak pin displacement (mm)	Pin displacement at plateau (mm)	Plug length at peak (mm)	Contact time (ms)	Time to peak force (ms)
0	8.5	0.0	0.5	0.0	15.8	80	54
5	48.5	11.2	2.6	0.7	13.4	206	80
10	107.0	69.3	5.8	3.9	11.6	250	99
14	158.4	118.7	8.6	6.5	10.4	280	117

apparent. Owing to the greater pin displacement with microcrystalline cellulose, work expended in compressing this substance was about double that used for compressing anhydrous lactose.

Further development of the same machine was carried out by Davar *et al.* (1997), who fitted each station with instrumented tamping pins. LVDTs were added to measure tamping pin penetration, pin displacement at peak pressure and the movement of the brass guide block. Using six formulations containing either lactose or microcrystalline cellulose, the relationships between compressing force and plug properties such as physical strength and density were investigated.

A totally different approach to instrumentation of tamping disk capsule-filling equipment has been described by Podczeck (2000, 2001). In the first of these papers, Podczeck pointed out that advances in instrumentation in tablet manufacture have led to feedback mechanisms that can be used to control tablet weight automatically (Chapter 11), but such developments in capsule filling had yet to be achieved. One possibility of changing fill weight was by altering the tamping distance of one or more of the tamping pins, but Podczeck and Newton (1999) demonstrated that only modest changes of a few milligrams could be achieved in this way. Larger changes in fill weight could be brought about by exchanging the springs inside the tamping fingers. However, changing springs and adjustment of penetrative depth can only be carried out when the equipment is stationary and so neither method lends itself to a feedback mechanism. For weight adjustment to take place when the machine was running, some form of electrical or electronic control was required.

Podczeck used a Bosch GKF 400S machine. In this equipment, there are five tamping stations each fitted with three tamping pins. In the one tamping block that was instrumented, the springs were removed and replaced with dashpots and a chamber filled with compressed air. The latter was in contact with a piezoelectric force transducer. By this arrangement, the point at which the springs were deflected could be altered continuously, and hence the volume available in the hole of the dosating disk was also changed. Force rapidly rose to a maximum, was maintained virtually constant for a time and then fell to a plateau level. It then remained constant for a further period before returning to zero. The change of force between maximum and plateau was attributed to deflection of the spring in the tamping pin, which in this case was simulated by air pressure. The plateau was not detected at tamping forces of less than approximately 60 N.

In practice, it was found that the pneumatic head was able to control fill weight but could only change it in small increments. If larger changes were needed, then the tamping pins had to be adjusted or the powder bed depth had to be altered. A significant finding in this work was that most of the powder that ultimately formed the plug entered the holes of the dosating disk by flow under gravity as the disk rotated. Only a minor portion of the plug came from powder pushed in by the tamping pins. It follows, therefore, that the ultimate plug weight is largely dependent on the flow properties of the powders rather than the force exerted by the tamping pins.

In a further study using this apparatus (Podczeck, 2001), the station bearing the instrumented head was varied so that the contribution of each station to plug formation could be assessed. It was found that the plug achieved its final length and density at station four, and so the best way to control plug weight would be to position the instrumented head at this station. Podczeck suggested that a feedback device could be achieved with the instrumented head at station four and a non-instrumented pneumatic head at station three. The internal pressure of the latter would be controlled by electrical signals from the former.

An important finding in this work was that, for any given plug, each successive tamp caused further densification. It will be recalled that Shah *et al.* (1986) reported that the density of each segment of the plug did not increase despite multiple applications of tamping force. Podczeck explained these contradictory results with reference to the consolidation mechanisms of the solids involved. Shah and colleagues had used lactose and dicalcium phosphate dihydrate in their studies. Both of these undergo consolidation by particle fracture, and since the forces involved in plug formation are well below the yield points of such substances, it is unlikely that these particles would undergo fragmentation. However, Podczeck used microcrystalline cellulose and pregelatinised starch. Both of these are ductile materials and are readily deformed by low forces. Hence progressive consolidation can be anticipated.

The simulation of dosating disk capsule-filling machines

A disadvantage of conventional tablet presses and capsule-filling equipment is that several hundred grams or even kilograms of particulate material may be needed for them to operate efficiently or even at all. In some circumstances, this quantity may not be available, and even if it were, considerable wastage would be unavoidable. It was primarily for this reason that tablet press simulators were introduced in the 1970s. These are hydraulic presses fitted with a die and two punches, the movement of the punches being precisely controlled. The die is usually filled manually and individual tablets are made and examined. A further advantage of tablet press simulators is that by adjusting the rate of movement of the punches to conform to a predetermined pattern, the speed of compression as well as the applied force can be controlled. Hence one versatile simulator can, in theory at least, imitate the mode of operation of any tablet press.

Though the advantages of economy and versatility can apply equally to simulation of capsule-filling equipment, there are other considerations that require a different emphasis. As already discussed, there are fewer suitable points of attachment for transducers to capsule-filling machinery without major modifications of the equipment. Hence a simulator can be designed so that it has the sufficiency of robust fixing points for the transducers that would not be available on the equipment itself. A further feature is that they are usually somewhat cheaper to construct and operate than their tablet counterparts. As mentioned above, tablet press simulators are hydraulic presses, and much of their expense arises from the need to move large volumes of hydraulic fluid rapidly and precisely at high pressures. Since capsule-filling simulators apply much lower forces than tablet presses, their control systems are less demanding and, therefore, cheaper. For example, the simulator described by Britten et al. (1995) is operated pneumatically from a commercial compressed air cylinder.

Any simulator must be capable of exerting the forces and reproducing the patterns of component movement of the equipment that it is designed to imitate. Since equations have been derived to predict punch movement of both rotary and eccentric presses (Chapter 9), a 'universal' simulator for tablet presses is at least theoretically possible. Not so with capsule-filling equipment, since each manufacturer has a different mechanism for inducing component movement. Consequently, movement is simulated either by using isolated parts of the equipment (e.g. Jolliffe et al., 1982) or by measuring component movement on an actual machine with a transducer and incorporating this knowledge into the design of the simulator (e.g. Britten et al., 1995). The latter method, of course, depends on the measuring device being properly sited, and lack of such siting points is one of the reasons for using a simulator in the first place.

The earliest attempts to simulate plug formation by tamping were not intended to imitate capsule filling per se, but to produce plugs under controlled conditions for dissolution studies. For example, Lerk et al. (1979) used a hand-operated press fitted with a plunger and die to produce plugs at a known constant force. This force was measured by a load cell fitted to the top of the plunger.

A device designed by Höfliger and Karg to select the correct dosating disk for a given formulation was used by Jones (1988, 1998) as a simulator. The device had one tamping pin, the force exerted by which could be measured with a load cell. Movement could also be detected, from which plug length could be calculated. Davar et al. (1997) used an Instron testing machine for the same purpose and confirmed their results by using the instrumented capsule-filling apparatus described above.

As stated above, the production of powder plugs by dosating disk machines is somewhat analogous to the compression of tablets. In their paper describing pin displacement measurements on dosating disk machines, Cropp et al. (1991) pointed out that if tamping pin displacement was known it should be possible to make a compaction simulator that could mimic the component movement and the low forces involved in plug formation. This development was reported by Heda et al. in 1999, using a Mand tablet compaction simulator so that both force and tamping pin movement could be independently controlled.

Since powder plugs for capsule fills have a greater height-to-diameter ratio than tablets, it was necessary to use a die that was much deeper than normal. A diameter of 5.71 mm was chosen, this being the same diameter as the tamping pin used to prepare a plug for a number 1 size capsule. In this, plugs of heights up to 12 mm could be prepared. Anhydrous lactose, microcrystalline cellulose and pregelatinised starch were used, and the die was lubricated by hand with a saturated solution of magnesium stearate in acetone. Forces up to around 400 N were used.

A feature of tablet press simulators is that they can be operated at a range of punch speeds, and Heda et al. (1999) studied plug formation at constant punch speeds of 1, 10 and 100 mm s^{-1}, using a saw-tooth waveform. The last speed is slightly greater than that of the tamping pins in a Bosch GKF 330 capsule-filling machine, and considerably higher than speeds encountered in a dosating piston machine such as the Zanasi LZ64.

These authors discovered that force transmission through the length of the plug was very dependent on plug length, as measured by the ratio of force detected at the lower punch to that applied by the upper punch. This they attributed to the large difference in packing densities between the two ends of the plug, which, in turn, leads to poor axial force transmission. Nevertheless, they found that their data could be fitted to the Shaxby–Evans equation (Shaxby and Evans, 1923), which predicts that force applied by the upper punch decays exponentially towards the lower punch at a rate dependent on plug dimensions and a constant that is substance specific. They also found that the Heckel (1961) and Kawakita (Kawakita and Ludde, 1970/71) equations, which have been applied to the study of tablet compression, apply equally well to the low-force environments of plug formation. Punch speed had no effect on the properties of plugs made from lactose, but with microcrystalline cellulose and pregelatinised starch, peak forces were at a maximum at a punch speed of 10 mm s^{-1}, again confirming findings in tablet preparation. As plug length was decreased, the forces that were generated also decreased through diminution of the total resistance to compression.

The authors pointed out that, though the information gained in this study is more obviously applicable to dosating disk machines, consolidation also occurs in dosating tube equipment and the same low-force powder physics could well apply to those machines too.

Instrumentation of dosating nozzle capsule-filling machines

The first published reports of instrumented capsule-filling equipment of the dosating nozzle type were made by Cole and May (1972, 1975), using a Zanasi LZ64 machine. Foil strain gauges (120 Ω) were mounted on flat surfaces ground on to opposing sides of the dosator piston shank, and the wiring from the strain gauges was led out through a hole drilled along the length of the piston to an amplifier and a recording oscillograph. Compression forces were thus measured along the axis of the dosator piston. Since the dosator on this type of machine constantly rotates during operation, Cole and May were faced with a problem similar to that encountered by users of instrumented rotary tablet presses, namely how to prevent twisting and rupture of the electrical cables leading to and from the transducers. They overcame this by fitting a planetary gear to the dosator head, which caused the dosator to make a complete clockwise rotation for each anticlockwise revolution of the dosator support arm.

Using this device, Cole and May were for the first time able to record the compression and ejection forces generated during plug formation and transfer for plugs made from lactose, microcrystalline cellulose and pregelatinised starch.

These powders were used either unlubricated or after the addition of 0.5% magnesium stearate. Cole and May noted that the low magnitude of the forces (typically 20–30 N) made measurement difficult, since a high degree of signal amplification was needed, with attendant problems caused by the signal-to-noise ratio. They reported that up to four regions could be distinguished on the oscillograph trace:

1. A force, tens of newtons in magnitude, represented the compression force generated as the dosator was pushed into the powder bed. It is worth noting that the compressive force in a tablet press would be of the order of tens of kilonewtons.
2. A force of a few newtons, termed the retention force, was detected while the dosator was being raised from the powder bed and positioned over the empty capsule shell. The presence of a retention force proved that the plug remained in contact with the face of the piston during transfer.
3. An ejection force occurred when the plug was pushed out of the dosator into the capsule shell. This force was very dependent on lubrication. For example, with unlubricated lactose, it could progressively rise to several hundred newtons, but the addition of 0.5% magnesium stearate virtually abolished it.
4. A 'drag force' resisted the full retraction of the dosator piston after ejection of the plug was complete, indicating that the dostor rod was in tension. The drag force, the magnitude of which was also dependent on the degree of lubrication, was attributed to particles lodging between the sides of the dosator rod and the inner surface of the nozzle. It was most marked with pregelatinised starch, which had the smallest particle size of the powders studied, and thus would be expected to show greatest penetration beyond the tip of the dosator piston.

Shortly after the full publication of the pioneering work of Cole and May, further work describing the instrumentation of a Zanasi LZ64 machine was published by Small and Augsburger (1977). This was the first part of a major body of work in the field of instrumented capsule-filling machines to come from a group led by Professor Augsburger at the University of Maryland.

One of the aims of Small and Augsburger was to modify the original equipment as little as possible. Four foil strain gauges (120 Ω) were bonded to flattened areas of the middle piston shank to give a complete Wheatstone bridge. They faced the same problem as Cole and May in getting the electrical supply to the transducer and the signals out while the dosator heads was rotating. They solved this by using a mercury contact swivel between the instrumentation and the amplifier. Signals were then fed into an oscilloscope or a recorder.

Small and Augsburger (1977) detected the compression, retention, ejection and drag forces reported by Cole and May. Additionally, they showed that the compression event itself could be divided into two stages. The first stage occurred as the dosator plunged into the powder bed. No force was detected until the dosator had penetrated to a depth equal to the height of the piston in the dosator. Then, a force built up as the dosator continued downwards, the maximum force coinciding with maximum penetration. This they termed pre-compression. Then the main compression event took place, caused by downward movement of the piston inside the dosator, the body of the dosator remaining stationary. The adjustable movement of the piston is a feature of the Zanasi design but was not commented upon by Cole and May in their work.

Retention forces were not observed with lubricated powders, and the authors surmised that this was because the lubricant had allowed the plug to slip inside the dosator, contact between the plug and the piston tip thereby being lost. The negative force after ejection that been reported by Cole and May was also noted by Small and Augsburger, and attributed to the same cause.

Thus from the work of Cole and May (1975) and Small and Augsburger (1977), the sequence of changes in force that occur when a plug is formed in a dosating nozzle machine can be envisaged (Figure 10.4). The instrumentation consisted of force transducers mounted on the piston shank. It must be borne in mind that Figure 10.4 shows all the events that can occur. The magnitude of these events will depend on the formulation, especially the degree of lubrication, and some events might not be detectable at all.

Point A. The dosating nozzle descends into the powder bed. Powder enters the tube and comes into contact with the piston tip, where a force is detected.

Point B. The dosating nozzle continues to descend, and hence an increasing force is detected at the piston tip. There is no *relative* movement between the piston tip and the end of the dosating nozzle. At B, the nozzle descends no further, and a constant force – the pre-compression force – is detected.

Points C–E. The piston now moves down the dosating nozzle, compressing the powder in it. Force increases to a maximum at D, after which it decreases rapidly as the dosator is drawn out of the powder bed. However, force does not drop to zero if the plug remains in contact with the piston tip (point E).

Between points E and F. The dosator assembly is rotated so that the nozzle containing the plug is positioned over the body of an empty capsule shell.

Points F and G. The piston moves downwards, thereby ejecting the plug from the dosator nozzle. The magnitude of the ejection force (point G) is very dependent on the concentration of lubricant in the formulation.

Point H. After the plug has been ejected, there is now no contact between it and the piston tip. Hence force should fall to zero, but in fact it may fall below zero. This means that an extensive stress rather than a compressive stress is detected by the transducers. This has been termed the 'drag force' and is attributed to powder on the inside of the dosator tube preventing full retraction of the piston.

In a subsequent study, Small and Augsburger (1978) examined the effects of powder bed height, piston height, lubricant type, lubricant concentration and compression force on the force needed to eject the plug from the dosator, using three fillers (microcrystalline cellulose, pregelatinised starch and anhydrous lactose). As might be expected, ejection force increased with compression force. However, ejection force was also directly proportional to powder bed height and piston height. As these two factors are increased, the plug length is also increased and hence there is a greater area of contact between the sides of the plug and the inside of the dosator tube. It follows from this work that con-

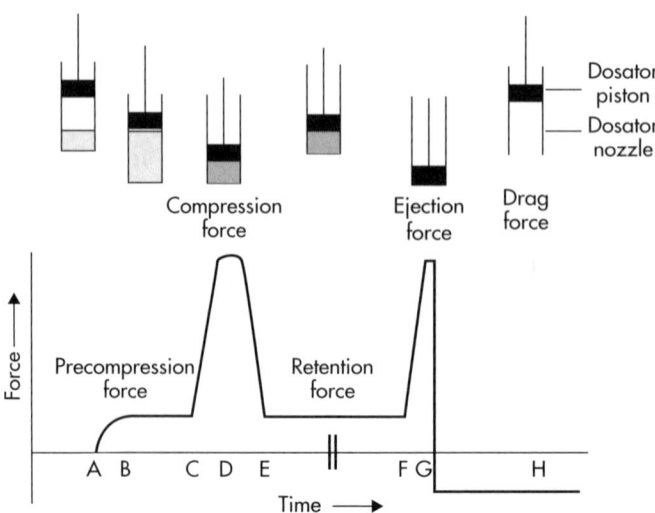

Figure 10.4 Force–time profile from instrumented dosating nozzle capsule-filling equipment. See text for details.

sistency in both powder bed height and piston settings is necessary for reproducible plug properties.

Ejection force minima were achieved with 1% magnesium stearate for anhydrous lactose, 0.5% for microcrystalline cellulose and 0.1% for pregelatinised starch. It is interesting to note that, despite the much lower forces used for compression, the ejection forces were comparable to those encountered in tablet presses and the required levels of lubricant were very similar to those needed for tablet formulations containing these three diluents.

Properties of the powder plug such as physical strength and release characteristics that are dependent on the compression force were investigated by Mehta and Augsburger (1981a), using the instrumented machine described above. They studied the effect of magnesium stearate concentration on plug strength and dissolution of active ingredient (hydrochlorothiazide) at a constant compressing force of 15 kg (approximately 150 N). They measured the physical strength of plugs by a three-point bending test and found that this could be correlated with dissolution rate. With microcrystalline cellulose, the strength of the plugs decreased markedly as the magnesium stearate concentration was increased. A similar reduction has been noted with microcrystalline cellulose tablets lubricated with magnesium stearate (Bolhuis and Hölzer, 1996).

The effect on release characteristics was more complex. At low lubricant concentrations, the reduction in physical strength of the plugs permitted easier water penetration, but as the level of lubricant was raised, increased hydrophobicity inhibited drug release. With lactose, strength was not significantly reduced, and only the retarding effect on dissolution was noted. This too has parallels in tablet formulation.

The instrumented Zanasi LZ64 filling apparatus was modified by Botzolakis et al. (1982), who replaced the potentially hazardous mercury swivel contact with slip-rings, an approach that had been used with rotary tablet presses by Ridgway Watt and Rue (1979). They studied the effect of disintegrating agents on capsule fills containing hydrochlorothiazide and paracetamol (acetaminophen) and pointed out that the capsules had been hand-filled in many previous studies on the release of active ingredient from hard-shell capsules, with a resulting high porosity. It was therefore, not surprising that disintegrating agents appeared to have little effect, since there was no structure for them to press against, and hence wettability and water penetration would be the more important factors. Botzolakis et al. (1982) were able to keep piston height, powder bed height and compression force constant while they examined the effects of a range of disintegrating agents on capsule fills made from dicalcium phosphate dihydrate lubricated with magnesium stearate. All disintegrating agents improved the release of active ingredient, with cross-linked sodium carboxymethyl cellulose being the most effective and cross-linked polyvinylpyrollidone being the least effective.

In a more elaborate study, Botzolakis and Augsburger (1984) used hydrochlorothiazide mixed with either dicalcium phosphate dihydrate or anhydrous lactose in a three-factor, two-level factorial design, the three factors being disintegrant concentration, lubricant concentration and compression force. The responses were physical strength, disintegration time and dissolution rate. They found that disintegration times did not always have the same rank order as dissolution rates, but that all three factors and their interactions had significant effects, the magnitude of which differed according to the solubility of the filler in water.

Fitting LVDTs to dosating nozzle capsule-filling equipment was first reported by Mehta and Augsburger (1980). They monitored movement of the instrumented piston described by Small and Augsburger (1977), a task complicated by the rotation of the piston during operation. This was overcome by threading a spring-loaded rod on to the core of the LVDT, which, in turn, was kept in contact with a bracket fixed to the dosator housing. Cables for all transducers were connected to the mercury swivel assembly described above.

The force and displacement traces obtained by Mehta and Augsburger (1980) confirmed suggestions made earlier. When the dosator enters the powder bed, pre-compression force develops without piston movement. After the dosator has descended fully, the piston moves downwards, exerting the compression force. Mehta and Augsburger reported that the maximum compression force preceded the point of maximum piston displacement by approximately 40 ms and attributed this to the action of the overload spring. However, a similar non-coincidence of force and displacement maxima has been noted in the force–displacement curve for tablets and this has been linked to viscoelastic behaviour of the powder particles (Ho and Jones, 1988). At the ejection stage, downward movement of the piston results in a rise in the force detected at the piston tip, and this rapidly falls away as the frictional forces holding the plug in the nozzle are overcome. Though the authors signified their intention to calculate work expenditure during plug formation by calculating areas under the force–displacement curves (Mehta and Augsburger, 1981b), few results seem to have been published. Since forces are so low, the work expended will consequently also be low. The accuracy of the results would be highly dependent on accurate measurement of displacement and this would undoubtedly be complicated by the presence of the return spring in the dosator.

Teams of workers other than Augsburger's group have fitted instrumentation to dosating nozzle capsule-filling equipment. For example, Mony et al. (1977) fitted piezoelectric load washers to the ends of the pistons of a Zanasi RV59 machine. With this arrangement, a force can only be detected during the compression and ejection events when the piston is being depressed and so pre-compression, retention and residual forces cannot be studied. These workers investigated the effect of magnesium stearate and talc on compression and ejection forces. A similar study was carried out by Maury et al. (1986), using load washers mounted not on the pistons but on the compression and ejection platens. Again only compression and ejection can be studied. Rowley et al. (1983) attached a load washer to the ejection knob of a Zanasi LZ64, and this arrangement can only be used to study the ejection event.

An instrumented Zanasi LZ64 was used by Hauer et al. (1993) to examine the formulation variables of a mixture of microcrystalline cellulose – a viscoelastic material – and anhydrous lactose – a brittle material. They found that the better the powder flow the more variable the fill weight, as the mixture was more difficult to densify. Magnesium stearate proved to be a superior lubricant to stearic acid, but the concentration was shown to be critical in each case.

The simulation of dosating nozzle capsule-filling equipment

An important development in capsule machine instrumentation came with the publication by Jolliffe et al. in 1982 of details of the construction of a dosating tube simulator based on an mG2 model G36 machine. The problems of connecting electrical wiring to a rotating component have been discussed above. In conventional mG2 machines, the filling turret rotates and the powder hopper beneath it is stationary. In this simulator, these roles were reversed so that the turret to which the transducers were connected was stationary and the powder hopper rotated around the dosator. There was no relative movement between the feed tray and the nozzle at the moment when the dosator entered the powder bed. Four semiconductor strain gauges were mounted on the dosator piston in a Wheatstone bridge configuration to measure stress, and displacement transducers monitored the vertical movements of the piston and the dosator nozzle. In this way, the movement of the whole dosator and the relative movement of dosator and piston could be followed. A force could be applied in two ways. A pre-compression force was exerted by adjusting the height of the piston in the nozzle, and this was found to be particularly useful to consolidate beds of low bulk density. Compression force was exerted by movement of the piston when the nozzle was in the powder bed and was altered by raising or lowering the compression cam, the precise position of which was recorded by the piston movement transducer.

A considerable body of work carried out on this simulator has been published. Newton and his colleagues were particularly interested in elucidating those factors that contributed to uniformity of plug weight. They found that fine lactose particles gave acceptable uniformity over a wide range of compression settings, whereas the larger the particles, the smaller the range over which satisfactory filling was achieved (Jolliffe and Newton, 1982). Fine cohesive powders gave the best results because they underwent greater volume reduction on compression than coarser particles.

Jolliffe and Newton (1978, 1980) had shown theoretically that a stable arch had to be formed at the outlet of the nozzle in order for a powder to be retained within the dosator nozzle during transfer from powder bed to capsule shell. This was related to the flow properties of the powder, in that cohesive powders would require a lower degree of compression for the arch to form. They also surmised that arch formation would depend on the nature of the surface of the inside of the nozzle, which would, in turn, govern the frictional forces between the nozzle and powder. The surface could be affected either by roughness of the metal or by a coating of powder. They prepared nozzles with a range of surface textures and confirmed that there is an optimum degree of surface roughness needed to ensure powder retention in the nozzle (Jolliffe and Newton, 1983a). These findings were confirmed when they used an mG2 G36 production machine, thereby validating their original approach of using a simulator (Jolliffe and Newton, 1983b). They found that fine cohesive powders gave acceptable fill weight uniformity over a wide range of compression settings, but this range was reduced with more free-flowing powders.

A series of papers by Tan and Newton (1990a–d) extended this work using the same simulator, which was now connected to a computer to capture and manipulate data. Using five common capsule diluents, the relationship between uniformity of fill weight and a range of parameters related to powder flow were investigated. Particle size, morphology, bulk density and compressive force were found to be important. They found that there was no correlation between uniformity of weight and measures of friction such as angle of internal flow and angle of effective friction (Tan and Newton 1990a). After each filling cycle, the dosator was weighed, and hence information was gained on the build-up of powder on the inner surface. It was found that lactose was particularly prone to binding.

They found that the texture of the inner wall of the dosator had no significant influence with powders with low binding affinity such as microcrystalline cellulose and pregelatinised starch (Tan and Newton, 1990b).

A later paper (Tan and Newton, 1990c) showed that fill weight variability also depended on powder bed density. The most uniform weights were achieved when no compressing force was applied during the filling process. As compression was increased, fill weight decreased. This was attributed to coating of the wall of the nozzle and loss of powder as particles were forced behind the tip of the piston, which in extreme cases led to the piston jamming in the nozzle. Tan and Newton (1990d) then compared the observed plug densities calculated from plug dimensions with predicted values based on knowledge of powder bed density and piston position. Correlation was poor because of weight variation, which was greatest with fine powders at high compression settings.

Another simulator based on the dosating nozzle principle was constructed by Britten and colleagues (Britton and Barnett, 1991; Britton et al., 1995). In this pneumatically driven apparatus that simulated the Macofar MT13-2 machine, there were no rotating components at all. In a conventional Macofar machine, the dosator nozzles are plunged into the powder bed, a plug is withdrawn and then ejected. In this simulator, the dosator mechanism was stationary and the powder brought to it by upward vertical movement of the powder bowl. A pre-compression force could be exerted, followed by a compression force that was applied by a downward movement of the piston. Once the plug was formed, the dosator nozzle ascended out of the powder bed, and the plug was ejected by means of further downward motion of the piston. No attempt was made to eject the plug into an empty capsule shell.

Compression force was measured by semiconductor strain gauges fitted to the dosator piston and arranged in a Wheatstone bridge conformation. An additional development on this apparatus was to fit strain gauges to the outer surface of the dosating nozzle in order to measure axial stresses brought about by the presence of the plug in the dosator. These strain gauges were positioned 6 mm from the tip of the dosator. Since direct contact between the piston and the LVDT was not feasible, a small arm, fitted to the piston shank and in contact with the LVDT was used to determine the position of the piston within the dosator. Vertical movement of the powder bowl was also determined by LVDT. The output of all transducers was fed into a computer and manipulated with a spreadsheet program.

This simulator could be set to operate in a variety of modes, all at a range of bowl and piston speeds:

- pre-compression simulation, when the powder plug was formed solely by the dosator plunging into the powder bed
- constant displacement simulation, when an additional tamp was applied to each plug by the dosator piston moving a predetermined distance
- constant pressure simulation, when the piston was allowed to travel as far as possible until the resistance of the powder to undergo further consolidation equalled the applied compression pressure.

Two-factor, two-level factorial designs were used to study variation in plug weight and density in relation to compression pressure, pre-compression velocity, compression velocity and ejection velocity using plugs made from pre-gelatinised starch and lactose (Britten et al., 1996). It was possible to form plugs of starch without lubrication, but addition of 1% magnesium stearate was necessary for lactose. The rate of ejection had no effect on plug weight or density. However, an increase in the pre-compression speed caused a fall in plug weight. At higher speeds, powder was pushed ahead of the nozzle rather than entering it, and there was also less consolidation. A similar observation of reduced consolidation was made during tablet compression, especially with pregelatinised starch (Armstrong and Palfrey, 1989).

When the simulator was run in constant pressure mode, the effect of pre-compression velocity disappeared, and there was no evidence that higher pressures had a significant effect on plug density. Consequently, a relatively high tamping pressure is indicated if reproducible and predictable plug weights are required. However, this may cause an increase in the physical strength of the plugs, which may, in turn, delay drug release (Mehta and Augsburger, 1981a). Hence for any given formulation, an optimum pressure must be sought. Britten et al. (1996) noted that no plugs fell out of the dosating tube before active ejection by the piston despite the radial pressures being as low as 0.01 MPa. It followed that, from the point of view of plug retention, high compression pressures are not required, a view also expressed by Tan and Newton (1990d).

A more elaborate study on lactose using the same simulator was carried out by Tattawasart and Armstrong (1997), who studied the effects of lubricant concentration, dosator pressure and dosator piston height on plug properties by means of a three-factor, three-level Box Behnken design followed by stepwise multiple regression. While pressure and piston height had significant effects on plug properties, lubricant concentration did not, and it was concluded that the lowest concentration of magnesium stearate examined (0.5%) was more than adequate.

References

Armstrong NA, Palfrey LP (1989). The effect of machine speed on the consolidation of four directly compressible tablet diluents. *J Pharm Pharmacol* 41: 149–151.

Bolhuis GK, Hölzer AW (1996). Lubricant sensitivity. In Alderborn G, Nystrom C (eds), *Pharmaceutical Powder Compaction Technology*. New York: Marcel Dekker, pp 517–560.

Botzolakis JE, Augsburger LL (1984). The role of disintegrants in hard-gelatin capsules. *J Pharm Pharmacol* 36: 77–84.

Botzolakis JE, Small LE, Augsburger LL (1982). Effect of disintegrants on drug dissolution from capsules filled on a dosator-type automatic capsule filling machine. *Int J Pharm* 12: 341–349.

Britten JR, Barnett MI (1991). Development and validation of a capsule filling machine simulator. *Int J Pharm* 71: R5–R8.

Britten JR, Barnett MI, Armstrong NA (1995). Construction of an intermittent-motion capsule filling machine simulator. *Pharm Res* 12: 196–200.

Britten JR, Barnett MI, Armstrong NA (1996). Studies on powder plug formation using a simulated capsule filling machine. *J Pharm Pharmacol* 48: 249–254.

Cole GC, May G (1972). Instrumentation of a hard shell encapsulation machine. *J Pharm Pharmacol* 24: 122P.

Cole GC, May G (1975). The instrumentation of a Zanasi LZ/64 capsule filling machine. *J Pharm Pharmacol* 27: 353–358.

Cropp JW, Augsburger LL, Marshall K (1991). Simultaneous monitoring of tamping force and pin displacement (F–D) on an Höfliger-Karg capsule filling machine. *Int. J. Pharm* 71: 127–136.

Davar N, Shah R, Pope DG, Augsburger LL (1997). The selection of a dosing disk on a Höfliger-Karg capsule-filling machine. *Pharm Technol* 21: 32–48.

de Blaey CJ, Polderman J (1970). Compression of pharmaceuticals. 1: The quantitative interpretation of force–displacement curves. *Pharm Weekblad* 105: 241–250.

Hauer B, Remmele T, Züger O, Sucker H (1993). Rational development and optimisation of capsule formulations with an instrumented dosator capsule filling machine. 1: Instrumentation and influence of the filling material and the machine parameters. *Pharm Ind* 55: 509–515.

Heckel RW (1961). Density-pressure relationships in powder compression. *Trans Metal Soc AIME* 221: 671–675.

Heda PK, Muller FX, Augsburger LL (1999). Capsule filling machine simulation. 1: Low force powder compaction physics relevant to plug formation. *Pharm Dev Tech* 4: 209–219.

Ho AYK, Jones TM (1988). Punch travel beyond peak force during tablet compression. *J Pharm Pharmacol* 40: 75P.

Jolliffe IG, Newton JM (1978). Powder retention within a capsule dosator nozzle. *J Pharm Pharmacol* 30: 41P.

Jolliffe IG, Newton JM (1980). The effect of powder coating on capsule filling with a dosator nozzle. *Acta Pharm Technol* 26: 324–326.

Jolliffe IG, Newton JM (1982). An investigation of the relationship between particle size and compression during capsule filling in an mG2 simulator. *J Pharm Pharmacol* 34: 415–419.

Jolliffe IG, Newton JM (1983a). The effect of dosator wall texture on capsule filling with the mG2 simulator. *J Pharm Pharmacol* 35: 7–11.

Jolliffe IG, Newton JM (1983b). Capsule filling studies using an mG2 production machine. *J Pharm Pharmacol* 35: 74–78.

Jolliffe IG, Newton JM, Cooper D (1982). The design and use of an instrumented mG2 capsule filling machine simulator. *J Pharm Pharmacol* 34: 230–235.

Jones BE (1988). Powder formulations for capsule filling. *Mfg Chem* 59: 28–30, 33.

Jones BE (1998). New thoughts on capsule filling. *STP Pharmacol* 3: 777–783.

Jones BE (2001). The filling of powders into two-piece hard capsules. *Int J Pharm* 227: 5–26.

Kawakita K, Ludde KH (1970/71). Some considerations on powder compression equations. *Powder Technol* 4: 61–68.

Lerk CF, Lagas M, Lie-A-Huen L et al. (1979). In vitro and in vivo availability of hydrophilised phenytoin from capsules. *J Pharm Sci* 68: 634–638.

Maury M, Heraud P, Etienne A et al. (1986). Measurement of compression during the filling of capsules. In *Proceedings of the 4th International Conference on Pharmaceutical Technology*, Paris, pp 384–388.

Mehta AM, Augsburger LL (1980). Simultaneous measurement of force and displacement in an automatic capsule filling machine. *Int J Pharm* 4: 347–351.

Mehta AM, Augsburger LL (1981a). A preliminary study of the effect of slug hardness on drug dissolution from hard gelatin capsules filled on an automatic capsule-filling machine. *Int J Pharm* 7: 327–334.

Mehta AM, Augsburger LL (1981b). Quantitative evaluation of force–displacement curves in an automatic capsule-filling operation. In *30th National Meeting of the American Association of Pharmaceutical Scientists*, St Louis.

Mony C, Sambeat C, Cousins C (1977). The measurement of compression during the formulation and filling of capsules. In *Proceedings of the 1st International Conference on Pharmaceutical Technology*, Paris, pp 98–108.

Podczeck F (2000). The development of an instrumented tamp-filling capsule machine. 1: Instrumentation of a Bosch GKF 400S machine and feasibility study. *Eur J Pharm Sci* 10: 267–274.

Podczeck F (2001). The development of an instrumented tamp-filling capsule machine. 2: Investigations of plug development and tamping pressure at different filling stations. *Eur J Pharm Sci* 12: 515–521.

Podczeck F (2004). Dry filling of hard capsules. In Podczeck F, Jones BE (eds), *Pharmaceutical Capsules* 2nd edn. London: Pharmaceutical Press, 119–138.

Podczeck F, Newton JM (1999). Powder filling into hard gelatine capsules on a tamp filling machine. *Int J Pharm* 185: 237–254.

Ridgway Watt P, Rue PJ (1979). The design and construction of a fully instrumented tablet machine. In *Proceedings of the International Conference of Pharmacy Technology and Product Manufacture*, Copenhagen.

Rowley D, Hendry R, Ward MD, Timmins P (1983). The instrumentation of an automatic capsule filling machine for formulation design studies. In *Proceedings of the 3rd International Conference on Pharmaceutical Technology*, Paris, pp 287–291.

Shah KB, Augsburger LL, Small LE, Polli GP (1983). Instrumentation of a dosing disk automatic capsule filling machine. *Pharm Tech* 7: 42–54.

Shah KB, Augsburger LL, Marshall K (1986). An investigation of some factors influencing plug formation and fill weight in a disk-type automatic capsule filling machine. *J Pharm Sci* 75: 291–296.

Shah KB, Augsburger LL, Marshall K (1987). Multiple tamping effects on drug dissolution from capsules filled on a dosing disk type automatic capsule filling machine. *J Pharm Sci* 76: 639–645.

Shaxby JH, Evans JC (1923). The variation of pressure with depth in columns of powders. *Trans Faraday Soc* 19: 60–72.

Small LE, Augsburger LL (1977). Instrumentation of an automatic capsule-filling machine. *J Pharm Sci* 66: 504–509.

Small LE, Augsburger LL (1978). Aspects of the lubrication requirements for an automatic capsule filling machine. *Drug Dev Ind Pharm* 4: 345–372.

Tan SB, Newton JM (1990a). Powder flowability as an indicator of capsule filling performance. *Int J Pharm* 61: 145–155.

Tan SB, Newton JM (1990b). Capsule filling performance of powders with dosator nozzles of different wall texture. *Int J Pharm* 66: 207–211.

Tan SB, Newton JM (1990c). Influence of compression setting ratio on capsule fill weight and weight variability. *Int J Pharm* 66: 273–282.

Tan SB, Newton JM (1990d). Observed and expected powder plug densities obtained by a capsule dosator nozzle system. *Int J Pharm* 66: 283–288.

Tattawasart A, Armstrong NA (1997). The formation of lactose plugs for hard shell capsule fills. *Pharm Dev Technol* 2: 335–343.

Further reading

Augsburger LL (1988). Instrumented capsule filling machines. Methodology and application to product development. *STP Pharma* 4: 116–122.

Hardy IJ, Fitzpatrick S, Booth SW (2003). Rational design of powder formulations for tamp filling processes. *J Pharm Pharmacol* 55: 1593–1599.

11

Automatic control of tablet presses in a production environment

Harry S Thacker

Introduction

In early research into the compaction process using multi-station rotary presses, two methods were used to measure the force applied to the material in the die. One method was to use strain gauges fitted to a pair of punches and the other was by fitting the strain gauges to a stationary part of the machine. The former method has the advantage of making the measurement as close as possible to the place where the compression is occurring and is, therefore, less likely to be subject to errors. It has two disadvantages. It requires a means, such as radio-telemetry, of transmitting the output from the gauges to a stationary receiver and it is only capable of taking measurements from one station of the press. The second method has the disadvantage of the strain gauges being placed further away from the point where the compression occurs, with possible loss of accuracy, but the measurements can be taken from every station of the press. A machine was built using both systems and Wray et al. (1966) reported good agreement between the two systems. However, this would only hold true if the machine is in good mechanical order. If the pressure roll was not rotating freely on the roll-pin, the value of the force measurement would be increased. An increase would also be observed if any of the pivot points in the linkage between the roll carrier and the site of the gauges was tight (Knoechel et al., 1967a; Figure 11.1). Although the early work was designed to study the compaction process, its potential as a means of controlling the process was soon realised.

One of the first examples of the use of compression force measurements to control tablet weight on a rotary tablet press in a production situation was by Knoechel et al. (1967a,b) of the Upjohn Company at Kalamazoo, USA. They used both a Stokes 540-35 press and a Stokes BB2-27

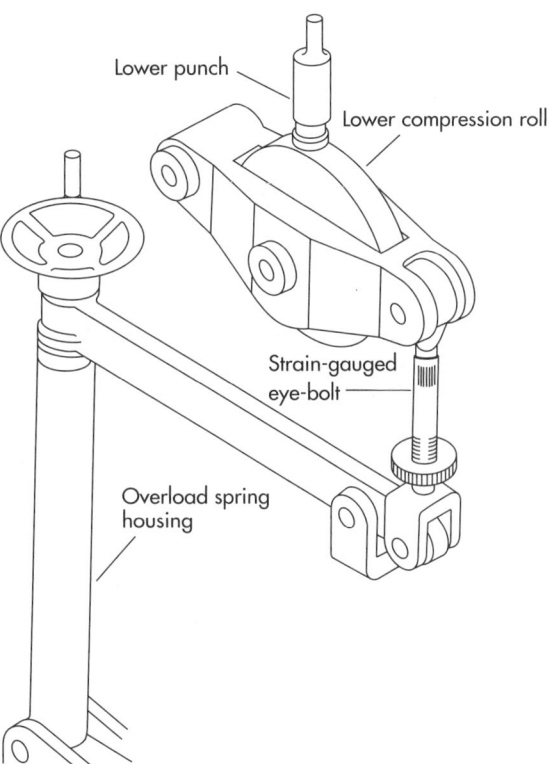

Figure 11.1 Pivot points between the pressure roll carrier and the strain gauge site. This arrangement is simple to fit but its response depends upon the condition of the pivots. (Knoechel et al., 1967a).

press, both fitted with compaction force measuring systems. The strain gauges were fitted to parts of the overload systems of the press and the signal obtained was displayed on an oscilloscope. By observing the trace on the oscilloscope and adjusting the die fill accordingly, the press operators obtained improved tablet weight consistency when compared with the conventional system of sampling and weighing at fixed time intervals. It was a comparatively short step from this system to the development of an automatic method for controlling tablet weight.

The force–weight relationship

If the force used to make a tablet is to be the basis of a system for controlling the tablet weight, the relationship between compression force and tablet weight must be considered. The force used to compress material into a tablet is proportional to the density of the tablet. For the variations in that force to be proportional to the variations in tablet weight, the volume of each tablet must be constant. This is not normally true for tablets made on a conventional tablet press for the following reasons.

1. The punches are not all of the same length, and the varying lengths will give rise to variations in the thickness of the tablet. Even for a new set of tooling, the lengths will vary within manufacturing tolerances. This variation will increase with use, as the punches are unlikely to wear at the same rate. However, the amount of wear that will occur over the time needed to produce a batch of tablets is not normally significant unless the material being compressed is highly abrasive.
2. In the case of punches that are not flat faced, the geometry of the punch tip will vary for the same reasons as those given for the variation in punch lengths.
3. The size of the die bore and consequently the diameter of the tablets will vary for the same reasons as the variations in the punches.
4. The pressure rolls on the machine are unlikely to be perfectly concentric and this will affect the thickness of the tablets. The thickness will vary depending upon which part of the roll is in contact with the punch head when the tablet is compressed. In addition, wear in the bearings in the centre of the roll will alter the effective radius of the roll, giving further variations in tablet thickness.

These factors only vary within fixed limits. The punch lengths and the other factors are not continually changing, at least not during the production of a single batch of tablets. However, this does indicate that if machine is fitted with a tablet weight control system based on compaction force, it is good practice to check the tooling for wear after each batch of tablets is produced and to ensure that the machine is in good mechanical condition.

A pair of punches does not move from one die to another and, therefore, all possible combinations of punch length, tip geometry and die diameter will have been encountered in one revolution of the turret. After a few further revolutions, all combinations of the longest and shortest punches with the minimum and maximum radius of the pressure roll will also have occurred. This means that even though the forces used to compress individual tablets are not necessarily proportional to the weight of the tablet, the mean compaction force is proportional to the average tablet weight.

When the tablet is at the point of compression and the force to compress it is measured, the amount of material in the die and hence the tablet weight is already fixed. This is another reason why it is not possible to control the weight of individual tablets by force measurements. However, adjusting the die fill for subsequent tablets in response to changes in the mean compaction force can control the mean tablet weight, as was shown by Knoechel et al. (1977b).

Other factors that affect the force–weight relationship are tight punches (i.e. where there is excessive friction between the punch and its guide) and changes in the bulk density of the feed material. These should also be taken into account when a tablet weight-control system is in use. Excessive friction between the punch and its guide can result from insufficient lubrication or from fine particles of the material being compressed when entering the punch guide and mixing with the lubricant. This increased frictional force will be measured in addition to the compression force, thereby giving a false reading. In a more sophisticated control system, the frictional force on the punch can be monitored and the press stopped when friction rises above a preset value. Variations in the bulk density of the feed material are quite rare, but where material is being fed from a bulk hopper, slight vibration of the hopper can cause segregation of its contents with the smaller particles percolating to the bottom. As the hopper empties, the mixture being fed to the press contains a greater proportion of coarse particles and has a lower bulk density. Modern control systems can normally cope with this problem, but it can be observed that the control system gradually increases the die fill towards the end of the batch to maintain tablet weight.

A further factor that needs careful attention is the overload system that is fitted to most rotary tablet presses. It is designed to protect the punches from being used at forces greater than their design will accommodate. The overload system is normally set at or below the maximum allowable force for the set of tooling being used, so that in the event of the die containing an excess of material, resulting in an abnormally high compression force, the tooling will not be damaged by the machine attempting to compress it to the set thickness. Overload systems for rotary presses are shown in Figures 11.1 and 11.2. If the press is operating with compression forces very near to overload limit, it is possible that system will not be rigid enough to allow accurate force measurements to be obtained and the control system will not function correctly.

All the above factors are important in ensuring that the control system operates efficiently, but many modern computer-based systems are able to monitor these factors and either report variations or stop the press if they are likely to result in loss of control.

Monitoring systems

The simplest form of control system is one that monitors the compaction force, indicates the variations in the mean compaction force and stops the machine if the mean force exceeds either the upper or the lower preset limit. As there is no direct correlation between the compaction force and the tablet weight, all control systems based on force measurement need to be set up by first adjusting the machine to make tablets of the correct weight, thickness and hardness or crushing strength. When these parameters are within the required limits, the mean force can be set. The unit will then indicate any variation in the mean force. On early electro-mechanical control systems, this was by means of a centre zero meter. The meter was set to zero when the press was making tablets of the correct weight, thickness and crushing strength, and any increase or decrease in the mean compaction force was indicated by the pointer on the meter moving away from its zero position. By adjusting the die fill until the meter again indicated zero, the operator was able to achieve a tablet weight consistency that was better than that achieved with the normal method of sampling and weighing tablets at fixed intervals.

One reason for the success of these simple systems was that small increases in the weight of the tablets give rise to quite large increases in the compaction force. The ratio of the increase in force to the increase in tablet weight will vary from one formulation to another, but a 1% increase in tablet weight can typically give rise to a 10% increase in compression force.

Control systems

Linking a motor to the tablet weight adjustment enabled simple automatic control systems to be developed. These used a feedback loop to adjust the die fill in order to maintain the compression force within a set range. As computer control gradually replaced electromechanical control, it became possible to display the values of the forces on a screen, and a better system of control based on the individual force measurements could then be devised. Although the force measurements from individual tablets do not relate directly to individual tablet weights, the individual measurements can be used to give

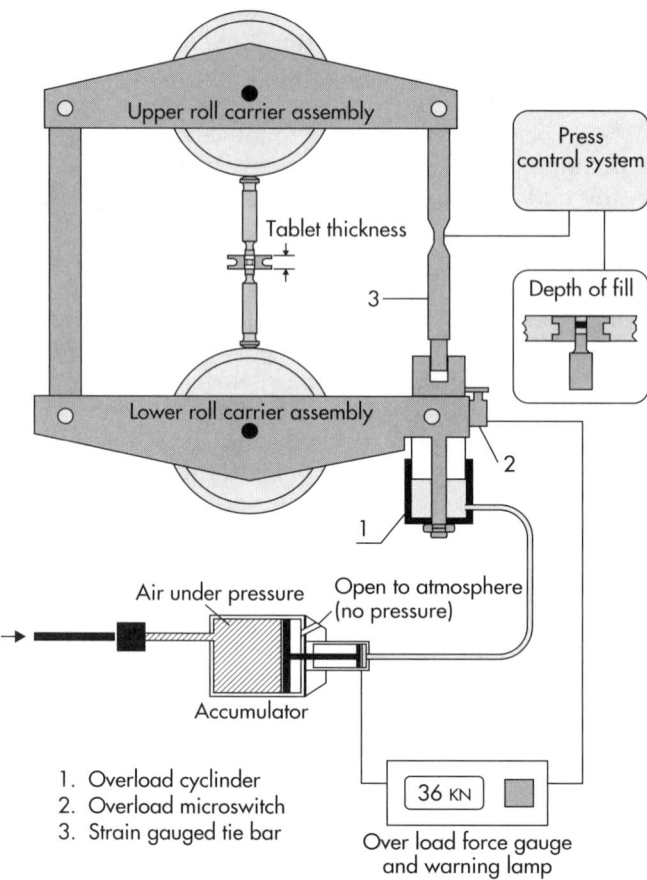

Figure 11.2 The overload system of a rotary press. Manesty control system.

better control than that obtained from a system that simply returns the mean force to a fixed value every time it deviates. A more rapidly reacting method of control is shown in Figure 11.3. The screen shows a graph where the crosses represent the peak force from individual tablets. The lines show the required mean force, the control limits and the reject limits. The system is set by arranging the control limits so that approximately an equal number of tablets exceed the upper and the lower limit. Any change in the mean force will cause more tablets to give rise to forces exceeding one limit and to fewer exceeding the other. The control system simply counts the number of tablets exceeding each limit and adjusts the die fill to restore the balance. This system keeps the mean force at the required level rather than between limits.

Reject systems

It is possible to reject an individual tablet if the force required to compress it is abnormally high or low. This does not guarantee that all rejected tablets are outside a particular weight limit. However, an abnormally high or low compaction force means that the weight of that tablet is at least suspect and it could be rejected on those grounds. The limits could be set to guarantee that all accepted tablets were within

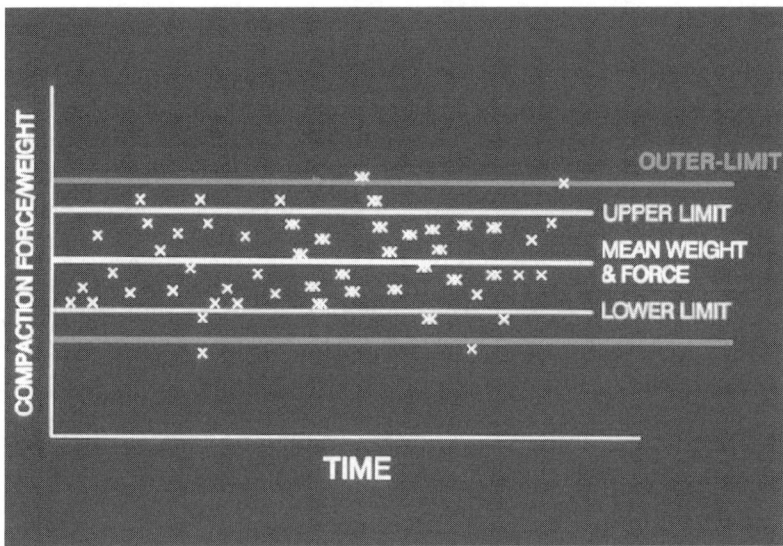

Figure 11.3 A control system for monitoring the weight of individual tablets. X, peak force from individual tablets; upper and lower limits, control limits; outer limits, reject limits.

the set weight limits but this, in turn, would mean that some of the rejected tablets may also be of acceptable weight. If the product were of low value, the rejected tablets could be discarded; if the product had a high value, then the rejects could be weighed individually and the tablets that were within specification recovered.

As the compaction force is measured when the tablet is in the die at the compression position, a device is required that 'remembers' the suspect tablet and operates a reject mechanism when that tablet reaches the ejection position. A reject system of this type can also be used to take sample tablets for weighing.

Weighing systems

As the relationship between compaction force and tablet weight is not perfect, a system based on weighing tablets as well as measuring compaction force is likely to be more reliable. It is not possible with present technology to weigh every tablet. However, a system that weighs sample tablets, calculates the mean tablet weight and uses this information to adjust the force-measuring system to compensate for variations in the force–weight relationship has advantages over a system based on force measurements alone. In this method, every tablet is checked by force measurement, suspect tablets are rejected and samples of the good tablets are checked for weight. Such a system was made possible by the development of a high-speed reject gate that can remove one known tablet from a machine producing thousands of tablets per minute and also with the introduction of automatic balances.

Various sampling routines can be used, but one in which samples are successively taken from each station ensures that every station is monitored. The fastest rate at which tablets can be weighed occurs when the balance calls for another tablet as soon as it has stabilised and recorded the weight of the previous tablet. Using this method continuously would remove too many tablets from the batch.

Most pharmacopoeias have essentially the same standard for uniformity of tablet weight. A sample of 20 tablets is taken; the tablets are weighed individually and the mean weight is calculated. Two limits are set. The weight of not more than 2 tablets in a sample of 20 is permitted to deviate from the mean by a given percentage, the size of which is governed by the mean weight. This is often termed the 'A' limit. No tablets are permitted to deviate in weight from the mean by more than double this percentage. This is the 'B' limit.

Individual companies will often set more stringent in-house limits than those published in pharmacopoeias. In these, the permitted percentage deviations from the mean weight are smaller, but the overall structure of the test is the same.

One method of automatic sampling weighs 20 tablets from successive stations and then waits for a fixed period before weighing the next group of 20. The weights of the 20 tablets can be then be checked against the relevant 'A' and 'B' limits. If any tablet is found to be outside the 'B' limit, or more than two tablets are outside the 'A' limit, the press is stopped. If one or two tablets are outside the 'A' limits, the press is allowed to continue but the time, station number and the tablet weight are reported. This system can also collect weight data for batch records.

Because the tablets forming the sample are separated from the main batch as each individual tablet leaves the die, they will not pass through the tablet-dedusting system on their route to the balance. To obtain an accurate tablet weight, the system should incorporate some form of dedusting for the individual tablet, this being situated between the press and the balance.

The display for the weighing system normally shows the weights of tablets being produced station by station. This display can be either graphic or numeric. In the graphical form, the display shows station number on the x axis and weight on the y axis (Figure 11.4), and is similar to the force display. Each tablet that has been weighed is represented by a dot and the dots gradually build up into a line for each station. The length of the line represents the range of tablet weights being produced by a particular station of the press. An individual dot appearing on the line represents the weight of the last tablet sampled from that station. The upper and lower limits can be displayed on the graph as horizontal lines, so it can be observed whether or not tablets are well within the limits or if any station is producing tablets near to one of the limits. In the numerical mode (Figure 11.5), the display shows a list of, say, 10 stations (or whatever number is convenient to display on the screen). After each station number, the display shows the mean weight and the maximum and minimum weight of the tablets weighed from that station. The remaining stations can be displayed by either scrolling the display or on a screen-by-screen basis.

Weighing systems can be extended to include automatic measurement of tablet thickness and crushing strength. Feedback to the press becomes more complicated when all three measurements are used, as the parameters are interrelated and one cannot be altered while the others remain constant. Most control systems use feedback for only one parameter.

Computer-controlled systems

Current computer-based control systems are capable of providing the machine operator with very detailed information on the operation of the press and of the performance of the product. Modern press design, which must provide for easy cleaning, has resulted in the elimination of manual controls. As the press can now be set up and operated remotely, the process control system and the press operation system have been combined into one unit. A modern control unit can set up the press, monitor and control the process, record data and events, and produce records.

A control system of this type will be described with reference to that used to control Manesty presses (Figure 11.6). Force measurement is normally made using a strain-gauged roll-pin.

The system is operated via a range of computer screens. Data can be entered either via the touch sensitive screen or via a keyboard. The main screen (Figure 11.7) is used for setting the operating parameters.

The following parameters can be reset from the main screen.

1. The mean compression force. This can be for the main compression event or for pre-compression or both, depending upon what is available on the machine and the requirement of the product. If two stages are used,

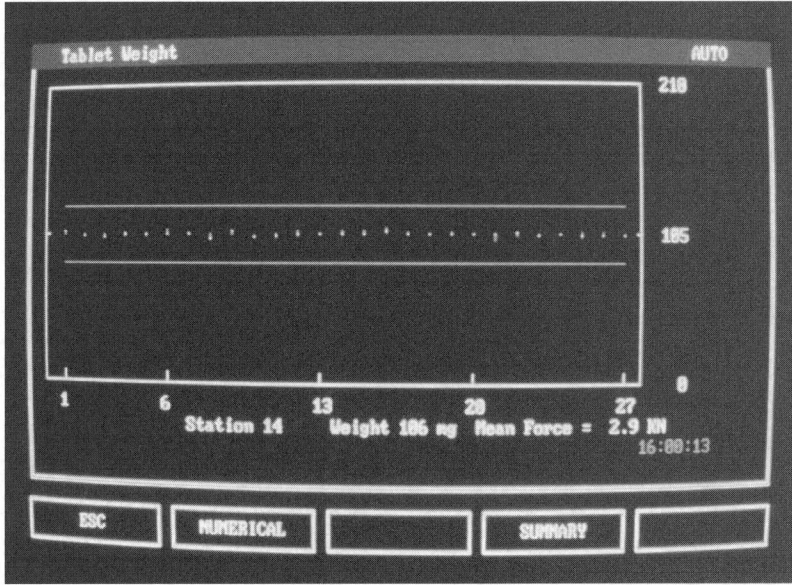

Figure 11.4 A graphical representation of tablet weights.

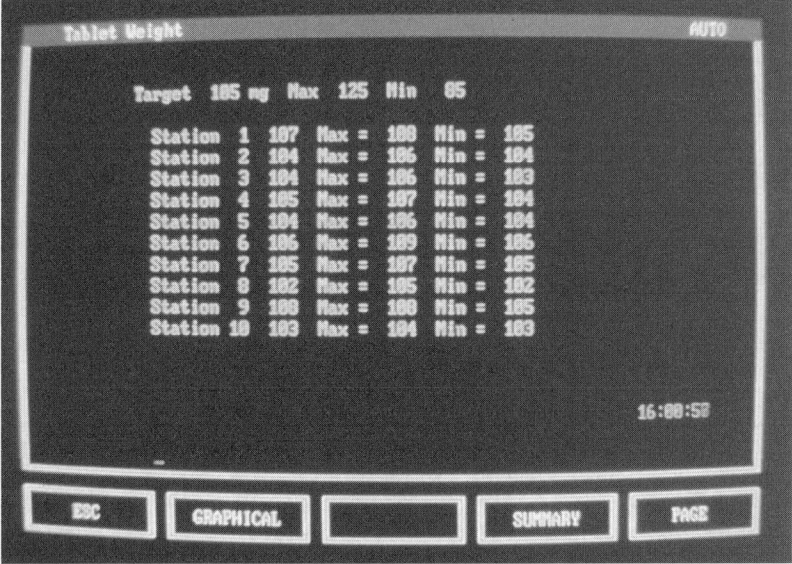

Figure 11.5 A numerical representation of tablet weights.

either one can be controlled but the other is only monitored.

2. The tablet thickness at each stage of the compression. This is shown as the distance separating the upper and lower punch tips. For tablets made with concave punch tips, the true tablet thickness will be greater because of the depth of the concavities.

3. The depth of fill. This is the distance of the tip of the lower punch from the top of the die at

230 Tablet and capsule machine instrumentation

Figure 11.6 The Manesty system for the control of tablet weight.

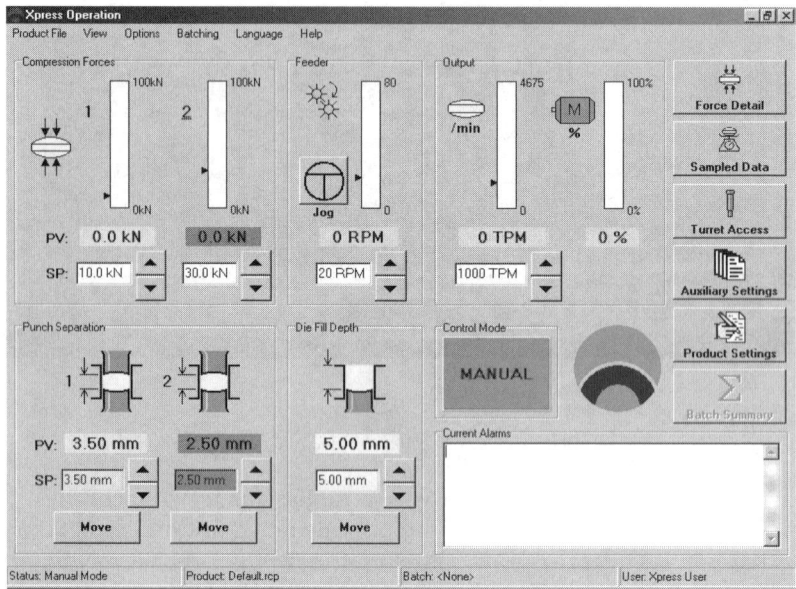

Figure 11.7 The main display screen of the Manesty control system.

the position where the excess of fill material is scraped off.
4. The speed of the feeder. This is given by the revolutions per minute of the main paddle.
5. The machine speed. This is the output in tablets per minute.

When the machine is started, the actual values of the peak compression force for the first and second stages of compression, the feeder paddle speed, the tablet output and the percentage of full load power being used by the motor are all indicated both graphically and numerically. For a product that has not been previously made using the control system, the die fill and tablet thickness can be varied until a satisfactory tablet is obtained. The present value (PV) readings that are shown on the screen at this time can be used to enter the set points (SP). Although the values for the set points are entered on the screen, these are not activated until the 'move' button is selected.

When the move button is selected, the punch positions for the tablet thickness and depth of fill are adjusted by means of stepping motors that are linked to rotary encoders. The feedback loop between the motor and the encoder ensures very accurate positioning of the punches.

When the machine is switched to automatic mode, the control system will adjust the press to maintain the compaction force and hence the mean tablet weight. Any tablets outside specification will be rejected and the machine stopped if the set criteria are exceeded. The criteria that can be set for stopping the machine are either a fixed number of reject tablets per revolution of the press or a fixed number from a particular station.

Once the operating conditions for a particular product have been established, they can be stored for future use. Additional information such as the description of the tooling and the target values for the weight, thickness and hardness (crushing strength) for the product can be entered via an additional screen (Figure 11.8). The data stored on the product can be accessed at any time and modified if necessary by appropriately authorised personnel (Figure 11.9). Recalling this file and using it to reset the machine for a further batch of that product will give exact reproducibility of the operating parameters. This, in turn, will give a much more consistent product than can be achieved by manually setting up the machine for each batch. It will also reduce the machine set-up time.

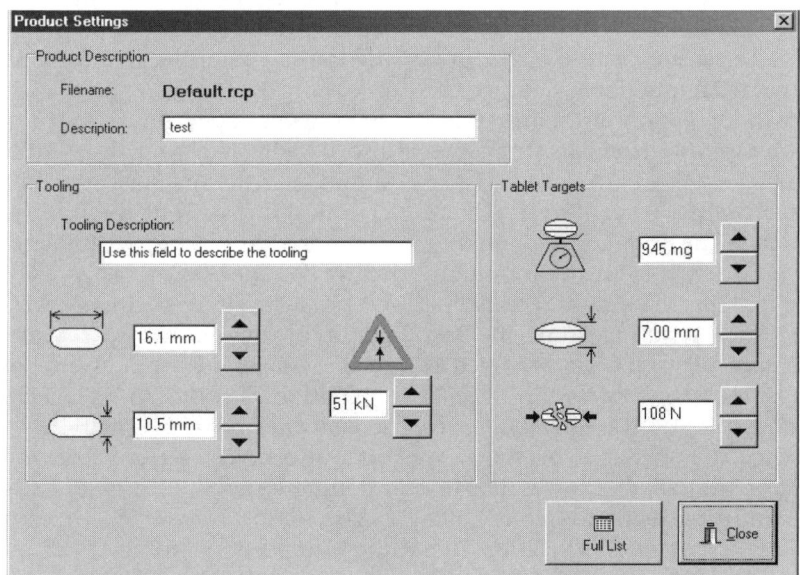

Figure 11.8 The product setting screen display of the Manesty control system.

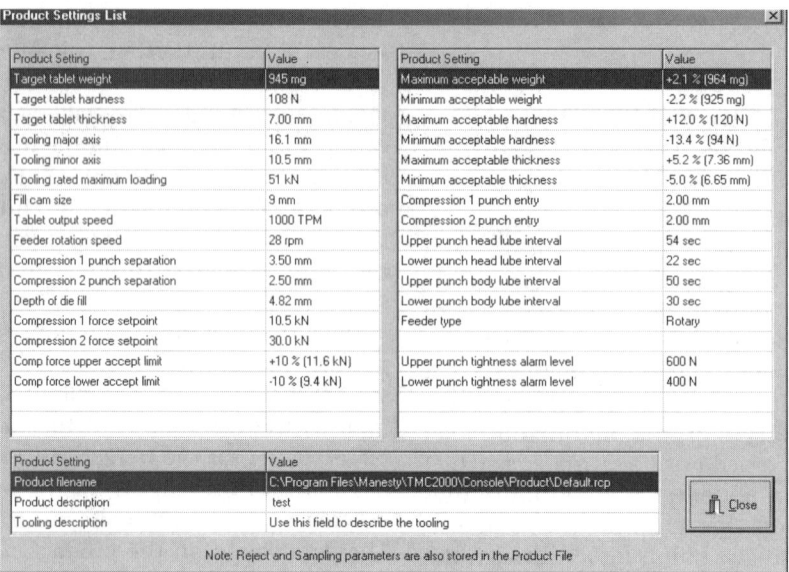

Figure 11.9 The product setting list display of the Manesty control system.

Further screens giving more detailed information on the operating conditions can be accessed from the main screen via speed buttons or drop down menus.

Force screens

There are three force screens available that give details of the forces from individual stations, the mean force for each revolution of the turret and details of the force profile from an individual compression.

In the Individual Station Forces Screen (Figure 11.10), each station of the press is represented by a vertical line that extends from the minimum acceptable force to the maximum acceptable force. The peak force resulting from the last tablet made on a particular station is displayed as a small square. If the tablet is within specification, the square will be on the line and leaves a history trail when it is replaced by the next square. This enables the range of forces actually occurring to be compared with the acceptance limits. If the tablet is out of specification, the square will be off the line and the tablet will be rejected if the system is in automatic mode.

The screen also gives a numerical display of the maximum, minimum and mean forces for the last revolution of the turret together with the coefficient of variation (CV), which is the standard deviation as a percentage of the mean. The coefficient of variation gives an indication of the consistency of the die filling. The optimum feeder paddle speed can be established by varying the speed until a minimum value of the CV is obtained. The acceptance limits, which are expressed as a percentage of the mean force, can also be adjusted from this screen. Figure 11.10 shows only the first stage of compression in the graphical display, but the mean values for both stages are shown in the numerical panel at the bottom of the display.

The Mean Force History Screen displays a bar graph of the mean force from the first and second stages of compression for a number of revolutions of the turret (Figure 11.11). The screen will update automatically, but there is an option to hold a set of values for more detailed examination.

The Station Force Profile Screen gives details of the compaction profile of an individual tablet (Figure 11.12). Up to 70 force readings are taken during the compression of a tablet and this

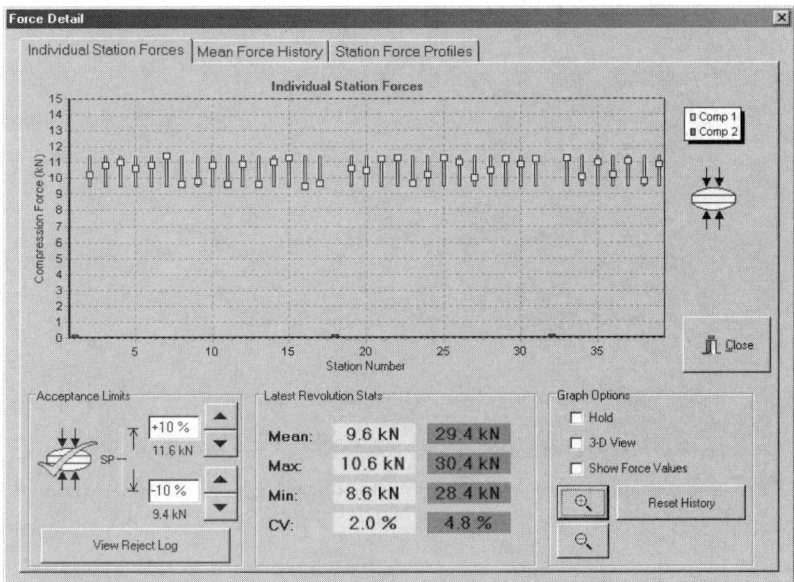

Figure 11.10 The individual punch force display of the Manesty control system.

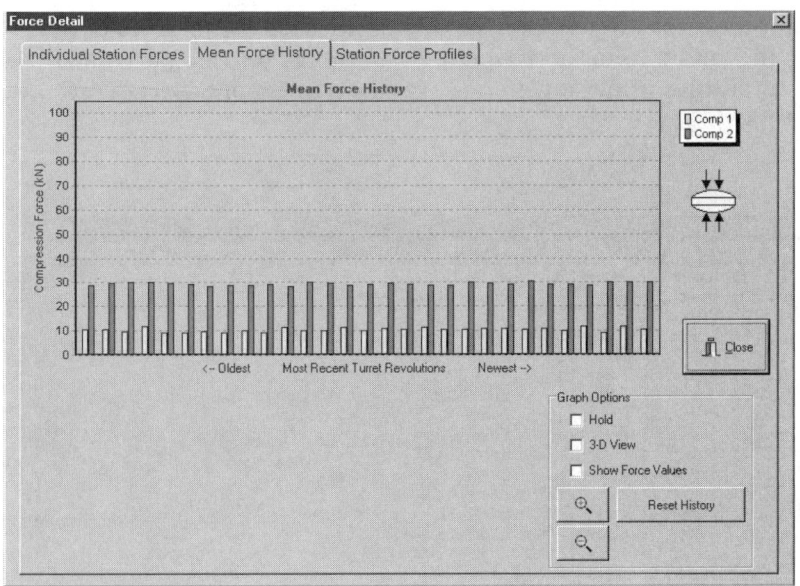

Figure 11.11 The mean force history display of the Manesty control system.

enables the force profile to be studied in detail. The ascending side can give an indication of the degree to which the compaction is occurring by brittle fracture or by plastic flow, and the descending side an indication of the extent to which the tablet is expanding after compression. Figure 11.12 is a simulated profile and does not show the true shape of the curve.

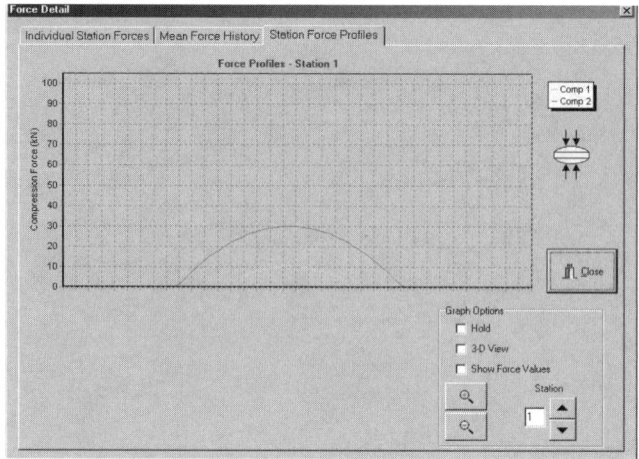

Figure 11.12 The force–time waveform of an individual tablet, as displayed on the Manesty control system.

The system keeps a log of any tablets that have been rejected because the compaction force was outside the acceptance limits (Figure 11.13). It shows the time, station number, compaction force and the error (i.e. the percentage deviation from the mean value and whether this deviation was positive or negative).

Information screens

Additional screens are available if a unit to measure weight, thickness, hardness (WTH unit) is attached to the press (Figure 11.14). The unit must first be configured and this includes setting the number of tablets in each group of tablets to

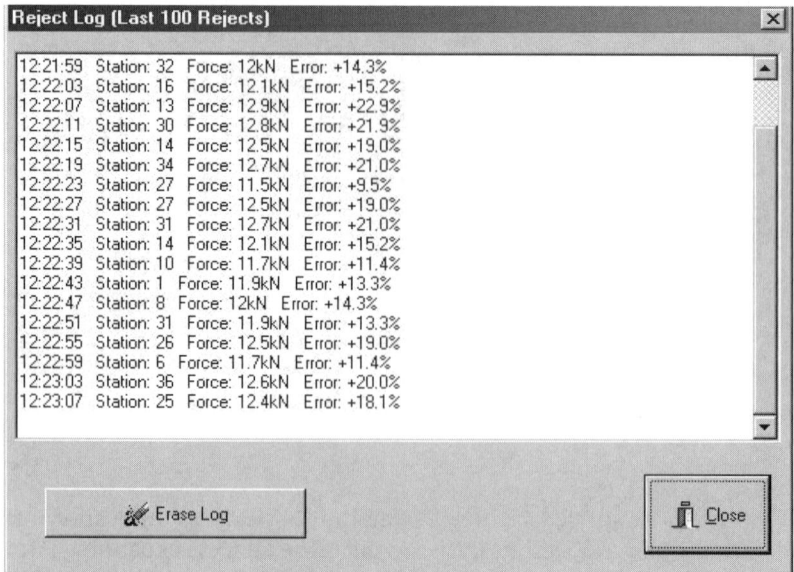

Figure 11.13 The reject log display of the Manesty control system.

be sampled for testing. It is also necessary to set a sampling time interval to reduce the number of tablets being taken from the batch. All the tablets in a group will have their weights and thicknesses measured, but they are not all automatically checked for hardness. If all the tablets in the group are not checked for hardness (a destructive test), some or all of the remaining tablets can be saved as a sample of the batch or for additional analysis. When the WTH unit is installed, screens for display of weight, thickness and hardness become available on the control unit.

Tablet weight screens

There are two weight screens (Figures 11.15 and 11.16). The first shows the mean weight of each group of tablets and this is superimposed on a bar that represents the range of weights within the group. It also shows the acceptance limits and these can be modified from this screen. The second screen shows the weights of individual tablets both graphically and numerically. It also shows the maximum, minimum, mean and CV for the group being weighed. A panel at the side of the screen shows the progress of the weighing system as it moves through the group. When the unit is in automatic mode, the press will be stopped if the weights are outside set criteria.

Tablet thickness screens

The thickness screens are very similar to the weight screens. The first shows data from the groups of tablets tested, with facilities to vary the acceptance limits. The second screen shows data from individual tablets within a group together with the maximum, minimum, mean and coefficient of variation for the group.

Tablet hardness screens

These again are similar to the weight and thickness screens displaying hardness data and statistics.

Figure 11.14 A unit for measuring tablet weight, thickness and hardness for attachment to a tablet press.

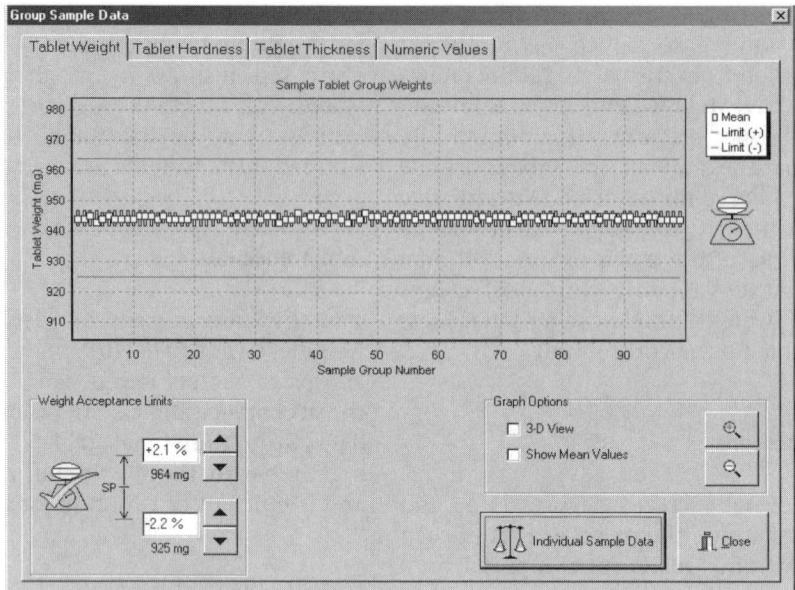

Figure 11.15 The tablet weight display of the Manesty control system, showing data for a group of tablets.

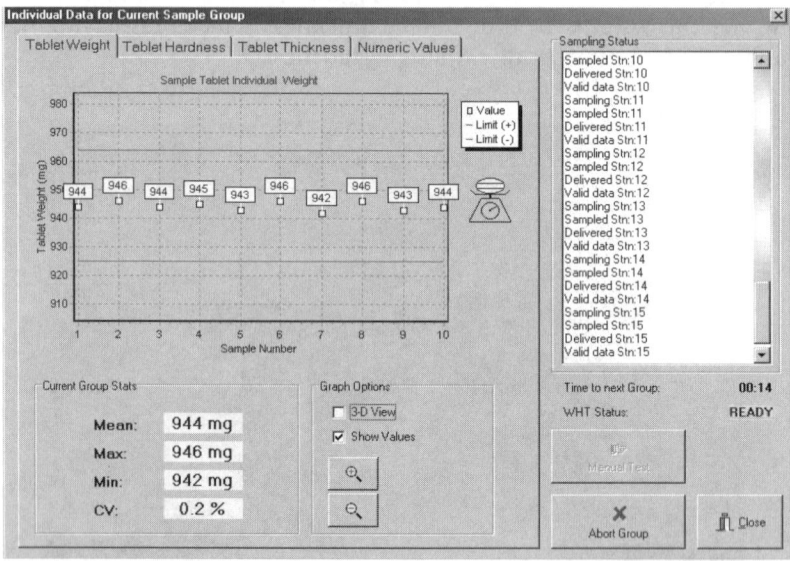

Figure 11.16 The tablet weight display of the Manesty control system, showing data for individual tablets.

Screen associated with the weight, thickness, hardness unit

A final screen associated with the WTH unit (Figure 11.17) shows numerical data from the group of samples being tested.

In cases where a WTH unit is not in use, the control system has facilities for taking samples that can be tested manually or on an automatic test unit not linked to the machine. This enables either individual tablets or groups of tablets to be sampled. In the individual mode, the samples

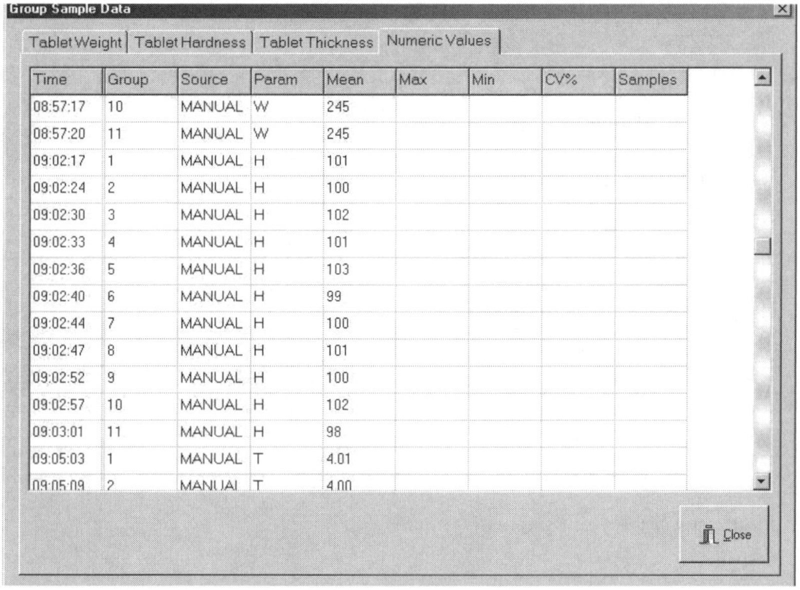

Time	Group	Source	Param	Mean	Max	Min	CV%	Samples
08:57:17	10	MANUAL	W	245				
08:57:20	11	MANUAL	W	245				
09:02:17	1	MANUAL	H	101				
09:02:24	2	MANUAL	H	100				
09:02:30	3	MANUAL	H	102				
09:02:33	4	MANUAL	H	101				
09:02:36	5	MANUAL	H	103				
09:02:40	6	MANUAL	H	99				
09:02:44	7	MANUAL	H	100				
09:02:47	8	MANUAL	H	101				
09:02:52	9	MANUAL	H	100				
09:02:57	10	MANUAL	H	102				
09:03:01	11	MANUAL	H	98				
09:05:03	1	MANUAL	T	4.01				
09:05:09	2	MANUAL	T	4.00				

Figure 11.17 The tablet weight display of the Manesty control system, showing numerical data for individual tablets.

can be taken from any particular station of the press and a time interval can be set between samples. In the group mode, the number of tablets in the group can be set, together with the time interval between groups. As the measuring system is not linked to the press, the data must be entered manually.

When the WTH unit is available, data are stored as the batch of tablets progresses. A series of summary screens then becomes available, covering the compression force, weight, thickness and hardness. The summary screen for compression force is shown in Figure 11.18, and the screens for the other parameters are similar. The system has extensive data storage facilities. The data are normally stored on a batch by batch basis, but sub-batches are allowed and can be used for collecting samples at intervals throughout the batch. At the end of the batch or sub-batch, a printed report can be generated containing any of the data that have been recorded. The amount of data stored electronically is very large and it is normal to print only those parts that are specifically required. The remaining information can be more efficiently stored on disk. The information contained in the report is selected from the report generator screen. This will consist of a basic batch report plus a number of other sets of information e.g. initial settings, system events, user adjustments, data relating to rejected tablets, and many others. An example of the first page of a report is shown in Figure 11.19.

The system has a number of other useful features. These include event logs that automatically record all actions taken during the production of a batch of tablets. The event log can be printed but to ensure accuracy of the record, it cannot be amended (Figure 11.20). Another feature is that instead of all the transducers, motors, encoders, etc. being wired to the control unit, the system uses a computer bus to link all the components. This simplifies the wiring of the system. Facilities are provided to calibrate the position encoders (Figure 11.21) and for removing any zero offset on the load cells (Figure 11.22). Two other features that are essential in a system of this type are an uninterruptible power supply to ensure all data are saved in the event of a power failure and that the method of checking signatures is 21CFR part 11 compliant.

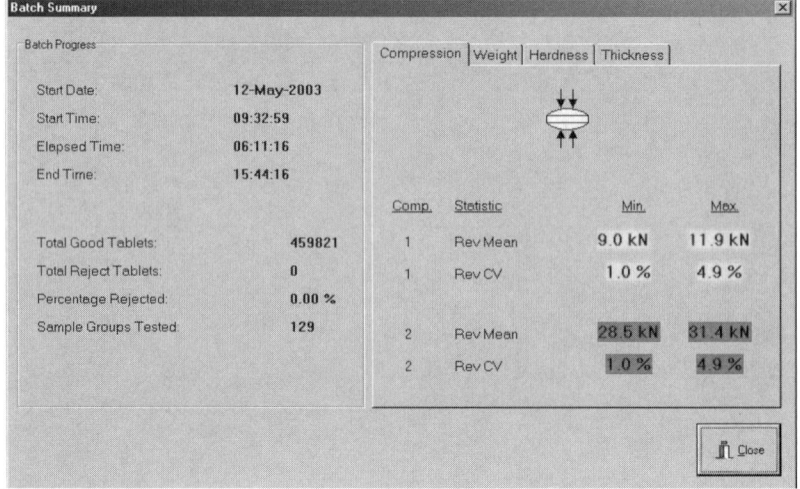

Figure 11.18 Batch summary display of compression force.

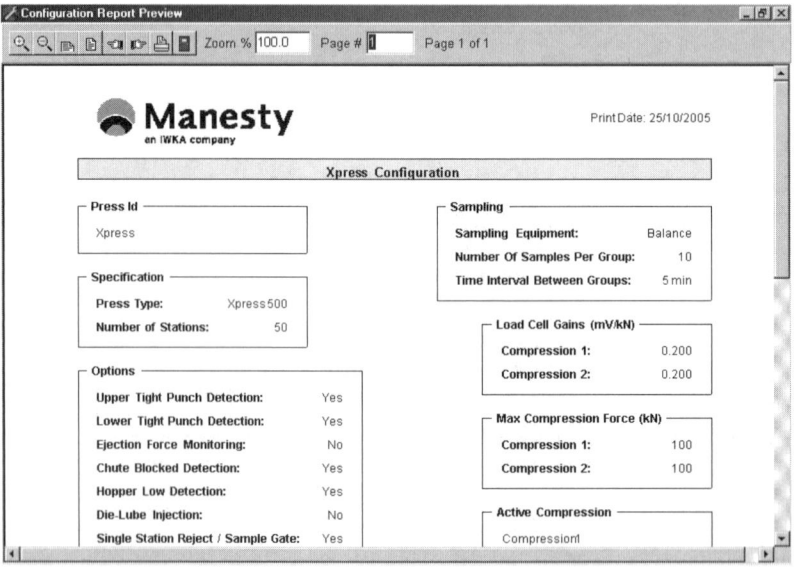

Figure 11.19 The first page of a sample report.

There have been considerable developments in the instrumentation of presses over recent decades and it will be interesting to see what the future holds. At the present time, tablet press output is restricted by the ability of the materials to undergo compression at high punch speeds, rather than being limited by the mechanical speed of the press. Using the

Time	Event	Description	Data 1	Data 2	Data 3	Data 4
09:48:42	1163	Update: Last rev force CV (comp;%)	2	2.1		
09:53:42	1164	Update: Last rev upper TPD force (N)	59			
09:53:42	1165	Update: Last rev lower TPD force (N)	71			
09:53:42	1167	Update: Turret motor load (%)	42			
09:53:42	1160	Update: Good tablets so far (side;tablets)	1	74138		
09:53:42	1161	Update: Reject tablets so far (side;rejects)	1	0		
09:53:42	1166	Update: Last rev ejection force (side;kN)	1	0.5		
09:53:42	1168	Update: Current fill-depth (side;mm)	1	5.00		
09:53:42	1162	Update: Last rev mean force (comp;kN)	1	10.8		
09:53:42	1163	Update: Last rev force CV (comp;%)	1	3.9		
09:53:42	1162	Update: Last rev mean force (comp;kN)	2	30.9		
09:53:42	1163	Update: Last rev force CV (comp;%)	2	1.0		
09:58:42	1164	Update: Last rev upper TPD force (N)	54			
09:58:42	1165	Update: Last rev lower TPD force (N)	66			
09:58:42	1167	Update: Turret motor load (%)	35			
09:58:42	1160	Update: Good tablets so far (side;tablets)	1	80338		
09:58:42	1161	Update: Reject tablets so far (side;rejects)	1	0		
09:58:42	1166	Update: Last rev ejection force (side;kN)	1	0.4		
09:58:42	1168	Update: Current fill-depth (side;mm)	1	5.00		
09:58:42	1162	Update: Last rev mean force (comp;kN)	1	9.4		
09:58:42	1163	Update: Last rev force CV (comp;%)	1	1.6		
09:58:42	1162	Update: Last rev mean force (comp;kN)	2	30.0		
09:58:42	1163	Update: Last rev force CV (comp;%)	2	2.6		

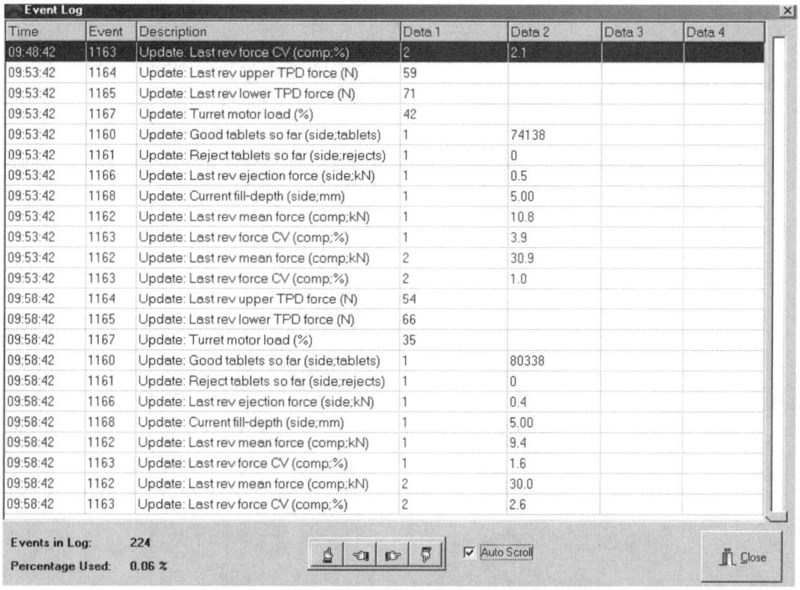

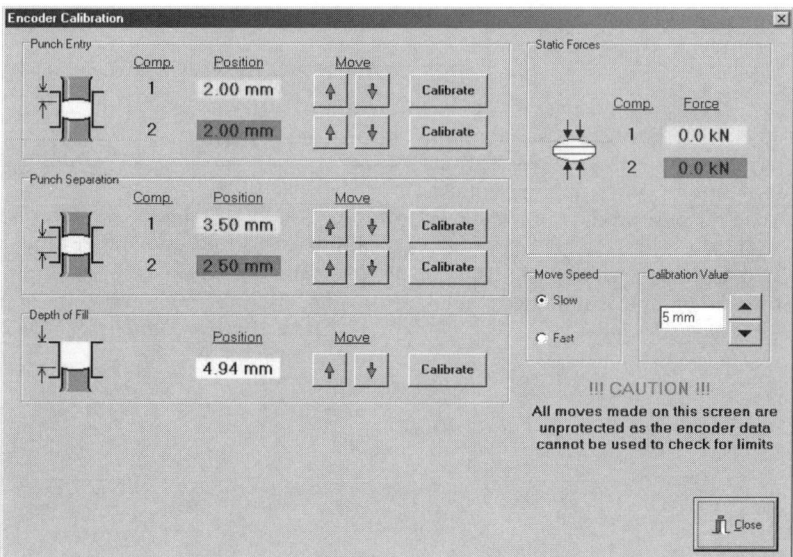

Figure 11.20 An events log.

Figure 11.21 The encoder calibration display.

latest instrumetation systems, the mechanism of the compaction process can be studied under actual production conditions and this may lead to advances in the formulation of tablets that will allow even higher outputs to be obtained.

References

Knoechel EL, Sperry CT, Ross HE, Lintner CJ (1967a). Instrumented rotary tablet machines. 1: Design, construction and performance as pharmaceutical research and development tools. *J Pharm Sci* 56: 109–115.

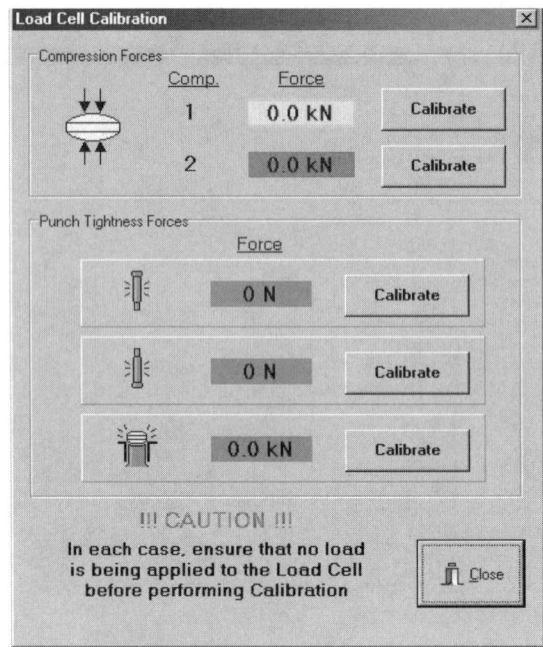

Figure 11.22 The load cell calibration display.

Knoechel EL, Sperry CT, Lintner CJ (1967b). Instrumented rotary tablet machines. 2: Evaluation and typical applications in pharmaceutical research, development and production studies. *J Pharm Sci* 56: 116–130.

Wray PE, Vincent JG, Moller SW, Jackson GJ (1966). Design and construction of an instrumented rotary tablet press. *A Ph A Annual Meeting*, Dallas.

Further reading

Murray FJ (1996). Tablet press automation: a modular approach to fully integrated production. *Drug Dev Ind Pharm* 22: 35–43.

Wray PE (1992). The physics of tablet compaction revisited. *Drug Dev Ind Pharm* 18: 627–658.

Appendix

Suppliers of materials and services

Some suppliers of equipment and services for the measurement of force and displacement are listed below, but this is by no means a definitive list.

Suppliers of force-measuring equipment

ABB Ltd
Daresbury Park
Warrington WA4 4BT
UK
+44 (0)1925 714111
www.abb.co.uk

Amber Instruments Ltd
Dunston House
Dunston Road
Chesterfield
Derbyshire S41 9QD
UK
+44 (0)1246 260250
www.amberin.clara.net

Applied Measurements Ltd
3 Mercury House
Calleva Park
Aldermaston
Berkshire RG7 8PN
UK
+44 (0)118 9817339
www.appmeas.co.uk

Control Transducers Ltd
North Lodge
25 Kimbolton Road
Bedford MK40 2NY
UK
+44 (0)1234 217704
www.controlt.co.uk

Entran Ltd
4 Thames Park
Lester Way
Wallingford
Oxfordshire OX10 9TA
UK
+44 (0)1491 893999
www.entran.co.uk

Graham and White Instruments Ltd
137a Hatfield Road
St Albans
Herts AL1 4LZ
UK
+44 (0)1727 844323
www.sensorsuk.com

Hottinger Baldwin Messtechnik (HBM)
HBM United Kingdom Ltd
1 Churchill Court
58 Station Road
North Harrow
Middlesex HA2 7SA
UK
+44 (0)208 5156100
www.hbm.com

Kistler Instruments Ltd
13 Murrel Green Business Park
London Road
Hook
Hants RG27 9GR
UK
+44 (0)1256 741550
www.kistler.com

Kulite Sensors Ltd
Kulite House
Stroudley Road
Kingsland Business Park
Basingstoke
Hants RG24 0UG
UK
+44 (0)1256 461646
www.kulite.co.uk

LCM Systems Ltd
Unit 15 Newport Business Park
Barry Way
Newport
Isle of Wight PO30 5GY
UK
+44 (0)1983 249264
www.lcmsystems.com

L H Transducers Ltd
25 Perry Hill
Sandhurst
Berkshire GU47 8HS
UK
+44 (0)1252 890900
www.acamltd.co.uk

Microsensors Division of Howard A Schaevitz Technologies Inc
7300 US Route 130 North
Building 22
Pennsauken
NJ 08110–1541
USA
+1 856 662 8000
www.microsensors.com

National Instruments Corporation (UK) Ltd
Measurement House
Newbury Business Park
London Road
Newbury
Berkshire RG14 2PS
UK
+44 (0)1635 523545
www.ni.com

Proctor & Chester Measurements Ltd
Chester House
Dalehouse Lane
Kenilworth
Warwickshire CV8 2UE
UK
+44 (0)1926 864444
www.pcm-uk.com

RS Components Ltd
Birchington Road
PO Box 99
Corby
Northants NN17 9RS
UK
+44 (0)845 8509922
rswww.com

Scientific Electro Systems Ltd
1 Rose Way
Purdeys Industrial Estate
Rochford
Essex SS4 1LY
UK
+44 (0)1702 530174
www.sesystems.co.uk

Strainstall UK Ltd
9–10 Mariners Way
Cowes
Isle of Wight PO31 8PD
UK
+44 (0)1983 203600
www.strainstall.com

Straintek Precision Services
Parsonage Side
Otterhampton
Bridgwater
Somerset TA5 2PT
UK
+44 (0)1278 652062
www.straintek.com

Stresscoat Inc
1334 North Benson Avenue
Unit A
Upland
CA 91786
USA
+1 909 981 7444
www.stresscoat.com

Techni Measure Ltd
Alexandra Buildings
59 Alcester Road
Studley
Warwickshire B80 7NJ
UK
+44 (0)1527 854103
www.techni-measure.co.uk

Thames-Side-Maywood Ltd
2 Columbus Drive
Summit Avenue
Southwood
Farnborough
Hants GU14 0NZ
UK
+44 (0)1252 555811
www.thames-side-maywood.com

TSM Ltd
Sensor House
Wrexham Technology Park
Wrexham
Flintshire LL13 7YP
UK
+44 (0)1978 291800
www.esrtec.com

Vishay Measurements Group UK Ltd
Stroudley Road
Basingstoke
Hants RG24 8FW
UK
+44 (0)1256 462131
www.measurementsgroup.co.uk

Strain gauge application services

Proctor & Chester Measurements Ltd
Chester House
Dalehouse Lane
Kenilworth
Warwickshire CV8 2UE
UK
+44 (0)1926 864444
www.pcm-uk.com

Straintek Precision Services
Parsonage Side
Otterhampton
Bridgwater
Somerset TA5 2PT
UK
+44 (0)1278 652062
www.straintek.com

The Strain Gauging Company Ltd
Rookery Lodge
College Lane
Ellisfield
Hants RG25 2QE
UK
+44 (0)1256 381006
www.tsgcltd.co.uk

TSM Ltd
Sensor House
Wrexham Technology Park
Wrexham
Flintshire LL13 7YP
UK
+44 (0)1978 291800
www.esrtec.com

Vishay Measurements Group UK Ltd
Stroudley Road
Basingstoke
Hants RG24 8FW
UK
+44 (0)1256 462131
www.measurementsgroup.co.uk

Suppliers of displacement transducers and accelerometers

Analog Devices Inc.
One Technology Way
Norwood
MA 02062
USA
+1 781 329 4700
www.analog.com

Automatic Systems Laboratories Ltd
40 Tanners Drive
Blakelands
Milton Keynes
Bucks MK14 5BN
UK
+44 (0)1908 440666
www.aslltd.co.uk

Entran Ltd
4 Thames Park
Lester Way
Wallingford
Oxfordshire OX10 9TA
UK
+44 (0)1491 893999
www.entran.co.uk

Graham and White Instruments Ltd
137a Hatfield Road
St Albans
Herts AL1 4LZ
UK
+44 (0)1727 844323
www.sensorsuk.com

IDM Electronics Ltd
Unit 30
Suttons Park Avenue
Suttons Industrial Park
Reading
Berkshire RG6 1AW
UK
+44 (0)118 9666044
www.idmelectronics.co.uk

Kistler Instruments Ltd
13 Murrel Green Business Park
London Road
Hook
Hants RG27 9GR
UK
+44 (0)1256 741550
www.kistler.com

Micro-Epsilon UK Ltd
Dorset House
West Derby Road
Liverpool L6 4BQ
UK
+44 (0)151 2609800
www.micro-epsilon.co.uk

Midori Inc
2555E Chapman Avenue State 400
Fullerton
CA 92831
USA
+1 714 449 0999
www.midoriamerica.com

Monitran Ltd
Monitor House
33 Hazelmere Road
Penn

High Wycombe
Bucks HP10 8AD
UK
+44 (0)1494 816569
www.monitran.co.uk

Penny and Giles Controls Ltd
15 Airfield Road
Christchurch
Dorset BH23 3TJ
UK
+44 (0)1202 409409
www.pennyandgiles.com

Positek Ltd
Manor Farm Industrial Estate
Andoversford
Cheltenham
Gloucestershire GL52 4LB
UK
+44 (0)1242 820027
www.positek.com

Precision Varionics International
Sensor House
Langley Road
Swindon
Wiltshire SN5 5WB
UK
+44 (0)1793 879879
www.pvi.co.uk

RS Components Ltd
Birchington Road
PO Box 99
Corby
Northants NN17 9RS
UK
+44 (0)845 8509922
rswww.com

Sensors (UK) Ltd
137a Hatfield Road
St Albans
Herts AL1 4LZ
UK
+44 (0)1727 844323
www.sensorsuk.com

Sensortronic Ltd
45 Leaver Road
Henley on Thames
Oxfordshire RG4 1UW
UK
+44 (0)1491 412055
www.sensortronic.com

Techni Measure Ltd
Alexandra Buildings
59 Alcester Road
Studley
Warwickshire B80 7NJ
UK
+44 (0)1527 854103
www.techni-measure.co.uk

Variohm EuroSensor
Williams Barn
Tiffield Road
Towcester
Northants NN12 6HP
UK
+44 (0)1327 351004
www.variohm.com

Vishay Measurements Group UK Ltd
Stroudley Road
Basingstoke
Hants RG24 8FW
UK
+44 (0)1256 462131
www.measurementsgroup.co.uk

Suppliers of instrumentation packages

Astech Electronics Ltd
Forge Industrial Estate
The Street
Binsted
Alton
Hants GU34 4PF
UK
+44 (0)1420 22689
www.astechelectronics.com

Copley Scientific Ltd
Colwick Quays Business Park
Nottingham NG4 2JU
UK
+44 (0)115 9616229.
www.copleyscientific.co.uk

Instron Ltd
Coronation Road
High Wycombe
Bucks HP12 3SY
UK
+44 (0)1494 456815
www.instron.co.uk

Korsch AG
Breitenbach Strasze
13509 Berlin
Germany
+49 30 435760
www.korsch.deu

Metropolitan Computing Corporation
6 Green Meadow Lane
East Hanover
NJ 07936
USA
+1 973 887 7800
www.mcc-online.com

Puuman Oy
Sammonkatu 6
70501 Kuopio
Finland
+358 17 288 1222
www.puuman.com

SMI Incorporated
PO Box 356
Lebanon
NJ 08833
USA
+1 908 534 1500
www.smitmc.com

Miscellaneous

Addlink
Adcon Telemetry GmbH
Inkustrasze 24
Klosterneuberg
A-3400
Austria
+43 2243 382800
www.adcon.at

Amplicon Liveline Ltd
Centenary Industrial Estate
Hollingdean Road
Brighton
Sussex BN2 4AW
UK
+44 (0)1273 570215
www.amplicon.uk

Analog Devices Inc
One Technology Way
Norwood
MA 02062
USA
+1 781 329 4700
www.analog.com

Data Translation Inc
100 Locke Drive
Marlboro
MA 01752 1192
USA
+1 508 481 3700
www.data_translation.com

Mantracourt Electronics Ltd
The Drive
Farringdon
Exeter EX5 2JB
UK
+44 (0)1395 232020
www.mantracourt.co.uk

Metglas Inc
440 Allied Drive
Conway
SC 29526
USA
+1 800 581 7654
www.metglas.com

Microchip Technology Inc
2355 West Chandler Boulevard
Chandler
AZ 85224 6199
USA
+1 480 792 7200
www.microchip.com

Newport Spectra Physics Ltd
Light House
Newbury Business Park
London Road
Newbury
Berkshire RG14 2PZ
UK
+44 (0)1635 521757
www.newport.com

Texas Instruments Inc
12500 TI Boulevard
Dallas
TX 75243 4136
USA
+1 800 336 5236
www.ti.com

Index

Page numbers in *italic* refer to figures, those in **bold** refer to tables.

abrading of surface for strain gauges, 52–53
abrasives, 62
AC waveform, 93
AC–AC transducers, 68
accelerometer, 79–81, 125
    components, 80
    piezoelectric, 81
    punch movement, 133
    servo-force balance, 81
    suspension, 80–81
acrylic varnish, 58, 63
actuators, 72
adhesion force, punch, 131–132, 202
adhesive bonding
    inspection, 58–59
    strain gauges, 53–56, 58–59
adhesives, 61
    cold cure, 54, 55–56
    heat cure, 54, 56
    instant, 53, 54–55
    storage, 54
    types, 53–54
air abrading, 53
aliasing, 147–148
ammonia, weak solution, 53, 61
amplifiers, integral charge, 43
analogue multiplexer, 154–155
analogue-to-digital cards, commercially available, 161–162
analogue-to-digital converter (ADC), 152–153, 155–157
    flash devices, 156
    multiplexer, 154–155
    parallel devices, 156
    parallel output, 155
    sub-system, 155
    voltage reference, 157
annealing, 25
auxetics, 13

base units, 4–5
bending, bridge circuit, 36–37
bolts, strain-gauged, 41, 108, 130
    ejection ramp, 127
bonded coefficient, 36
British Society for Strain Measurement (BSSM), 64

brittle lacquer technique, 30
Brockedon press, 1
Butterworth filters, 154
calibration
    die, 197, 198
    die-wall stress, 142, 197, 198
    direct, 140
    displacement measurement, 142–145
    displacement transducers, 144–145
    encoder, 237, *239*
    force, 140–142
    fundamental standards, 139
    indirect, 140–141
    load cells, 102, 103
        display, 237, *240*
        force, 140, *141*
        indirect, 140
    micrometer gauge, 143–144
    practical standards, 139–140
    problems, 145–146
    punch, 140, 141
    secondary standards, 140
    signal problems, 145–146
    strain gauges, 59
    transducer systems, 139–146
capacitance changes, 48
capacitative velocity transducers, 82
capacitors, 89
    linear variable, 69
    piezoelectric load cells, 43
capsule(s), 207
    history, 1
capsule-filling machinery, **3**
    compression force, 207
    dosating disk machines, 208
        instrumentation, 209–214
    dosating nozzle machines, 208–209
        instrumentation, 214–220
    instrumentation, 3–4, 7–8, 207–220
        packages, 8
    parameters for measurement, 3–4
    types, 208–209
casting polymers, 24
Chebychev filters, 154
cleaning materials, 62

clevis pins, 108
coefficient of variation (CV), 232
cold cure adhesives, 54, 55–56
cold curing, 54
compact disks (CDs), 163–164
compaction
  data storage for tablet presses, 95
  power of, 195
  punch velocity, 190–191
compaction force measurement, 227
compaction simulators, 134–136
  die-wall stress, 200
complimentary metal oxide semiconductor (CMOS), 155
composite gauges, 35
composite materials, 61
compression
  behaviour of solids, 167
  capsule-filling machinery, 207
  dosating disk machines, 210
  energy state of tablet, 190
  equations, 183–187
  events, 182, 190
  force–displacement curve, 188–189
  phase of force–time profile, 181
  punch shortening, 175–176
  stages, 215
  tablet press distortion, 175–176
  triaxial system, 200
compression curves, 197–198
  tablet presses, 103
compression cycles, 197–198
  axial force, *198*, 200
  plastic deformation, 199
  predictive use, 200
compression force
  capsule-filling machinery, 207
  computer-controlled systems, 231, *238*
  dosating nozzle machinery, 214–215, 217–218, 219–220
  feedback loop to maintain with die fill, 225
  relationship with tablet properties, 167
  static conditions, 43
  tablet weight control, 222–225
compression pressure measurement, 2
computer instrumentation, 8
computer interfacing, 157–162
  built-in, 158–159
computer switch-mode power supplies, 90
computer-controlled tablet presses, 228–229, *230*, 231–239, *240*
  data storage, 237
  encoder calibration, 237, *239*
  event log, 237, *239*
  information screens, 234–235
  operating condition storage, 231, *232*
  weight, thickness, hardness unit, 234–239, *240*
conductive plastics film, temperature coefficient of resistivity, 71
consolidation process, dosating disk machines, 209, 212–213

Constantan, 14, 17
  gauge life, 19
  measuring range, 18
Cooper–Eaton equation, 185, 186
copper/copper alloys, 61
cosine error, 102, *103*
cutters, 63
cyanoacrylate adhesives, 53, 54–55

D connector, 158, *159*
data acquisition
  frequency response, 153
  Hall effect transducer, 97
  mains noise, 90
  signal display, 95–97
  tablet presses, 91–95
data handling, 147–164
  backup, 163–164
  computer interfacing, 157–162
  electronics sub-systems, 153–157
  embedded systems, 157
  sampling system theory, 147–148, *149–151*, 152–153
  software, 162–163
data-logging system, 157–158
DC–DC transducers, 68
decompression curves, 179, 197–198
degreasing of surface for strain gauges, 52, 61
Denisyuk method, 27
densification, dosating disk machines, 212–213
die
  calibration, 197, 198
  cutaways, 197, 199
  cylindrical housing, 200
  dosating disk machines, 214
  flexible, 200
  force required to eject tablet, 202
  layered, 199
  segmented, 197–200
  strain gauge attachment, 197
die fill, feedback loop to maintain compression force, 225
die table, 171, *172*, 173
die-wall stress, 196–201
  calibration, 142, 197, 198
  strain gauges, 198–200
  wall thickness/width, 199
digital signal processor chip, 157
digital versatile disks (DVDs), 164
diodes, photosensitive, 77–78
displacement, 65
  dosating nozzle machinery, 217–218
  force relationship, 187–190
  rate of change, 190–191
displacement measurement, 65–84
  accuracy, 121
  calibration, 142–145
  dynamic, 79–82
  errors, 189
  flash photography, 83–84

laser speckle photography, 82–83
machine distortion, 121–124
punch, 121
punch tilting, 134
tablet presses, 119–127
  rotary, 125–127
transducers, 65–79
displacement transducers, 65
  analogue output, 65–76
  calibration, 144–145
  digital output, 76–79
  fitting to dosating disk machines, 210–211
  LVDT, 122–124
  non-contact, 146
  positioning, 122, **123**
displacement–time profiles
  eccentric tablet presses, 168–171
  Heckel equation, 185–186
displacement–time waveform, 135
dissolution, dosating disk machines, 210
dosating disk machines, 208
  compression, 210
  consolidation process, 209, 212–213
  densification, 212–213
  dies, 214
  dissolution, 210
  fill weight control, 212
  force–displacement curves, 211–212
  force–time relationships, 211
  instrumentation, 209–214
  LVDT fitting, 210–211
  peak maximum force, 211
  plug formation, 210, 213–214
  simulation, 213–214
  tamping pins, 209–210, 212
dosating nozzle machines, 208–209
  compression force, 214–215, 217–218, 219–220
  displacement, 217–218
  ejection force measurement, 215, 216–217, 218
  fill weight, 219
  force, 217–218
  instrumentation, 214–220
  LVDT attachment, 217, 220
  plug
    density, 220
    formation, 215–216
  powder bed density, 219
  pre-compression force, 217–218, 219
  simulation, 218–220
double-weighing, 11
dual-slope integrating devices, 155

eccentric presses, 7–8, 99–111
  compression, 170
  displacement measurement, 119–125
  displacement–time profiles, 168–171
  force–time profiles, 177–178, 180–181
  machine distortion, 121–124
  punch, 122–124
    upper, 168–169

velocity, 191
sample rate choice, 148, 152
scale length, 120–121
ejection event, 148
ejection force measurement, 127–129
  dosating nozzle machines, 215, 216–217, 218
  force calibration, 142
ejection ramp, 127–129
  cantilever beams, 129
  modified designs, 129
  non-contact, 146
  piezoelectric load cells, 128
  shear force load cell mounting, 129
  split, 129
elastic deformation, powder consolidation, 176
elastic expansion, 188–189
electromagnetic interference suppression ferrites, 90
electronics sub-systems, 153–157
  amplification, 153–154
  analogue multiplexer, 154–155
  analogue-to-digital converters, 155–157
  dual-slope integrating device, 155
  low-pass filter, 154
  sample and hold, 155
  successive approximation register devices, 155–156
elliptic filters, 154
embedded systems, 157
encoder calibration, 237, *239*
energy
  SI unit, 5–6
  state of tablet after compression, 190
  tablet formation, 187
epoxy resins, 54, 55–56
epoxy-phenolics, solvent-thinned, 54, 56
Ethernet, 159
Excel spreadsheet program, 162
excitation voltage, 33, 38

fast Fourier transform (FFT), 152, 162–163
ferrites, 90
fiber optic transducers, 75–76
fibre optic strain gauges, 48–49
field effect transistor (FET), 43
FIFO memory, 162
files, 62
  abrading of surface for strain gauges, 52
fill depth, computer-controlled systems, 229, 231
fill weight
  control by dosating disk machines, 212
  dosating nozzle machines, 218, 219
filters
  design, 154
  mains, 90–91
  switched-capacitor, 154, *156*
finite element analysis, 22
  die-wall stress, 199–200
flash photography, 83–84
  three-colour, 84
floppy disks, 163
flow properties of powders, 218

force, 6
    accelerometer, 80–81
    axial, *198*, 200
    computer screens, 232–234
    dosating nozzle machinery, 217–218
    effective, 181
    measurement, 11–49
    peak
        maximum for dosating disk machines, 211
        punch penetration, 180, 189–190
    powder consolidation, 176–178
    rate of change, 190–191
    required to eject tablet from die, 202
    servo-driven balance, 81
    SI units, 5
    stress analysis, 31
    sweep-off, 202
    transmission between particles, 179–180
    see also compression force; punch force
force calibration, 140–142
    error sources, 141–142
    telemetry, 142
force transducers, 140–141
force–displacement curves, 121, 187–190
    area under, 188–190, 195
    dosating disk machines, 211–212
force-measuring equipment suppliers, 241–243
force–porosity relationships, 182–186, *186*
    compression equations, 183–187, *186*
force–time profiles, 176–187
    compression phase, 181
    dosating disk machines, 211
    dwell time, 181
    eccentric tablet presses, 148, 152, 177–178, 180–181
    phases, 181
    relaxation phase, 181
    rotary tablet presses, 178–179, 181
    shape, 179–182
    waveforce, 148, 152, *234*
force–weight relationship, 224–225
Fourier transform (FT), 152, 162–163
frictional force, 224–225
frictional resistance of sliding components, 102

gauge blocks, transducer calibration, 142–143
gauge factor, 13
    temperature variation, 17
    variation, 35
gauged punches, force transducers, 140
gauging transducers, 67, 121
    fitting, 126–127
gratings, multiple-track, 77
grinders, 62
grit blasting, 53
grit paper, 62
    abrading of surface for strain gauges, 52–53

half-bridge, 33, 35
    completion resistors, 37–38
Hall effect transducer, 70–71, 97

hard drives, 163
health and safety, strain gauges, 51
heat curing, 54, 56
Heckel coefficients, 185
Heckel equation, 183, 186
holographic camera, 27
holographic interferometry, 26–27
    real-time, 27, 83
holography, 26–28
    white light, 27
Hooke's law, 12
HP-IB parallel bus, 160
hydraulic presses, 132–133
hysteresis
    eccentric tablet press, 104
    loop area, 200–201

inductive transducers, 78
    remote, 125
inductive velocity transducers, 81–82
Infrared Data Association (IrDA) standard, 157
infrared port, 160
instrumentation of capsule-filling machinery, 8, 207–220
    dosating disk machines, 209–214
    dosating nozzle machines, 214–220
instrumentation of presses, 3–4, 6–7, 8
    applications, 167–203
    die wall stress, 196–201
    force calibration, 142
    force–displacement curves, 187–190
    force–porosity relationships, 182–186, *186*
    force–time profiles, 176–182
    lubrication studies, 201–202
    packages, 8, 132–136
    punch displacement–time profiles, 168–171, *172*, 173–176
    punch velocity, 190–196
instrumentation packages
    suppliers, 245–246
    tablet presses, 8, 132–136
instrumented systems, 2–4
    components, *4*
integral charge amplifiers, 43
integrated micro electro mechanical systems (iMEMS), 81
interferometers
    displacement measurement, 142
    Michelson, 82
    optical, 79
interferometry, holographic, 26–27
    real-time, 27, 83
IrDA port, 160
isopropanol, 61

K-alloy, 15, 17
Kawakita equation, 184–185, 186

laser
    holography, 27

moiré grid technique, 28
laser Doppler velocimetry, 65, 82
  tablet punches, 124–125
laser Doppler-shift measurement, 82
laser speckle photography, 82–83
laser speckle strain gauges, 49
layout marking tools, 63
lead zirconate titanate (PZT), 41
leadwire(s)
  intra-bridge, 57, 58
  main, 57
  tablet presses, 91–92, *94*
leadwire attachment to strain gauges, 57–58, 59, 62
  high-temperature installations, 61
  unbonded, 60
length, base unit, 4–5
light-emitting diode (LED), 95
linear diode array, 77–78
linear inductive displacement transducer, 68–69
linear magnetoresistive transducer, 69–70
linear regulators, 88–89
linear variable capacitor, 69
linear variable resistor, 71–72
linear variable-differential transformer (LVDT), 65–68
  construction, 65
  demodulation, 68
  dimensions, 67–68
  displacement transducer, 122–124
  dosating disk machines, 210–211
  dosating nozzle machines, 217, 220
  eccentric tablet press, 120
  excitation, 68
  gauging transducers, 67
  load cell, 44, *45*, 46
  machine distortion, 121–123
  output signal, 65
  punch, 126
  repeatability, 66
  scale length, 120–121
  spring-loaded, 121, 127
  thermal effects, 67
  zero setting, 66
load cells, 32, 38–46
  advantages, 44–45
  based on strain gauges, 38–39
  calibration
    against, 103
    comparison for indirect, 140
    display, 237, *240*
    force, 140, *141*
    against transfer standard, 102
  concentric element, 39, *40*
  disadvantages, 45–46
  ejection ramp mounting, 129
  force calibration, 140, *141*
  force transfer, 105
  linear variable-differential transformer, 44, *45*, 46
  lower punch holder, 109
  magnetoelastic, 43–44
  piezoelectric, 41–43, 105, 106, 109

accelerometer, 81
  ejection ramp, 128
  force calibration, 141
resolution, 45
rigidity, 46
rotary tablet presses, 115, 116
shear force, 129
shear-beam, 44
  strain-gauged, 105, 106
  strain-gauged metal cylinder, 109
  temperature compensation, 45
load testing, strain gauges, 59
load washers, 42
  piezoelectric, 43, 116, 218
  strain-gauged, 43
lubricants, actions, 201–202
lubrication, tablet press instrumentation, 201–202

machine distortion
  eccentric tablet presses, 121–124
  punch force measurement, 122–123
magnetic sensors, 74
magneto optical (MO) disks, 164
magnetoelastic cells, 43–44
magnetoresistive effect, 69–70
magnetorestrictive transducer, 73–74
magnetostrictive strain gauges, 48
mains filters, 90–91
mains noise, 89–91
  transient suppressors, 91
  upper arm, 102–103
Manesty control system, *226*, *230*, *231*, *232*, *233*, *234*, *236*, *237*
Manesty F3 press, 2
masked aperture camera, 28
mass, base unit, 5
mercury seals, tablet presses, 92
Michelson interferometer, 82
microcrystalline wax, 58
micrometer gauge, 143–144
modulus compensation, 17, 34–35
Mohr's bodies, *198*, 198
moiré grid technique, 27–28
monitoring systems, tablet presses, 225
multiple-track grating, 77
multiplexing, 95

necking down, 112, 114
neutralising solutions, 53, 61
null-modem cable, 158
Nylatron, 130
Nyquist frequency, 154

optical grading transducers, 76–77
optical interferometers, 79
optical linkage, tablet presses, 95
optical potentiometers, 76
optical transducers, 75–76
optical velocity transducers, 81
optical wedge, 76

oscillographs, 97
oscilloscopes, 95–97
  digital storage, 96–97
  mixed signal, 97
  rotary tablet press, 114
output of equipment, 190
overload system, 225, *226*

particle density, 182
particle fragmentation, powder consolidation, 176
PCI analogue card, 161–162
  (intelligent), 162
PCI interfaces, 159
Perspex, 24, 25
photoelastic film measurement, 25
photoelastic models, 22–24, 25
piezoceramic materials, 41
piezoelectric accelerometer, 81
piezoelectric effect, 41–42
piezoelectric load cells, 41–43
  accelerometer, 81
  ejection ramp, 128
  force calibration, 141
  lower punch holder, 109
  upper punch holder, 105, 106
piezoelectric load washers, 43, 116, 218
piezoelectric transducers, 41–43
  rotary tablet presses, 116–117
plastic deformation, 198
  compression cycles, 199
  powder consolidation, 176
plastics, 61
  annealing, 25
  commercial sheet, 24, 25
plug
  density, 220
  length, 214, 215
  properties with dosating nozzle machinery, 217, 220
  weight, 220
plug formation
  dosating disk machines, 210, 213–214
  dosating nozzle machines, 214–215, 215–216
  packing density, 214
  punch speed, 214
Poisson effect, 35
Poisson gauge, 16, 33–34
Poisson ratio, 13, 16
polarised light, stress analysis, 23
polysulphide rubber, 58
polyurethane varnish, 58, 63
pore fraction, 182–186
porosity
  compression equations, 183–187
  force relationships, 182–186
  measurement, 121
  out-of-die, 182
  powder consolidation, 176–177
  punch velocity, 194–195
  in situ, 183

Portable Press Analyser, 95, 108, 133–134
potentiometers
  linear, 71
  optical, 76
  resistive, 71–72
powder, flow properties, 218
powder bed density, 219
powder consolidation, 176–177
  eccentric tablet presses, 177–178
power, SI unit, 6
power supplies, 87, 88–95
  battery, 91
  computer switch-mode, 90
  mains noise, 89–91
  tablet presses, 91–95
power supply unit (PSU), 88–89
  linear regulators, 88–89
  non-regulated, 88
pre-compression force, dosating nozzle machinery, 217–218, 219
pre-compression rollers, 43
pressure, 6
  conversion of normal to lateral, 197
  SI unit, 5
pressure roll, 171, *172*
  rotary tablet presses, 117–118, 173
pressure–time curve asymmetry, 182
pre-triggering, 96
printer port, parallel, 158–159
product setting, 231, *232*
programmable input–output controllers, 157
punch, 106–108
  adhesion force, 131–132, 202
  adhesion measurement, 202
  adjustable, 130
  calibration, 140, 141
  dimension measurement, 144
  displacement
    measurement, 121
    time, 174
  displacement–time profiles, 168–171, *172*, 173–176
  eccentric tablet presses, 122–124
  gauged, 140
  gauging methods, *119*
  instrumentation packages, 133
  laser Doppler velocimetry, 124–125
  lower, 111
    force ratio to upper, 202
    movement, 174
    tip, *170*
    tip position, 176
  maximum penetration and peak force, 180, 189–190
  points of contact, 107–108
  positions, 231
  rotary tablet presses, 116, 118–119, 171, *172*, 173–175
  shortening during compression, 175–176
  strain-gauged, 116, 127, 141
  tightness, 130–131

tilting, 134
upper
  force ratio to lower, 202
  force–displacement curve, 187–188, 189
  movement, 173–174, 191
  tip, 169–171
  tip velocity, 191–195
velocity, 169, 190–196, 214
  eccentric tablet presses, 191
  horizontal, 191
  porosity, 194–195
  rotary tablet presses, 191–194
  sensitivity of materials, 196
  tablet formulation dependency, 194–196
  tablet press simulators, 214
  vertical, 191–194
punch face adhesive forces, 131–132
punch force
  ratio to upper to lower, 202
  strain-gauge output, 198
punch force measurement, 99
  accuracy, 99
  linear variable-differential transformer, 120
  lower, 108–111, 112, *113*, 120
  machine distortion, 122–123
  pull-up/pull-down, 129–131
  upper, 100–108, 112, *113*, 120
punch force transducer, 140
punch heads
  dwell time, 174
  rotary tablet press, 115, 171, *172*, 173
punch holder, lower, 108–111
  materials, 110–111
  two-piece, 109–110
punch holder, upper, 104–106
punch movement, 65, 169, *170*, 173–175
  accelerometer measurement, 133
  examination, 135
  predicted, 176
  stages, 169, 170
punch securing screw, 105
punch tip
  deformation, 175
  lower, *170*, 176
  measurement error, 176
  upper, 169–171

quarter-bridge, 33, 34
quartz crystals, 41

radiofrequency interfaces, 160–161
radio-telemetry, tablet presses, 93–95
Recommended Standard (RS) 232 card, 158
Recommended Standard (RS) 422 card, 159–160
Recommended Standard (RS) 485 card, 160
reed switch and magnet, 83–84
reject log, 234
reject systems, tablet presses, 226–227
remote capacitative transducer, 73
remote inductive transducer, 72–73, 125

repeatability, linear variable-differential transformer, 66
resistivity, temperature coefficient, 36, 37
resistors, linear variable, 71–72
roll-pins
  rotary tablet presses, 111, 115–117
  strain-gauged, 40–41, 45
rosette gauges, 35, 101
rosin solvent, 61
rotary presses, 8, 111–112, *113*, 114–119
  action, 111
  die table, 171, *172*, 173
  displacement measurement, 125–127
  displacement–time profile, 171, *172*, 173–175
  distortion during compression, 175–176
  force measurement, 114–115
  force–time profiles, 178–179, 181
  load cells, 115, 116
  machine frame, 112, *113*, 114–115
  measurement errors, 126–127
  piezoelectric transducers, 116–117
  pre-compression rollers, 43
  pressure roll, 117–118, 171, *172*, 173
  punch, 118–119, 171, *172*, 173–175
  punch force, 112, *113*
  punch heads, 115, 171, *172*
  punch velocity, 191–194
  roll carriers, 112
  roll-pins, 111, 115–117
  sample rate choice, 148
  strain gauges
    semiconductor, 114–115
    siting, 112, 114
  strain-gauged punch, 116
  tie rods, 112
  wheel eye-bolts, 112, *113*

sample and hold, 155
sample rate
  aliasing, 148
  choice, 148, *149–151*, 152
sampling system theory, 147–148, *149–151*, 152–153
  aliasing, 147–148
  resolution, 152–153
  sample rate choice, 148, *149–151*, 152
self temperature compensating (STC) gauges, 18, 34–35
  curvature compensation, 37
semiconductor(s), 20–21, 35–36
semiconductor bridges, 35–36
semiconductor strain gauges, 20–21, 106, 130
  rotary tablet presses, 114–115
sensors, waveform digitisation, 152
Shaxby–Evans equation, 179–180
shear beams, 39, *40*
shear force load cells, ejection ramp mounting, 129
shear-beam load cells, 44
sheet plastics, commercial, 24
  annealing, 25
SI units, 5, 139

signal clipping, 145–146
signal display, tablet presses, 95–96
silicon, doped, 20–21, 35–36
silicon carbide papers, 62
silicone rubber, 58, 63
slip gauges, 142–143, 145
slip-ring gear, 92
    tablet presses, *93*, *94*
soldering
    flux removal, 59, 61
    leadwire attachment to strain gauges, 57–58, 59, 62, 63
speckle pattern analysis, 28–29
    see also laser speckle photography; laser speckle strain gauges
sputtered thin-film technology, 21
S-sensor, 47–48
steel, elasticity, 110
strain, 175
    apparent, 18, *19*
    definition, 12
    measurement, 11–32
strain gauges, 2, 11–12, 13–32
    abrading of surface, 52–53
    abrasives, 62
    AC bridge configuration, 100–101
    active, 33
    adhesive bonding, 53–56, 58–59, 61
    application services, 243–245
    attachment, 2
        to die, 197
    backing, 15
    bonding with adhesive, 53–56
    bridge configurations, 33
    calibration, 59
    capacitance, 48
    clamping, 60
    compensation, 18
    composite, 35, 101
    conditioning of surface, 53
    configuration, 15–17
    construction, 14–15
    degreasing of surface, 52, 61
    die-wall stress, 199–200
    eccentric tablet presses, 102, 103–104
    electrical characteristics, 17
    electrical resistance, 13–14, 32–33
    electrical testing, 59
    encapsulated, 46
    excitation level, 87–88
    fibre optic, 48–49
    grid elements, 16–17
    handling, 54, 62
    health and safety considerations, 51
    high-temperature installations, 60–61
    installation, 51–64, 107, 108
        accessories, 61–63
        kits, 63
        professional assistance, 63–64
        protection, 58
    insulation, 15
    internal mounting, 108
    investigative, 25–26
    laser speckle, 49
    leadwire attachment, 57–58, 62
        high-temperature installations, 61
        inspection, 59
    life, 19
    linearity of response, 17
    load testing, 59
    location lines, 53
    lower punch holder, 108–109
    magnetostrictive, 48
    materials, 14, 15, 17, 18
    measuring range, 18–19
    metal foil, 14, *15*, 18, *19*, 106
    modulus compensation, 17, 34–35
    neutralising surface, 53, 61
    operating cycles, 19
    output relationship with punch force, 198
    passive, 33
    positioning, 60
    protective coating, 58, 63
    quarter-bridge configuration, 33, 34
    resistance, 20, 33
        non-linearity, 17
    resistive, *16*
    resonating sensors, 47–48
    rosette, 35, 101
    segmented die attachment, 197
    semiconductor, 20–21, 106
        punch force measurement, 130
        rotary tablet presses, 114–115
    shear-sensitive, 41
    signal, 93
    siting, 22–32, 107, 108
        rotary tablet presses, 112, 114
    soldering, 57–58, 59, 61, 62, 63
    special materials, 61
    specialist applications, 59–61
    sputtered thin-film, 21
    stress analysis, 22–30
    sub-assemblies, 31–32
    surface
        curvature, 37
        preparation, 51–53, 62
    tablet press automatic control, 223–224
    temperature coefficient, 17–18
    terminals, 62
    thermal drift, 34–35, 38, 43
    tools, 61, 63
    unbonded and leadwire attachment, 60
    vibrating element, 47–48
    weldable, 46–47, 61
    zero drift, 34–35, 43
strain rate sensitivity, 194–195
strain shunts, 112, 114
strain-gauge conversion transducer, 72
strain-gauged bolts, 41, 108, 130
    ejection ramp, 127

strain-gauged load washers, 43
strain-gauged punch, 116, 127, 141
strain-gauged roll-pins, 40–41
stress, 175
   mechanical, 12
   time-dependent decay, 43
   see also die-wall stress
stress analysis, 22–30
   annealing, 25
   brittle lacquer, 30
   direct methods, 25–30
   force, 31
   holography, 26–28
   modelling methods, 22–25
   moiré grid technique, 27–28
   resins for modelling, 24–25
   speckle pattern, 28–29
   thermoelastic, 29–30
stress pattern analysis by thermal emission (SPATE), 29–30
stretching, bridge circuit, 36–37
stroboscopic systems, 84
   aliasing, 147
successive approximation register devices, 155–156
superglue *see* cyanoacrylate adhesives
super-linear variable capacitor (SLVC), 69
surface conditioner, 61
surface strain measurement, 11–12
sweep-off force, 202
switched-capacitor filters, 154, *156*
switch-mode regulators, 89

tablet(s)
   adhesion to punch measurements, 202
   compression force relationship with properties, 167
   compression pressure, 2
   computer-controlled systems, 229, 231, 235–239
   forces, 2
      required to eject from die, 202
   hardness, 235–239
   history, 1
   particulate mass forming, 167
   porosity, 182–186
   properties, 1–2
   punch face separation force, 202
   speed dependency of formulations, 194–196
   thickness, 229, 231, 235–239
tablet presses, 1, 2, **3**
   applied force, 101
   automatic control, 223–239, *240*
   calibration, 102, 103
   compaction data storage, 95
   compaction simulators, 8
   compression characteristics, 128
   compression curves, 103
   compression cycle simulation, 134–136
   computer-controlled systems, 228–229, *230*, 231–239, *240*
   control systems, 225–226, *227*
   cosine error, 102, *103*
   crossheads, 133
   cycle, 148, 170
   data acquisition, 91–95
   displacement measurement, 119–127
   drive belt, 100
   force–displacement characteristics, 111
   force–time characteristics, 111
   force–weight relationship, 224–225
   frame distortion, 101
   frictional resistance of sliding components, 102
   gauged arm, 103
   hydraulic, 132–133
   instrumented, 99–136
   leadwires, 91–92, *94*
   machine frame, 100–101
   main shaft, 100
   mercury seals, 92
   monitoring systems, 225
   motor, 100
   optical linkage, 95
   output, 190
   overload system, 225, *226*
   parameters for measurement, 3–4
   power supplies, 91–95
   radio-telemetry, 93–95
   reject systems, 226–227
   rotation speed, 128–129, 169–170
   signal display, 95–97
   simulators, 214
   slip-ring gears, 92, *93*, *94*
   thermal errors, 102–103
   upper arm, 101–104
      errors in gauging, 102–104
   waveform, 136
   weighing systems, 227–228
   see also computer-controlled tablet presses; eccentric presses; instrumentation of presses; punch; rotary presses
tablet weight
   computer-controlled systems, *229*, *230*, 235–239, *236*
   control
      automatic, 167
      using compression force measurement, 223–225
   uniformity, 227–228
   weighing systems, 227–228
tamping pins, 209–210
   instrumented, 212
tape drives, 163
tapes for handling strain gauges, 62
telemetry, force calibration, 142
temperature, thermodynamic, 5
temperature coefficient of resistivity, 36, 37
   conductive plastics film, 71
temperature compensation, load cells, 45
terminals, strain gauges, 62
thermal drift, 34–35, 38, 43
thermal output, 18, *19*
thermoelastic stress analysis, 29–30
tie rods, 112, 114

time, SI unit, 5
titanium, 61
transducers, 3, *4*
   AC–AC, 68
   calibration, 139–146
      gauge blocks, 142–143
   capacitative, 73
   DC–DC, 68
   displacement, **123**
      linear inductive, 68–69
   fiber optic, 75–76
   force indirect calibration, 140–141
   gauging, 67, 121
      fitting, 126–127
   Hall effect, 70–71
   inductive, 72–73, 125
   inductive element, 78
   linear inductive displacement, 68–69
   linear magnetoresistive, 69–70
   magnetorestrictive, 73–74
   non-linear signals, 146
   optical, 75–76
   optical grading, 76–77
   optical velocity, 81
   piezoelectric, 41–43
   remote capacitative, 73
   remote inductive, 72–73, 125
   short-scale, 120
   signal, 93
   strain-gauge conversion, 72
   ultrasonic, 74–75
   variable capacitance, 69
   velocity, 81–82, 125
   see also displacement transducers
transformer
   variable turns-ratio, 69
   see also linear variable-differential transformer (LVDT)
transient suppressors, solid-state, 91

triaxial compression system, 200
tweezers, 63

ultrasonic transducers, 74–75
ultraviolet recorders, high-speed, 95–96, 97
units of measurement, 4–6
   derived, 5–6
universal serial bus (USB) interface, 159
USB analogue converter, 161
USB interface, 159

variable turns-ratio transformer, 69
velocity, 6
velocity transducers, 81–82, 125
   capacitative, 82
   inductive, 81–82
   optical, 81
vernier scale, 144

waveform fuzziness, 102
Weibull function, 182
weighing
   force measurement, 11
   tablet press systems, 227–228
   see also tablet weight
Wheatstone bridge circuit, 32–38
   bending, 36–37
   completion resistors, 37–38
   excitation level, 87–88
   excitation voltage, 33, 38
   output linearisation, 36
   quarter-bridge configuration, 33, 34
   stretching, 36–37
   zero drift, 34–35
work, SI unit, 5–6

Young's modulus, 12–13, 110, 111, 175

zero drift, 34–35, 43
zip drives, 163